Advances in Intelligent Systems and Computing

Volume 309

T0138174

Series editor

Janusz Kacprzyk, Polish Academy of Sciences, Warsaw, Poland
e-mail: kacprzyk@ibspan.waw.pl

About this Series

The series "Advances in Intelligent Systems and Computing" contains publications on theory, applications, and design methods of Intelligent Systems and Intelligent Computing. Virtually all disciplines such as engineering, natural sciences, computer and information science, ICT, economics, business, e-commerce, environment, healthcare, life science are covered. The list of topics spans all the areas of modern intelligent systems and computing.

The publications within "Advances in Intelligent Systems and Computing" are primarily textbooks and proceedings of important conferences, symposia and congresses. They cover significant recent developments in the field, both of a foundational and applicable character. An important characteristic feature of the series is the short publication time and world-wide distribution. This permits a rapid and broad dissemination of research results.

More information about this series at http://www.springer.com/series/11156

Lakhmi C. Jain · Srikanta Patnaik
Nikhil Ichalkaranje
Editors

Intelligent Computing, Communication and Devices

Proceedings of ICCD 2014, Volume 2

 Springer

Editors
Lakhmi C. Jain
Faculty of Education, Science, Technology
  and Mathematics
University of Canberra
Canberra, ACT
Australia

and

University of South Australia
Mawson Lakes, SA
Australia

Srikanta Patnaik
Department of Computer Science and
  Engineering
SOA University
Bhubaneswar, Odisha
India

Nikhil Ichalkaranje
Department of Premier and Cabinet
Office of the Chief Information Officer
Adelaide, SA
Australia

ISSN 2194-5357   ISSN 2194-5365 (electronic)
ISBN 978-81-322-2008-4 ISBN 978-81-322-2009-1 (eBook)
DOI 10.1007/978-81-322-2009-1

Library of Congress Control Number: 2014944718

Springer New Delhi Heidelberg New York Dordrecht London

Preface

The Organizing Committee is delighted to present the high-quality papers presented in the first International Conference on Intelligent Computing, Communication and Devices (ICCD 2014) being organized by SOA University during April 18–19, 2014. The title was chosen as this converges three upcoming technologies for the next decade. In recent time, "Intelligence" is the buzz word for any discipline and many scholars are working in these areas.

In simple definition, "intelligence" is the ability to think and learn. Looking back to its origin and development, report says, since 1956 artificial intelligence was formally found and has enjoyed tremendous success over the past 60 years. During the 1960s, the subject was dominated by traditional artificial intelligence that follows the principle of physical symbolic system hypothesis to get great success, particularly in knowledge engineering. During the 1980s, Japan proposed the fifth generation computer system (FGCS), which is knowledge information processing forming the main part of applied artificial intelligence. During the next two decades, the key technologies for the FGCS was developed such as VLSI architecture, parallel processing, logic programming, knowledge base system, applied artificial intelligence and pattern processing, etc. The last decade is observing the achievements of intelligence in the mainstream of computer science and at the core of some systems such as Communication, Devices, Embedded Systems, Natural Language Processor, and many more.

ICCD 2014 covers all dimensions of intelligent sciences in its three tracks, namely Intelligent Computing, Intelligent Communication, and Intelligent Devices. Intelligent Computing track covers areas such as Intelligent and Distributed Computing, Intelligent Grid and Cloud Computing, Internet of Things, Soft Computing and Engineering Applications, Data Mining and Knowledge Discovery, Semantic and Web Technology, Hybrid Systems, Agent Computing, Bio-Informatics, and Recommendation Systems.

At the same time, Intelligent Communication covers communications and networks technologies, including mobile broadband and all optical networks, which are the key to the groundbreaking inventions to intelligent communication technologies. This covers Communication Hardware, Software and Networked

Intelligence, Mobile Technologies, Machine-to-Machine Communication Networks, Speech and Natural Language Processing, Routing Techniques and Network Analytics, Wireless Ad Hoc and Sensor Networks, Communications and Information Security, Signal, Image and Video Processing, Network Management and Traffic Engineering.

The Intelligent Device is any type of equipment, instrument, or machine that has its own computing capability. As computing technology becomes more advanced and less expensive, it can be built into an increasing number of devices of all kinds. The Intelligent Device covers areas such as Embedded Systems, RFID, RF MEMS, VLSI Design and Electronic Devices, Analog and Mixed-Signal IC Design and Testing, MEMS and Microsystems, Solar Cells and Photonics, Nano-Devices, Single Electron and Spintronics Devices, Space Electronics, and Intelligent Robotics.

The "Call for Paper" for this conference was announced in the first week of January 2014 and due to shortage of time we have to keep a very tight deadline for paper submission, i.e., March 15. But to our surprise, we have received 324 papers, which were considered for review and editing. Out of these 324 papers, 163 papers were accepted for the presentation and publication whereas 147 papers were registered, which are covered in this proceeding.

I am sure the participants must have shared a good amount of knowledge during the two days of this conference. I wish all success in their academic endeavor.

Srikanta Patnaik

Contents

ICCD-2014 Conference Committee

Chief Patron

Prof. Manojranjan Nayak
President, Siksha 'O' Anusandhan University, India

Patron

Prof. R.P. Mohanty
Vice Chancellor, Siksha 'O' Anusandhan University, India

General Chair

Prof. Chitta Ranjan Das, Penn State University, USA

Programme Chair

Prof. Srikanta Patnaik, Siksha 'O' Anusandhan University, India

Programme Co-Chair

Prof. Kwang Baek Kim, Silla University, South Korea

Finance Chair

Prof. Manas Kumar Mallick, Siksha 'O' Anusandhan University, India

Convener

Dr. Alok Kumar Jagadev, Siksha 'O' Anusandhan University, India

Co-Convener

Dr. Ajit Kumar Nayak, Siksha 'O' Anusandhan University, India

International Advisory Committee

Prof. Florin Popentiu Vlădicescu, UNESCO Chair in Information and
    Communication Engineering, City University, London
Prof. Ishwar Sethi, Oakland University, USA
Prof. Reza Langari, Texas A&M University, USA
Prof. Rabi Mahapatra, Texas A&M University, USA
Prof. Kazumi Nakamatsu, University of Hyogo, Japan
Prof. Ma Maode, Nanyang Technological University, Singapore
Prof. Bruno Apolloni, Università degli Studi di Milano, Italy
Prof. Yadavalli Sarma, University of Pretoria, South Africa
Prof. DeSouza, Guilherme N., West University of Missouri-Columbia, Missouri
Prof. Zbigniew Michalewicz, School of Computer Science, Australia
Prof. Akshya Kumar Swain, The University of Auckland, New Zealand
Prof. Zhihua cui Taiyuan University of Science and Technology, China
Prof. Rajib Mall, IIT Kharagpur, India
Prof. Ashish Ghosh, ISI Kolkata, India
Prof. P.K. Dash, Siksha 'O' Anusandhan University, India
Prof. R.K. Mishra, Siksha 'O' Anusandhan University, India
Prof. P.K. Nanda, Siksha 'O' Anusandhan University, India
Prof. G.C. Bose, Siksha 'O' Anusandhan University, India
Prof. R.K. Hota, Siksha 'O' Anusandhan University, India

Programme Committee

Dr. Xiaolong Li, Indiana State University, USA
Dr. Yeon Mo Yang, Kumoh University, Korea
Dr. Sugam Sharma, Iowa State University, USA

Dr. Arturo de la Escalera Hueso, Intelligent Systems Lab, Spain
Dr. Debiao He, Wuhan University, China
Dr. Nadia Nouali-Taboudjemat, Research Centre on Scientific and Technical Information, Algeria
Prof. Doo Heon Song, Yong-in SongDam College, South Korea
Prof. Baojiang Zhong, Soochow University, China
Prof. Nitaigour Mahalik, California Sate University, Fresno, CA
Prof. Guangzhi Qu, Oakland University, USA
Prof. Peng-Yeng Yin, National Chi Nan University, Taiwan
Dr. Yaser I. Jararweh, Jordan University of Science and Technology, Jordan
Dr. Jayanthi Ranjan, Information Management and Systems, Ghaziabad

Organizing Committee
Organizing Chair

Dr. Debahuti Mishra

Organizing Co-Chair

Badrinarayan Sahu
Sarada Prasanna Pati

Organizing Committee

Dr. Renu Sharma
Dr. Shazia Hasan
Trilok Nath Pandey
Sharmistha Kar
Sashikala Mishra
Saumendra Kumar Mohanty
Debasish Samal
Jyoti Mohanty
Shruti Mishra
Prabhat Kumar Sahoo
Priyabrata Pattnaik
Aneesh Wunnava
Sarbeswar Hota
Debabrata Singh
Kaberi Das

Publicity Chair

Dr. B.K. Pattanayak

Publicity Co-Chair

Bibhu Prasad Mohanty
Dr. D.B. Ramesh

Publicity Committee

Chinmaya Kumar Swain
Shrabanee Swagatika
Pandab Pradhan
Sandeep Kumar Satapathy
Subrat Kumar Nayak
Kulamala Vinod Kumar
Madhuri Rao
Meera Nayak
Susmita Panda
Jeevan Jyoti Mahakud

Technical Chair

Prof. Niva Das

Technical Co-Chair

Dr. Guru Prasad Mishra
B.M. Acharya

Technical Committee

Dr. Sukanta Sabut
Dr. Satyanarayan Bhuyan
Dr. Benudhar Sahu
Dr. Mihir Narayan Mohanty
Gyana Ranjan Patra
Smita Prava Mishra
Manoranjan Parhi

Minakhi Rout
Nibedan Panda
Ambika Prasad Mishra
Satya Ranjan Das
Sarita Mahapatra
Barnali Sahoo
Bandana Mahapatra
Alaka Nanda Tripathy

About the Editors

Prof. Lakhmi C. Jain is with the Faculty of Education, Science, Technology, and Mathematics at the University of Canberra, Australia, and University of South Australia, Australia. He is a Fellow of the Institution of Engineers, Australia. Professor Jain founded the KES International, a professional community for providing the opportunities for publications, knowledge exchange, cooperation, and teaming. Involving around 5,000 researchers drawn from universities and companies worldwide, KES facilitates international cooperation and generate synergy in teaching and research. KES regularly provides networking opportunities for professional community through one of the largest conferences of its kind in the area of KES. His interests focus on the artificial intelligence paradigms and their applications in complex systems, security, e-education, e-healthcare, unmanned air vehicles, and intelligent agents.

Prof. Srikanta Patnaik is presently a Professor of Computer Science and Engineering, SOA University, Bhubaneswar, India. He holds Doctor of Philosophy in Engineering from Jadavpur University, India. He has published more than 80 research papers and articles in international journals and magazines of repute. He has supervised 10 research scholars for their Ph.Ds. He has completed various funded projects as Principal Investigator from various funding agencies of India. He is an Editor-in-Chief of two international journals, namely International Journal of Information and Communication Technology and International Journal of Computational Vision and Robotics, published from Inderscience Publishing House, England, and also a Series Editor of Springer Book Series on Modeling and Optimization in Science and Technology (MOST).

Dr. Nikhil Ichalkaranje is currently serving as Senior ICT Strategist, Office of the Chief Information Officer (CIO), Department of Premier and Cabinet, the Government of South Australia. He holds Doctor of Philosophy in Computer Systems Engineering from University of South Australia (March 2006). He served in various capacities such as Assistant Director, Cyber Security Programs and Security Strategy, Senior Technology Advisor, Department of Broadband, Communications and Digital Economy (DBCDE), Government of Australia. He has produced useful and meaningful research outcomes from diverse methodologies, perspectives, and concepts to solve real life problems and support policymakers in forming evidence-based policy. He has a strong academic record of research publications in the computer science and telecommunication industry including journals, conference papers, and edited international books from renowned publishers.

A Survey of Security Concerns in Various Data Aggregation Techniques in Wireless Sensor Networks

Mukesh Kumar and Kamlesh Dutta

Abstract Achieving security in case of data aggregation in wireless sensor network (WSN) is a challenging task because of its limitations in terms of computation, resources, battery power, transmission capabilities, etc. This paper provides a comprehensive discussion on literature of security concerns in data aggregation in WSNs. Main contributions of this paper are describing the fundamentals of secure data aggregation in WSN, identifying the important parameters for the classification of secure data aggregation techniques for WSN, considering key characteristics of existing secure data aggregation techniques, comparing the existing secure data aggregation techniques, and introducing table of comparison based on various parameters such as security principles, prevention of attacks by protocols, aggregation function, and cryptographic techniques used.

Keywords Data aggregation · Security · WSN

1 Introduction

Wireless sensor network (WSN) is a heterogeneous system, consisting of tiny reasonably priced sensors, actuators, and general purpose computing elements. Sensors are spread over a specific geographical area, which are competent of sensing changes in parameters such as temperature, pressure, humidity, and noise level [1] of surrounding environment. These sensor nodes are scattered in an unattended environment (i.e., sensing field) to sense the physical world. They communicate with other devices over a specific area using transceiver and send

M. Kumar (✉) · K. Dutta
Department of CSE, National Institute of Technology, Hamirpur 177005, H.P., India
e-mail: mukeshk.chawla@gmail.com

K. Dutta
e-mail: kdnith@gmail.com

© Springer India 2015
L.C. Jain et al. (eds.), *Intelligent Computing, Communication and Devices*,
Advances in Intelligent Systems and Computing 309,
DOI 10.1007/978-81-322-2009-1_1

sensed information to a central location, so that central processing can be performed on it to achieve desired functionality such as environment monitoring, providing security at home or in public place. The most efficient model for WSN is cluster-based hierarchical model. WSN is like an ad hoc network, so it is also called "ad hoc wireless sensor network." Potential applications include monitoring factory environment such as instrumentation, pollution level, fire alerts, free way traffic, climate monitoring and control [2], medical monitoring and emergency response [3], monitoring remote or unfriendly habitat [4, 5], military target tracking [6, 7], natural disaster relief [8], wildlife habitat monitoring [9], forest fire prevention [10], military surveillance [10], and biomedical health monitoring [8, 9]. The preferred features of WSN are security, reliability, robustness, self-healing, self-organizing capabilities, dynamic network topology, limited power, node failures and mobility of nodes, short-range broadcast communication and multi-hop routing, and large scale of deployment and scalability [11].

1.1 Data Aggregation

Sensor networks are event-based system. Sensor node triggers when some particular event occurs in the environment for which it is deployed. Generally, there is redundancy in the event data. Individual sensor readings are of limit use. We require average, min, max, sum, count, predicate, quartile, etc., of some particular readings. If we forward data without aggregation, it will be too expensive as it consumes more energy as well as bandwidth. One of the best solutions to this problem is to combine the data coming from sensor nodes. Data aggregation techniques explore how the data are to be routed in the network as well as the processing method that is applied on the packets received by a node. Data aggregation techniques are used to reduce energy consumption of nodes, and thus, network efficiency is increased due to reduced number of transmission. Data aggregation [12] is defined as global process of gathering and routing information through a multi-hop network, processing data at intermediate nodes with the objective of reducing resource consumption, thereby increasing network lifetime. It will eliminate redundancy and minimize the number of transmissions. Thus, aggregation is the summarization of the data combined to answer some particular query.

1.2 Data Aggregation Security Concerns

- *Confidentiality* ensures that information is only disclosed to those who are authorized to see it. The confidentiality principle in case of WSN will be achieved when we are able to aggregate encrypted data

- *Integrity and freshness* ensures that information is accurate and complete and has not been altered in any way. There should not be alterations in aggregated data
- *Availability* ensures that a system can accurately perform its intended purpose and is accessible to those who are authorized to use it. Here, availability is limited to availability of data aggregator
- *Authentication* ensures correctness of claimed identity. Absence of proper source authentication invites Sybil attacks against data aggregation
- *Authorization* ensures permissions granted for actions performed by entity

2 Secure Data Aggregation Protocols

Secure data aggregation protocols broadly can be classified based on cryptographic techniques and based on non-cryptographic techniques.

2.1 Secure Data Aggregation Based on Cryptographic Techniques

Secure data aggregation protocols based on cryptographic techniques are further classified on the basis of which cryptographic technique it is using, e.g., symmetric key cryptography, asymmetric key cryptography, hash function, and MAC. There are further two categorizations of non-cryptographic solutions for secure data aggregation. First one is based on trust/reputation. Second category of solution is based on soft computing techniques as described in Fig. 1.

2.1.1 Secure Data Aggregation Based on Symmetric Key Cryptography

We start discussing the work proposed by Hu and Evans [13]. It is based on symmetric key cryptography and aggregation of unencrypted data. In this scheme, messages are not aggregated at the immediate next hop, are forwarded as it is over the first hop, and then are aggregated at the second hop. The proposed protocol ensures data integrity but does not provide data confidentiality. Another drawback of the scheme is that if both of parent node and its child are compromised nodes, then data integrity will not be assured.

The next protocol which falls under the category of aggregation protocol based on symmetric key cryptography is proposed by Yang et al. [14]. In this, nodes are dynamically divided in a tree topology by using probabilistic grouping technique. SDAP is based on the fact that more trust has to be placed on the high-level nodes

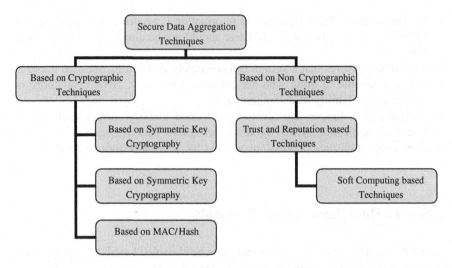

Fig. 1 Classification of various secure data aggregation techniques

(i.e., nodes closer to the root) as compared to low-level nodes during a normal hop-by-hop aggregation process in a tree topology. If a compromised node is closer to the base station, the false aggregated data produced by this compromised node will have a larger impact on the final result computed by the base station. SDAP dynamically separates the topology tree into multiple logical groups of similar sizes. SDAP provides data confidentiality, source authentication, and data integrity.

The next paper under this category was proposed by Ozdemir in 2007. In this paper, author argues that compromised nodes have access to cryptographic keys that are used to secure the aggregation process, and therefore, cryptographic primitives alone cannot provide a sufficient enough solution to secure data aggregation problem. The SELDA [15] is based on the concept that sensor nodes observe behavior of their neighboring nodes to build up trust levels for both the environment and the neighboring nodes. A web of trust is formed by exchanging trust level of nodes with each other that allows them to determine secure and reliable paths to data aggregators. Main contribution of SELDA is the detection of DoS attack by the use of monitoring mechanisms.

Westhoff et al. [16] proposed concealed data aggregation (CDA) protocol in which sensor nodes have a common symmetric key with the base station that is reserved secret from intermediate aggregators. In the proposed approach, data aggregators carry out aggregation functions that are applied to cipher texts. This scheme has main advantage that intermediate aggregators do not have to perform decryption and encryption operations at their ends. Therefore, data aggregators do not have to store a susceptible cryptographic key which ensures an unrestricted aggregator node election process during the WSN's lifetime.

A wide range of CDA based on homomorphic encryption has been proposed in the literature. In these techniques, aggregation function can be performed on the cipher text without decrypting it. Technique proposed by Sun et al. [17] also falls under this category. It provides data confidentiality as well as integrity of data. Moreover, it is resistant to eavesdropping attack. It also provides malicious node detection with some extra overhead. Main limitation of the scheme is that no source authentication is achieved, therefore resilient to Sybil attack. Another limitation is extra overhead due to malicious node detection.

Now, we discuss the work proposed by Castellucia et al. in 2009 in which authors proposed a protocol which achieves end-to-end confidentiality while achieving data aggregation. This protocol performs data aggregation without requiring the decryption of sensor data at aggregator node.

Next, we discuss SEEDA [18] and main feature of SEEDA is less communication overheard. In SEEDA instead of sending information of non-responding nodes, cipher text for non-responding nodes is considered as 0. This protocol has to compute average of the received data. Therefore, along with the cipher text count of number of nodes is added. Limitations of this scheme are that only confidentiality is achieved and very elementary method is used for achieving it. No integrity and authentication is achieved.

2.1.2 Secure Data Aggregation Protocol Based on Asymmetric Key Cryptography

In year 2007, Ozdemir [19] proposed CDAP which is an asymmetric key-based privacy homomorphic encryption technique. By using this technique, end-to-end data confidentiality and data aggregation can be achieved together. This protocol uses a number of more efficient sensor nodes called aggregator node for privacy homomorphic encryption and aggregation of the encrypted data. During the network deployment, each aggregator node shares pair-wise keys with its neighboring nodes. The purpose of sharing these keys is to enable neighboring nodes to send their sensor readings securely. Each neighboring node encrypts its data and sends the encrypted data to its aggregator node. After receiving, the data aggregator node decrypts all the data, aggregates the data, and encrypts the aggregated data using the privacy homomorphic encryption algorithm. The data which are encrypted with the privacy homomorphic encryption algorithm can only be decrypted with private key own by base station.

Another end-to-end data aggregation approach which employs homomorphic encryption on elliptic curve cryptography is proposed by Jacques et al. [20]. Elliptic curve cryptography facilitates nodes to generate comparatively smaller key size as compared to existing complex schemes. However, same security can be achieved using elliptic curve cryptography. It prevents the decryption phase at the aggregator's layers and saves energy. This scheme is based on the cryptosystem which has been proved safe and has not been crypt analyzed. Another important feature of this scheme is that for two identical messages, it generates two different cryptograms. It is resistant to known plain text attack, chosen plain text attack, and man in middle attack. This scheme provides only confidentiality not integrity and authentication.

2.1.3 Secure Data Aggregation Protocol Based on MAC/Hash Function

A witness-based data aggregation scheme for WSNs is proposed by Du et al. [21]. In this approach, the witness nodes of each data aggregator also perform data aggregation and compute MACs of the aggregated data. MAC of the aggregated data of every witness node is sent to the data aggregator. The data aggregator collects and forwards the MACs to the base station. These MACs are used by the base station to verify the integrity of the data send by data aggregators. The proposed protocol provides only integrity of the data. It does not provide confidentiality and authentication.

In 2010, Suat Ozdemir and Hasan Cam worked on false data detection and data aggregation and proposed a protocol. This protocol [22] supports data aggregation with false data detection. To achieve this, the monitoring nodes of every data aggregator also conduct the data aggregation as well as compute the corresponding MAC for data verification at their pair mates. To achieve confidentiality in data transmission, the sensor node between the successive data aggregators verifies the data integrity on the encrypted data rather than the plain data. The role of the data aggregator is selected on rotation basis among the sensor nodes based on their residual energy levels. This will lead to the balance of energy consumption.

Energy-efficient and high-accuracy scheme which is proposed by Hongjuan et al. [23] in 2011 can protect data privacy with moderate extra overhead; hence, this scheme has less bandwidth and energy consumption. The communication overhead of EEHA is reduced significantly, the amount of transmission is less, and hence, the chance of occurring collisions is also decreased which causes an improvement of aggregation accuracy and energy efficiency. To address privacy, authors adopt the "slicing and assembling" technique. This scheme is more suitable for applications that have relative loose requirements of privacy preservation, but place more emphasis on energy-efficiency and accuracy level [23].

The last protocol [24] under this category is proposed by Ozdemir and Yan. This technique is used for hierarchical data aggregation which uses different keys for encryption of data. It uses elliptical curve cryptography and homomorphic encryption for data integrity and confidentiality. In this protocol, network is virtually divided into several regions and a different public key is used for each region. For achieving data integrity, the MAC of each region is combined using XOR function resulting in a single MAC which is further verified by the base station. All messages are time stamped to prevent replay attack. This protocol is important when the base station may need to analyze the data for a certain region in a network. In this protocol, each sensor node encrypts its data using the public key of the region in which it resides and sends it to the data aggregator of the region. This protocol does not provide integrity protection for individual sensor reading. In that case, the base station would need the list of all sensor nodes that contributed to the aggregated data, thereby incurring too much communication overhead.

The technique proposed by Ozdemir and Xiao [25] is also known as FTDA which is based on locality-sensitive hashing which is used to represent sensor data compact. In this technique, recent m readings are encoded into LSH code of size *b*. LSH codes are forwarded to aggregator nodes, and data aggregator node checks the similarity between the LSH pairs. Now, the data aggregator node verifies from the neighboring node whether this outlier is due to any phenomenon in the neighboring nodes. Data aggregator will not include data of outlier nodes in the data aggregation. Further, duplicate data packets which have similar LSH codes are hereby removed in order to improve the bandwidth and energy efficiency. Those authors have mainly two contributions in this work: First, deletion of outlier and second, removal of redundant data.

2.2 Secure Data Aggregation Protocols Based on Non-Cryptographic Schemes

First protocol under this category is proposed by Roy et al. [26]. In this protocol, authors tried to overcome the limitations of existing aggregation framework called synopsis diffusion which combines multi-path routing schemes with duplicate-insensitive algorithms to accurately compute aggregated values in spite of message losses resulting from node and transmission failures. Also, this aggregation technique does not consider the problem of false sub-aggregate values generated by malicious nodes. These sub-aggregate values result in large errors in the final value computed at the sink. This is an important problem since sensor networks are highly susceptible to node compromise attack due to the unattended nature of sensor nodes and the lack of tamper-resistant hardware. In this protocol, authors make the synopsis diffusion approach secure against attacks in which compromised nodes contribute false sub-aggregate values. A novel lightweight verification algorithm is proposed by which the base station can determine whether the computed aggregated value includes any false data.

The next scheme [27] employs hierarchical inspecting of message integrity during aggregation. In this, result checking and aggregation are performed concurrently. Whenever attack is detected, sensor will send error report to the base station via different pre-decided routing paths. It does not require strong assumption, i.e., fixed network topology, topology knowledge security features. This scheme is secure against stealthy attack since the tampered results generated by the parent node can be detected by its child node. Main drawback of this scheme is that it cannot detect a compromised node which modifies its sensing reading, e.g., inserting an extra value cannot be prevented.

2.2.1 Secure Data Aggregation Protocols Based on Trust/Reputation

First scheme under this category [28] is based on defensible approach against insider attack. In this technique, trust value is assigned by the neighbor node by cross-checking. The readings are compared with the expected sensing result within the possible and legitimate sensing range. Authors assumed that there are some nodes called anchor nodes which know their location and orientation and are trustworthy nodes. Another assumption is that malicious node only tries to inject false data and do not increase the trust value of malicious nodes. In this, first of all, sensor network is partitioned into logical grids and unique identification number is assigned. Then, each sensor node estimates its location and grid number, and after that, each node evaluates trust worthiness of its neighbor nodes by cross-checking neighbor node's duplicate data with its own results. False data can be detected in this step. Finally, aggregator node computes the median of the data and transmits the aggregation result to the destination node.

In the next technique [29], trust worthiness/reputation of each sensor node is calculated by using information theoretic concept Kullback–Lieber (KL) distance. Uncertainty of the aggregation result is represented by an opinion, a metric of degree of belief. This opinion is routed by Josang's belief model. This scheme is resilient against node compromise attack and quantifies the uncertainty in the aggregated results. Statistical and signal processing techniques are used to measure the trustworthiness by comparing the readings with the reported data. In this technique, one cluster head and one aggregator for each cluster are appointed. Each sensor node sends its readings to the aggregator; then, aggregator classifies the sensor nodes into different groups based on reputation by comparing the data with aggregated results. Further, it calculates an opinion, metric of uncertainty to express trust worthiness of the sensor nodes. Cluster head receives report which includes aggregated result and opinion. Cluster head then combines all the reports calculated from different aggregators, recalculates opinion discount based upon aggregator's reputation, and sends to base station. In this scheme, new aggregator is elected on the basis of trust or reputation.

The last protocol [30] under this category is based on reputation system in which activity of every node is observed by neighbor's node in order to determine their trustworthiness. In this scheme, different nodes are considered for their different responsibilities such as sensing, and aggregation. However, this work is similar to that of Zhang' work [29] but it has overcome one important assumption that all nodes have a similar view of sensed event. Authors claimed that current work follows on practical implementation, while previous work was based on analytical work. Reputation system used here is called as beta reputation system which is able to describe the probability of binary event based on previous outcome of the event. This scheme follows Bayesian framework using the beta distribution for trustworthiness of sensor network.

2.2.2 Secure Data Aggregation Protocols Based on Soft Computing Techniques

According to the author [31], computational intelligence-based techniques can be widely used for secure data aggregation instead of traditional cryptographic techniques. Computational intelligence consists of neural network, fuzzy computing, swarm intelligence, and evolutionary computing. Data aggregation techniques consist of two kind of routing heuristic namely data centric and address centric. We can consider construction of aggregation tree as NP hard problem, but there exist some solutions which are based on computational intelligence.

First paper under the aforementioned category employs neural network. In this paper [32], authors have proposed a scheme to implement neural network into sensors and neural network is trained for false data detection. Neural network is trained with the data from neighbors. It compares the data of neural network with the sensor node and then evaluates error between the two data. The reason for devising neural network within the sensor node is that processing uses less energy as compared to the data transmission. This technique has two main contributions. First, it has increased the sensor network lifetime. Secondly, it is used to identify and detect false data during data aggregation.

In the second work [33], fuzzy logic is employed to test the trustworthiness of sensor nodes. This technique has basically three phases. In first phase, cluster formation takes place and cluster head is elected. Cluster head estimates the distance between the sensor node and itself. In second phase, cluster head determines the trust value with the help of spatial and temporal changes. In third phase, fuzzy logic is applied to test the trustworthiness of nodes. Fuzzy logic rules are constructed on the basis of trust level, power, and distance. There would be three output levels: best node, normal node, or worst node. Authors have not employed any cryptographic technique to detect malicious node.

3 Comparison of Secure Data Aggregation Protocols

In this section, comparison of various secure data aggregation protocols has been done on the basis of various parameters as shown in Table 1. These parameters are briefly explained below.

- Security principles: One parameter of comparison can be various security concerns achieved by the proposed techniques such as confidentiality, integrity, authentication, non-repudiation, and availability.
- Prevention of attacks: Other criteria for categorization can be the types of attacks which can be prevented by the existing literature such as DoS, replay, eavesdropping, Sybil, and false data injection.

Table 1 Comparison of secure data aggregation protocols

Protocol	Year	Data confidentiality	Data integrity	Source authentication	Node availability	Prevention of attacks	E-E security/ H-H security	Aggregation function	Techniques used
Hu and Evans [13]	2003	No	Yes	Yes	No	Eavesdropping	H-H	Sum	Symmetric key cryptography
Du et al. [21]	2003	No	Yes	Yes	No	Eavesdropping	H-H	Sum	MAC
Hur et al. [28]	2005	No	No	No	No	Malicious node	H-H	Median	Non-cryptographic
Zhang et al. [29]	2006	No	No	Yes	No	Comprised node, false data injection	H-H	Average, sum, histogram	MAC, trust based
SDAP [14]	2006	Yes	Yes	Yes	No	False data injection attacks	H-H	Sum	Symmetric key cryptography
CDA [16]	2006	Yes	No	No	No	Eavesdropping	E-E	Sum	Symmetric key cryptography
SELDA [15]	2007	No	Yes	Yes	Yes	DoS	H-H	Sum	Symmetric key cryptography
Ozdemir [19]	2007	Yes	No	No	No	Eavesdropping	E-E	Sum	Asymmetric key cryptography, homomorphic encryption
Sun et al. [17]	2008	Yes	Yes	No	No	Eavesdropping, malicious node detection	E-E	Sum	Homomorphic encryption, digital signature

(continued)

Table 1 (continued)

Protocol	Year	Data confidentiality	Data integrity	Source authentication	Node availability	Prevention of attacks	E-E security/ H-H security	Aggregation function	Techniques used
SEEDA [18]	2009	Yes	No	No	No	Eavesdropping	Both	Sum, avg	Homomorphic encryption
Jacques et al. [20]	2010	Yes	No	No	No	Known plain text attack, chosen plain text attack, and man in middle attack	E-E	–	Homomorphic encryption, elliptic curve cryptography
Ozdemir and Çam [22]	2010	Yes	Yes	Yes	No	Replay, false data detection, eavesdropping, Sybil	E-E	–	MAC, group key management
RDAS [30]	2010	No	No	No	No	Compromised node	H-H	–	Bayesian framework with beta reputation system
IPHCDA [24]	2011	Yes	Yes	No	No	Replay attack, eavesdropping, known plain text attacks, cipher text analysis, unauthorized aggregation	E-E	Sum	Homomorphic encryption, MAC
EEHA [23]	2011	Yes	Yes	No	No	Eavesdropping, replay attack.	E-E	Sum	MAC
Bahanfar et al. [32]	2011	No	No	No	No	False data injection	H-H	Sum	Neural network

(continued)

Table 1 (continued)

Protocol	Year	Data confidentiality	Data integrity	Source authentication	Node availability	Prevention of attacks	E-E security/ H-H security	Aggregation function	Techniques used
Hevin Rajesh et al. [33]	2012	No	No	No	No	Compromised node	H–H	–	Fuzzy computing
FTDA [25]	2012	No	No	No	No	False data injection	H–H	–	Locality sensitive hashing
Roy et al. [26]	2012	No	No	No	No	False data injection	E–E	Predicate count or sum	Non-cryptographic
SASHIMI [27]	2013	No	Yes	No	No	Stealthy attack	E–E	Average	Non-cryptographic

- Aggregation function: Generally, there are two approaches to perform secure data aggregation. Data aggregation can be done with size reduction and without size reduction.
- Technique used for security: Another criterion for comparison would be based on cryptographic techniques used like digital signature, symmetric key cryptography, asymmetric key cryptography, and MAC. Others techniques may be based on trust/reputation and soft computing based techniques.
- Hop-by-hop data aggregation: In this technique, the encryption of the data is performed by the sensing nodes and decryption by the aggregator nodes. The aggregator nodes aggregate the data and again encrypt the aggregation result. At the end, the sink node obtains the last encrypted aggregation result and decrypts it.
- End-to-end data aggregation: In this technique, the aggregator nodes in between do not contain any decryption keys and can only perform aggregation on the encrypted data.

4 Conclusion

In this paper, we started with brief introduction of WSN, its applications, characteristics, and limitations. We observed that after application of data aggregation, the number of message transfers reduced significantly. Further, security requirements increase when we apply aggregation on the data from sensor nodes .In literature, a number of secure data aggregation schemes have been proposed. We have given a comprehensive review of various secure data aggregation protocols with their advantages and limitations. This comparison is done on the basis of various parameters such as security principles, prevention of attacks by protocols, aggregation function, and cryptographic techniques used.

References

1. Pathan, A.-S.K., Islam, H.K., Sayeed, S.A., Ahmed, F., Hong, C.S.: A framework for providing e-services to the rural areas using wireless ad hoc and sensor networks. Appeared in IEEE ICNEWS (2006)
2. Chan, H., Perrig, A.: Security and privacy in sensor networks. Carnegie Mellon University, pp. 99–101
3. Welsh, M., Myung, D., Gaynor, M., Moulton, S.: Resuscitation monitoring with a wireless sensor network. In: Supplement to Circulation. J. Am. Heart Assoc. (2003)
4. Mainwaring, A., Polastre, J., Szewczyk, R., Culler, R.: Wireless sensor networks for habitat monitoring. In: First ACM International Workshop on Wireless Sensor Networks and Applications (2002)
5. Szewczyk, R., Polastre, J., Mainwaring, A., Culler, D.: Lessons from a sensor network expedition. In: First European Workshop on Wireless Sensor Networks (EWSN'04) (2004)

6. Duckworth, G.L., Gilbert, D.C., Barger, J.E.: Acoustic counter-sniper system. In: SPIE International symposium on Enabling Technologies for Law Enforcement and Security (1996)
7. Simon, G., Maroti, M., Ledeczi, A., Balogh, G., Kusy, B., Nadas, A., Pap, G., Sallai, J., Frampton, K.: Sensor network-based counter sniper system. In: Proceedings of the Second International Conference on Embedded Networked Sensor Systems (Sensys), Baltimore, MD (2004)
8. Castillo-Effen, M., Quintela, D.H., Jordan, R., Westhoff, W., Moreno, W.: Wireless sensor networks for flash-flood alerting. In: Proceedings of the Fifth IEEE International Caracas Conference on Devices, Circuits, and Systems, Dominican Republic (2004)
9. Gao, T., Greenspan, D., Welsh, M., Juang, R.R., Alm, A.: Vital signs monitoring and patient tracking over a wireless network. In: Proceedings of the 27th IEEE EMBS Annual International Conference (2005)
10. Lorincz, K., Malan, D., Fulford-Jones, T.R.F., Nawoj, A., Clavel, A., Shnayder, V., Mainland, G., Welsh, M., Moulton, S.: Sensor networks for emergency response: challenges and opportunities. In: Pervasive Computing for First Response (Special Issue), IEEE Pervasive Computing (2004)
11. Law, Y.W., Dulman, S., Etalle, S., Havinga, P.: Assessing security-critical energy efficient sensor networks. University of Twente, EA Enshede, Netherlands (2002)
12. Fasolo, E., Rossi, M., Widmer, J., Zorzi, M.: In-network aggregation techniques for wireless sensor networks: A survey. IEEE Wireless communication (2007)
13. Hu, L., Evans, D.: Secure aggregation for wireless networks. In: Proceedings of the Workshop on Security and Assurance in Ad Hoc Networks, Orlando, FL (2003)
14. Yang, Y., Wang, X., Zhu, S., Cao, G.: SDAP: a secure hop-by-hop data aggregation protocol for sensor networks. In: Proceedings of the ACM MOBIHOC'06, New York, USA, pp. 356–367 (2006)
15. S. Ozdemir.: Secure and reliable data aggregation for wireless sensor networks. LNCS, volume 4836, pp. 102–109 (2007)
16. Westhoff, D., Girao, J., Acharya, M.: Concealed data aggregation for reverse multicast traffic in sensor networks: encryption key distribution and routing adaptation. IEEE Trans. Mobile Comput. 5(10), 1417–1431(2006)
17. Sun, H.M., Lin, Y.H., Hsiao, Y.C., Chen, C.M.: An efficient and verifiable concealed data aggregation scheme in wireless sensor networks. In: Proceedings of ICESS08, pp. 19–26 (2008)
18. Castellucia, C., Chan, C.F., Mykletum, E., Tsudik, G.: Efficiently provably secure aggregation of encrypted data in wireless sensor networks. ACM Trans. Sensor Netw. 5(3) (2009)
19. Ozdemir, S.: Concealed data aggregation in heterogeneous sensor networks using privacy homomorphism. In: Proceedings of the ICPS'07: IEEE International Conference on Pervasive Services, Istanbul, Turkey, pp. 165–168 (2007)
20. Bahi, J.M., Guyeux, C., Makhoul, A.: Efficient and robust secure aggregation of encrypted data in sensor networks. In: proceedings of Fourth International Conference on Sensor Technologies and Applications, pp. 472–477 (2010)
21. Du, W., Deng, J., Han, Y.S., Varshney, P.K.: A witness-based approach for data fusion assurance in wireless sensor networks. In: Proceedings of the IEEE Global Telecommunications Conference (GLOBECOM'03), pp. 1435–1439 (2003)
22. Ozdemir, S., Çam, H.: Integration of false data detection with data aggregation and confidential transmission in WSN. IEEE/ACM Trans. Netw. 18(3) (2010)
23. Li, H., Lin, K., Li, K.: Energy-efficient and high-accuracy secure data aggregation in wireless sensor networks. Comput. Commun. 34, 591–597 (2011)
24. Ozdemir, S., Xiao, Y.: Integrity protecting hierarchical concealed data aggregation for wireless sensor networks. Comput. Netw. 55, 1735–1746 (2011). Elsevier
25. Ozdemir, S., Xiao, Y.: FTDA: outlier detection based fault tolerant data aggregation for WSN. In: Security and Communication Networks (2012)

26. Roy, S., Setia, S., Jajodia, S.: Secure data aggregation in WSN. IEEE Trans. Inf. Forensics **7**(3), 1040–1052 (2012)
27. Chen, C.M., Lin, Y.H.: SASHIMI: Secure aggregation via successively hierarchical inspecting of message integrity on WSN. J. Inf. Hiding Multimedia Signal Process. **4**(1) (2013)
28. Hur, J., Lee, Y., Hong, S., Yoon, H.: Trust-based secure aggregation in wireless sensor networks. In: Proceedings: The 3rd International Conference on Computing, Communications and Control Technologies: CCCT (2005)
29. Zhang, W., Das, S., Liu, Y.: A trust based framework for secure data aggregation in wireless sensor networks. In: Proceedings IEEE SECON'06 (2006)
30. Perez-Toro, C.R., Panta, R.K., Bagchi, S.: Reputation based resilient data aggregation. In: Proceedings of IEEE SECON (2010)
31. Farooq, M.U.: Computational intelligence based data aggregation technique for WSN. In: Proceedings of the World Congress on Engineering and Computer Science, vol. 2, 24–26 Oct, San Francisco, USA (2012)
32. Bahanfar, S., Kousha, H., Darougaran, L.: Neural network for error detection and data aggregation in WSN. IJCSI **8**(5–3) (2011)
33. Hevin Rajesh, D., Paramasivan, B.: Fuzzy based secure data aggregation technique in WSN. J. Comput. Sci. **8**(6), 899–907 (2012)

Secure and Dynamic IP Address Configuration Scheme in MANET

Poulami Choudhury, Koushik Majumder and Debashis De

Abstract Mobile ad hoc network (MANET) is an infrastructure-less network having dynamic topology and volatile nodes connected by wireless communication. Secure communication in MANET demands unique IP address assignment to ensure proper routing of packets. It is challenging for a decentralized network to have dynamic and unique IP addressing scheme. In this paper, we focus on the difficulties of secure message passing and assigning unique IP addresses to new nodes willing to join in MANET. The message passing is authenticated using both symmetric and asymmetric keys, and arrival conflict is diminished by time stamp. Each node being a proxy server can allocate unique IP address to new node. Each node maintains a unique tuple of own IP address, node ID, and MANET ID for efficient network merging and partitioning. This scheme offers a secure and efficient mechanism of configuring a MANET.

Keywords Mobile ad hoc network · IP address · Routing · IP address configuration · Proxy server · Symmetric and asymmetric key

1 Introduction

MANET has very vibrant characteristics which demand unique identification of each node in network for source destination communication. There is a manual configuration scheme for configuring IP addresses in an ad hoc network. It is well

P. Choudhury (✉) · K. Majumder · D. De
Department of Computer Science and Engineering, West Bengal University of Technology, Kolkata, India
e-mail: poulami.me.13@gmail.com

K. Majumder
e-mail: koushik@ieee.org

D. De
e-mail: dr.debashis.de@gmail.com

© Springer India 2015
L.C. Jain et al. (eds.), *Intelligent Computing, Communication and Devices*,
Advances in Intelligent Systems and Computing 309,
DOI 10.1007/978-81-322-2009-1_2

fitted in small-scale network but not for large network. Centralized server-based and fixed infrastructure network has a secure authenticated dynamic host configuration protocol (DHCP) server. As MANET is dynamic, the network might get partitioned at some times and might also merge. Therefore, manual configuration may lead to conflict of address. Centralized server DHCP cannot be used in a distributed network like MANET. For assigning an IP address, a standard IP addressing protocol should have the following objectives: dynamic IP address configuration, uniqueness, robustness, scalability, security, and adaptivity.

The rest of the paper is organized as follows: A brief note on the related research work in this area is given in Sect. 2. In Sect. 3, the proposed scheme is explained in detail. In Sect. 4, conclusions are presented.

2 Related Work Review

Address configuration schemes for MANET can be classified into three categories: neighbor-based schemes, decentralized schemes, and centralized schemes. The centralized scheme or leader-based allocations are DHCP, agent, or initiator-based allocation schemes. Most of the existing address allocation algorithms for a MANET use duplicate address detection (DAD) mechanism [1, 2] to resolve address conflict in the network. In neighbor-based schemes, a new node is configured by a neighbor node, so it does not suffer from network-wide flood or centralized control. In [3], a root node is also responsible for address reclamation and network merging. Therefore, if a root node fails, then the address configuration system may collapse. Ghosh et al. [4] proposed an ID-based address configuration scheme. The scheme assumes that in the initial state, the first new node in a MANET sets its address as 0. If more than one new node joins a MANET at the same time, then address conflict can occur.

In this paper, both the authentication and uniqueness are taken as a matter of concern for IP address allocation in distributed and dynamic MANET network.

3 Proposed Work

In this paper, we propose a secure and dynamic IP address configuration scheme. It is authenticated mutually, and uniqueness is guaranteed by using ID. We have given an improved solution to the problems that may arise due to host failures, message losses, mobility of the hosts, and network partitioning/merging.

Unique address allocation scheme: Our proposed address allocation algorithm has two parts: (a) for the new node (N_n) (*Algorithm* 1) and (b) for the proxy that assigns IP address (*Algorithm* 2).

A new node at first generates its public key using its hardware address and private keys using a random function and one-time session key. Then, it sends the

Discover message to 1 hop broadcast nodes. And a DiscTimer () is set on. If the DiscTimer expires and no existing nodes are present in MANET, then no Choice (CHOICE) message is received. In cases when such CHOICE message is not received or if the variable counter is greater than the threshold (value 2), then the new node calls self_configure (). This function sets the IP address of the node to 169.y.0.1 (for class B) and generates Node_id and MID using random function. On the other hand, if it receives multiple CHOICE messages with Priority (Pr_p) meant for generating IP address for child, then the highest priority proxy node is chosen by the new node. The priority is set depending on the available IP address in proxy recycle list and its father's recycle list. New node then sends request message REQ to that selected highest priority proxy node along with time stamp. This REQ message is encrypted using both symmetric key cryptography and asymmetric key cryptography. After receiving REQ message from proxy node, it will send a Reply (REP) message with unique IP address to the node N_n (Table 1).

Next, the N_n node generates its node ID using hash function on allocated IP address and public key and sets the MANET ID from father node. After that, the SELECT message is sent to the selected proxy node, and the REQTimer () is set off and acknowledgment timer (ackTimer ()) is set on (Table 2).

After receiving Select message from new node, the proxy n node sends an acknowledgment message along with the next IP value. And simultaneously, it broadcasts NextIP value to the nodes with same x.y.j.* (siblings). After getting the acknowledgment message from the selected proxy node, the new node is considered to be fully configured.

Algorithm 3 shows the steps for generating new unique IP address from IP address of Proxy, if available. Here, the range of IP address that the root proxy (169.y.0.1) can assign is from 169.y.1.1 to 169.y.254.1, and the range of IP address that other proxy can assign is from (i) 2 to 254 for each 169.y.j. as the NextIP (i) is broadcasted to the other nodes after each time of increase in NextIP (i). So there is no limitation for any proxy node to generate next IP. Proxy nodes have no limited range of IP addresses. This scheme requires synchronization between all the nodes in MANET. So the update of NextIP should be taken place consistently (Table 3).

Authentication scheme: Here, in combination of symmetric key cryptography and asymmetric key cryptography, the sender encrypts the message with the symmetric key (one-time symmetric key (K_s)) algorithm and transfers encrypted message and K_s. Sender also encrypts K_s with receiver's public key. This process is called key wrapping. Now, sender puts both the encrypted message and symmetric key together in digital envelop and sends the digital envelop to the receiver. Receiver receives, and after opening, the digital envelop uses the same asymmetric key algorithm as was used by sender and own private key to decrypt the logical box that contains the one-time symmetric key (K_s). Finally, receiver applies the same symmetric key algorithm as was used by sender and symmetric key (K_s) to decrypt the encrypted message. Each Request message has the priority, and the priorities of any two Request messages are compared according to the following algorithm: If the time stamp in one Request message T_n is earlier than the one in another Request message T_n', then T_n has higher priority than T_n'.

Table 1 Algorithm 1: IP address allocation for new node N_n

1. **begin**	26. Select OfferIP from REP message and check
2. Set threshold ← 2; begin ← true;config←	the TimeStamp $(T_n +1)$ nearest higher to its
false;counter ←1;	own TimeStamp;
3. Generate PublicKey← K_{n1};PrivateKey←	27. Generate node_id$_T$ = H(Pr_IP, K_{p1});
K_{n2};SymKey← K_s; // *public key is the hard-*	28. Generate SignN(REP, K_{p1});
ware address of the node	29. **if** SignN = = SignP and node_id$_T$ = =
4. **if** begin = = true and counter ≤ threshold **then**	node_id$_P$ **then**
5. DISC = {my_Hw_add, K_{n1}} and Send to 1	30. Generate node_id$_N$ =H(OfferIP, K_{n1}); //
hop broadcast;	*create node id using allocated IP address and*
6. Start DiscTimer( );	*public key*
7. begin ← false;	31. Set $MID_n = MID_p$;
8. **else**	32. Generate SignN (SELECT(node_id$_N$), K_{p1});
9. self_configure ( ); //*first node in MANET self-*	33. Send (SELECT(node_id$_N$) + SignN) mes-
configure its IP address with X.Y.0.1(class B)	sage to selected proxy;
10. Config = true;	34. Stop REQTimer( );
11. **end**	35. Start ackTimer( );
12. **if** more than one (SignP, CHOICE (Pr_p, PrIP,	36. **else**
K_{p1})) **then**	37. Select next highest probability proxy node
13. Set all the Pr_p in a queue in descending order;	from queue;
14. Select highest probability proxy from queue;	38. **goto step 15**;
15. Generate SignN= (CHOICE( ), K_{n2});	39. **end**
16. **if** (SignP = = SignN) **then**	40. **if** (ACK + SignP) is received from selected
17. Generate SignN(REQ, K_s);	proxy **then**
18. Generate K_{wrap} (K_s, K_{p1}); //*One time Key-*	41. Generate node_id$_T$ = H(IP,KPP);
wrap	42. Generate SignG(K_{P1}, ACK);
19. Send (K_{wrap}, SignN, T_n) message to the proxy	43. **if** SigG = = SigP and node_id$_T$ = = node_id$_P$
node with highest probability; //*$T_n$is the time*	**then**
stamp of new node	44. stop ackTimer;
20. Start ReqTimer( );	45. Config = true;
21. **else**	46. **end**
22. Select next highest probability proxy node	47. **if** timeout(DiscTimer) **then**
from queue;	48. begin ← true; counter ← counter +1;
23. **goto step 15**;	49. **if** timeout(REQTimer) **then**
24. **end**	50. begin ← true; counter ← counter +1;
25. **if** (REP+SignP+(T_n+1)) message from proxy	51. **if** timeout(ackTimer) **then**
node **then**	52. begin ← true; counter ← counter +1;

Departure of a node from MANET: A node in a MANET makes a request for graceful departure to its father node, and the father node allows the node to DEPART and update its recycle list by making the leaving IP address reusable. Dynamic topology, weaker wireless connection, and mobility of devices cause a node to be out of the radio range of MANET. Most of the time a node does graceless departure, which causes shortage of IP address. But by using HELLO message (message of AODV routing protocol) broadcast, a node gets the

Table 2 Algorithm 2: IP address allocation for existing proxy node

1. **Begin**	23. **if** (REP_F(OfferIP)+SignF) message is received
2. Set proxy_pub_key← K_{p1}; proxy_pri_key←	from father **then**
K_{p2};	24. Generate node_id$_T$= Hash(F_IP, K_{F1});
3. **if** DISC( ) message is received from N_n **then**	25. Generate SignP using REP_F(OfferIP) and
4. Set Prob_IP_alloc←Pr_p;	public key of Father;
5. Generate SignP(CHOICE (Pr_p, PrIP, K_{p1}),	26. **if** SignP = = SignF and node_id$_T$= =
K_{n1});	node_id$_F$then
6. Send (CHOICE, SignP) message to the new	27. Generate SignP(REP(OfferIP, con-
node;	fig_parameters), K_{n1});
7. **end**	28. Send (REP+SignP+(T_n+1)) message to N_n;
8. **if** (K_{wrap}, SignN, T_n) message is received from	29. **else**
node N_n**then**	30. Generate SignP (REFUSE, K_{n1});
9. Decrypt K_{wrap} using K_{p2} and get K_s;	31. Send (REFUSE + SignP) message to N_n;
//asymmetric key cryptography decryption	32. **end**
10. Generate SignP(REQ, K_s);	33. **else**
11. **if** SignP = = SignN and $T_n$is closest time-	34. Drop the (REQ, SignN, T_n) message;
stamp **then**	35. **end**
12. **if** free IP address is available in the Recyc-	36. **if** (SELECT(node_id$_N$) + SignN) is received
leList **then**	from N_n **then**
13. OfferIP = minimum IP from RecycleList;	37. Generate SignG(K_{n1}, SELECT(node_id$_N$));
14. Generate SignP (REP(OfferIP, con-	38. Generate node_id$_T$ = H(offerIP, K_{n1});
fig_parameters), K_{n1});	39. **if** SignG = = SignN and node_id$_T$ = =
15. Send (REP+SignP+(T_n+1)) message to N_n;	node_id$_N$ **then**
16. **else if** node is able to generate new IP **then**	40. Generate SignP (K_{P2} ,ACK);
17. OfferIP = generate_IP(my_IP); // unique	41. Send (ACK + SignP) message to N_n;
IP address generation from own IP address	42. Send (NextIP + SignP) to other nodes having
18. Generate SignP(REP(OfferIP, con-	the same x.y.j values;
fig_parameters), K_{n1});	43. **else** // unauthenticated node
19. Send (REP+SignP+(T_n+1)) message to N_n;	44. Drop the (SELECT(node_idN) + SigN)
20. **else**	message;
21. Send the REQ_F message to Father node	45. **exit;**
encrypted by pub_key of Father;	
22. **end**	

Table 3 Algorithm 3: unique IP address generation for new node N_n. **generate_IP(my_IP)**

1. **begin**	14. **else if** y!= 0 and j= =0 **then**
2. get my_IP←x.y.j.i; Set static count ← 1;	15. J=J+1;
3. Static J ← 0; Static Y ← 0; Static NextIP = 2;	16. **if** J≤ 254 **then**
4. **if** y= =0 and j= =0 **then**	17. **return** NEWIP x.y.J.i;
5. J=J+1;	18. **end**
6. **if** J≤ 254 **then**	19. **else if** y!= 0 and j!= 0 **then**
7. **return** NEWIP x.y.J.i;	20. i=NextIP;
8. **end**	21. NextIP++;
9. **else if** y= =0 and j!= 0 **then**	22. **if** i ≤ 254 **then**
10. Y=Y+1;	23. **return** NEWIP x.y.j.i;
11. **if** Y≤ 254 **then**	24. **else**
12. **return** NEWIP x.Y.j.i;	25. IP address is not available;
13. **end**	26. **end**

Table 4 Algorithm 4: *partition handler*

1. **begin**	8. config = false;
2. Set my_ip$\leftarrow$x.y.j.i;	9. call the address allocation algorithm as the new
3. Set my_nid$\leftarrow$node_id$_x$;	node N_n
4. Set my_MID$\leftarrow$MID$_x$;	10. **else**
5. **if** HELLO( ) message is received from other	11. Set my_MID$\leftarrow$MID$_y$;
partition with MID$_y$**then**	12. **else**
6. **if** MID$_x$<MID$_y$**then**	13. Set my_MID$\leftarrow$generate new MID$_x$ which is
7. **if** (number of neighbor of node (child node)	greater than MID$_x$;
with MID$_x$)<(number of neighbor of node	**end**
(child node) with MID$_y$)**then**	

information of the IP of the left node (i.e., when REPLY to the HELLO message is not received from any node that reflects that the node has left the MANET).

Partition and merging in MANET: Graceless/graceful departure or mobility of MANET leads to network partitioning. So, it generates another MID for it and broadcasts to the neighbors in the new range and make a new MANET network but with same IP address.

Each node in MANET is uniquely identified by a tuple which consists of (IP address, Node_id, MID). Then, the MID is checked, and if it is different, then partition handler (*Algorithm* 4) algorithm is followed. Let a MANET with MID$_x$ gets a HELLO () message from the other partition having MID$_y$. If the MID$_x$ is lesser than the value of MID$_y$, then the number of nodes in each MANET is compared. The MANET having lesser number of existing nodes will configure the nodes in it using address allocation algorithm or else set the MID with greater value (Table 4).

4 Conclusion

In our scheme, each node acts as proxy node capable of allocating and generating unique IP for a new node. So the DAD is not required. The calculation cost and overhead of each node are decreased as the highest priority proxy node only sends the unique IP address to the new node. By applying time stamp, arrival conflict is diminished. Every mobile node in MANET should maintain a table of IP address, status of IP address (allocated, free, father), and public key of allocated node in a MANET. This table increases the accessibility of MANET as each node is aware of state of the MANET nodes. This is indeed a low-cost addressing scheme and authenticated too. The security threats are avoided by the proposed authentication scheme. The authenticated node can only get the IP address while joining the network.

References

1. Vaidya, N.H.: Weak duplicate address detection in Mobile Ad Hoc Networks. In: Proc. ACM International Symposium on Mobile Ad Hoc Networking and Computing (MobiHoc02), pp. 206–216 (2002)
2. Weniger, K.: Passive duplicate address detection in Mobile Ad Hoc Networks. In: WCNC, (Florence, Italy) (2003)
3. Al-Mistarihi, M., Al-Shurman, M., Qudimat, A.: Tree based dynamic address auto configuration in mobile ad hoc networks. In: Elsevier, Computer Networks, vol. 55, pp. 1894–1908 (2011)
4. Ghosh, U., Datta, R.: A secure dynamic IP configuration scheme for mobile ad hoc networks, Ad Hoc Networks. In: Elsevier Journal Published, vol. 9(7), pp. 1327–1342 (2011)

A Novel Distributed Dynamic IP Configuration Scheme for MANET

Pritam Bag, Koushik Majumder and Debashis De

Abstract In mobile ad hoc network (MANET), unique address allocation is a major task in order to ensure proper routing. Most of the address configuration schemes are based on duplicate address detection (DAD) which requires lots of message passing. In this paper, we have proposed a novel address allocation technique based on simple mutual authentication where each node is capable of generating unique IP address for a new node. So, DAD is not required. This causes significant reduction in the number of message which ultimately results in fewer networks overhead. After a successful allocation, allocating node broadcasts some information about the new node and therefore, the new node is authenticated by the existing node. A meaningful ID is used for each node. Whenever a new node wants to join a MANET, it can choose the network based on its requirement depending on the ID. Our scheme also handles the network partitioning and merging efficiently.

Keywords Mobile ad hoc network · DHCP · DAD · IP address · Security · Uniqueness

P. Bag (✉) · K. Majumder · D. De
Department of Computer Science and Engineering, West Bengal University of Technology, Kolkata, India
e-mail: pritambag@yahoo.in

K. Majumder
e-mail: koushik@ieee.org

D. De
e-mail: dr.debashis.de@gmail.com

© Springer India 2015
L.C. Jain et al. (eds.), *Intelligent Computing, Communication and Devices*,
Advances in Intelligent Systems and Computing 309,
DOI 10.1007/978-81-322-2009-1_3

1 Introduction

Mobile ad hoc network (MANET) is an infrastructure less, self-configured network, where two or more devices can be connected through wireless medium. If two nodes in the network are situated in the radio range of each other, they can communicate directly and if not, then with the help of other nodes. In each network, unique identity is required, if two nodes communicate via other nodes and for unicast communication. Address allocation is done for that purpose. This requirement is also same for MANET. In traditional network, dynamic host configuration protocol (DHCP) is used, where a centralized DHCP server is responsible for address allocation and free address maintenance. But it cannot be used in MANET due to its infrastructure less nature. As MANET can be deployed anywhere, anytime and cannot use centralized schemes, address allocation is a challenging task to handle on. An ideal address configuration scheme for MANET should be designed with taking care of scalability, robustness, and authentication.

The rest of this paper is organized as follows. Section 2 defines and analyses the other works in this field. Our scheme is proposed and defined in Sect. 3, and conclusion is given in Sect. 4.

2 Literature Survey

Traditional address allocation schemes for MANET are based on duplicate address detection (DAD). In this technique, whenever a new node wishes to join the network, it chooses an IP address and sends a DAP message to the other nodes already in the network. If within a specified time, it receives an ACN from any node, then it chooses another IP and does the same until it does not receive any ACN within the specified time [1]. DAD has limitations that it requires a large number of message passing and if the replay fails to reach the destination, then the new node thinks that it has a unique IP and starts functioning. DHCP [2] is one scheme, where client–server architecture is used and a DHCP server is responsible for address allocation. Decentralized best effort allocation is based on DAD for checking the uniqueness of the address. Weak DAD [3] and passive DAD [4] are also variations of DAD, where duplicate address can be tolerated as long as packet reaches the correct destination. The second one uses periodic link state routing information to notify other nodes about their neighbours. It may cause broadcast storm problem [5]. Another scheme [1] is based on ID, where each node is capable of generating unique IP based on a simple formula. This scheme takes care of the small number of message passing, but a large computation is used for authentication purpose which may increase the delay. And if one side of the allocation tree increases, then the delay is prolonged because the allocating node has to contact its parent for addresses. Another variation of this scheme is MMIP [6]. In MMIP, MAC address is mapped with the allocated IP address.

3 Proposed Scheme

In our proposed scheme, whenever a new node wants to join the network, it can take its IP address from any existing node in the network because every existing node is capable of generating new IP address for the new node. And each node can generate a large number of addresses.

We have used a 20-bit ID which consists of four parts, MANET type, class of IP being used, MANET id, and node id. MANET type is a 2-bit id which is decided by the first node or the root node of the MANET whether it is a vehicular MANET (VANET) or an iMANET (MANET with connection to internet) or otherwise. 00 stands for otherwise, 01 for VANET, 10 for iMANET, and 11 for iVANET. Next 2 bits indicate what class of IP address is used. It is decided at the time of deployment, that is, when the first node creates the MANET. If class A IP is used, then 01 is assigned, and if class B, then 10 and 11 in case of class C. It is very helpful for unique address allocation. Next 8 bits stand for MANET id which is also decided by the root node of the MANET, and it is an 8-bit randomly generated decimal number ranging from 0 to 255, used for identification of the MANET. Last 8 bits are used for node id ranging from 0 to 255, a random number generated by individual nodes in the MANET, used for identification of each node in the network. And every node in the network is identified by a unique tuple <ID, IP>. We have also assumed that every node has a small amount of memory required for maintaining recycleLIST. The scheme works as following:

Joining of a New Node: According to our scheme, a new node can join a MANET in two ways—(a) initiated by itself and (b) initiated by any existing node.

For the first one, every existing node in the network sends an ADD message periodically after a given time stamp, which is an 1 hop broadcasted message to know that whether any node within that range wants to join the network or not. For the second one, new node sends a 1 hop broadcast of JOIN_REQ message requesting to start the process. After receiving the message, nodes which are already in the network send a JOIN message. After receiving JOIN or ADD message, the new node sends a JOIN_REP message seeking further communication. Then, the allocator node sends an ALLOCATION message which contains the IP address. After receiving that, new node replies with an ACK message confirming the end of the joining process. After a successful allocation, allocator node broadcasts an ALLOC_SUC message. This message contains the ID and IP of the new node, and therefore, the new node is authenticated by the existing node and further verification of the node is no longer needed. Every message is encrypted for maintain security (Tables 1 and 2).

Self-configuring of a node: Function configure() is used for self-configuring of a new node, i.e., the first node of the network. It is used when any ADD or JOIN message is not received. If any ADD message is not received, then the node sends JOIN_REQ message. If any reply to that JOIN_REQ is not received, then the node is the first node in the network. So it configures itself setting its ID and IP.

Table 1 Algorithm 1: algorithm for the allocating node already existing in the network

1.	Generate public_key ← Kpb, private_key ← Kpv;	18.	**If** free IP available on recycleLIST then
2.	Tsp ← Specified time;	19.	IP_ALLOC = minimum IP from the recyleLIST
3.	Timer T1 ← 0, T_JOIN ← 0;	20.	Else Generate IP_ALLOC = generate_IP();
4.	**If** T1 >Tsp then	21.	Set the final ID_NEWNODE of the node appending the 8 bit node_id with own ID's first 12 bit;
5.	Generate PL = (ID+Kpb);		
6.	(ADD+PL) message 1-hop broadcast;	22.	Encrypt IP_ALLOC and ID_NEWNODE with New nodes public key;
7.	Set T1← 0;		
8.	Else	23.	Generate PL1=(NKpb(IP_ALLOC+ID_NEWNODE));
9.	Increment T1;		
10.	**If** (JOIN_REQ) message received from new node	24.	Send (ALLOCATION+PL1) to the new node;
		25.	**If** (ACK+PL_N1) received then
11.	Generate PL = (ID+Kpb);	26.	Decrypt PL_N1 and get ID and IP
12.	Send (JOIN+PL) to the new node;	27.	**If** ok then goto 29;
13.	Start T_JOIN timer;	28.	Else Goto 24;
14.	If timeout (T_JOIN) then	29.	Generate ALLOC_SUC(IP_NEWNODE+ID_NEWNODE);
15.	End;		
16.	**If** (JOIN_REP+PL_N) message received	30.	Broadcast ALLOC_SUC;
17.	Decrypt new node's public key NKpb, and node_id with own private key and store;	31.	Call dep_child ( );
		32.	End;

Table 2 Algorithm 2: algorithm for the new node that wants to join the network

1.	Generate public_key ← NKpb, private_key ← NKpv;	15.	**If** multiple (JOIN+PL) received then choose any one
2.	Timer T_JOINR ← 0;	16.	Check first 12 bits of the ID
3.	**If** multiple (ADD+PL) received choose any one	17.	**If** ok thenGoto 8;
		18.	Else choose another one and goto 16;
4.	Generate ID from the message;	19.	**If** (ALLOCATION+PL1) received then
5.	Check first 12 bits of the ID	20.	Decrypt ID_NEWNODE and IP_ALLOC with NKpv;
6.	If ok then goto 8;		
7.	ElseIf choose another (ADD+PL) message and goto 4;	21.	Set IP = IP_ALLOC;
		22.	ID = ID_NEWNODE;
8.	Retrieve Kpb from the message;	23.	Call next();
9.	Generate random node_id;	24.	Generate PL_N1 = (Kpb(ID+IP));
10.	Generate PL_N = (Kpb,(node_id+NKpb));	25.	Send (ACK+PL_N1) to the allocating node;
11.	Send (JOIN_REP+PL_N) to the node;		
12.	End;	26.	End;
13.	Else send JOIN_REQ message 1-hop broadcast;	27.	**If** timeout(T_JOINR)
		28.	Configure();
14.	Start timer T_JOINR;	29.	End;

Graceful leaving of a node: If any node wants to leave the network, then it sends a LEAVE message to its allocator node. LEAVE message contains the ID, IP and public key of the node. After receiving the LEAVE message, allocator node decrypts the message by its private key and checks whether it is ok or not. After that, it sends an acknowledgement message back to the node. Allocator node saves the IP in its recycleLIST. After receiving the ACK_LEAVE message, corresponding node gracefully leaves the network. Thus, using this mechanism, IP of the leaving node can again be allocated to any other new node.

Unique Address Generation: Address uniqueness is maintained by our proposed scheme, as each node is capable of generating new unique IP for the new node. First node or the root node decides which class of IP address it is going to use and sets the address class bits accordingly.

Suppose, it decides to use class A IP address, then it configures itself accordingly and also decides its own IP. For class A network, the root node will self-configure its IP as a.0.0.0 according to the algorithm. The root node is capable of generating IP addresses in the range a.1.0.0–a.254.0.0. The nodes in the next level of the address tree will get their IP address as a.b.0.0. They are capable of generating IP addresses from a.b.1.0 to a.b.254.0. And the nodes in the next level are capable of generating IP addresses from a.b.c.1 to a.b.c.254. Although they are allocating in the same range, the conflict can be avoided by the AllocIP variable, which contains the last bit of IP address to be allocated next. The variable AllocIP is updated periodically after an ALLOC_SUC message is received from any node having the same allocation zone (Table 3).

Network Merging and Partitioning: According to our algorithm, after a successful address allocation, allocating node sends ALLOC_SUC message to every node in the network. That means it is a broadcasted message. If ALLOC_SUC message is not received for a long time, then the node thinks that it is partitioned from the network. In this situation, it sends the ALIVE message, which is a 1 hop broadcasted message. After receiving ALIVE message, every node, situated in the radio range, sends ALIVE_REP message to the corresponding node. After receiving ALIVE_REP message, that node is assured that it is not partitioned from the network. If it does not receive any ALIVE_REP message, then it thinks that it is partitioned from the network and self-configures itself as the first node of the network (Table 4).

In case of merging, root of the network sends a MERGE message to the root of the other network. MERGE message contains the ID of the network that wants to be merged. After receiving the MERGE message, root node checks the first 2 bits of the ID. If it is same, then the next 2 bits are checked. Then, again next 8 bits are checked. If any of this checking is failed, then it sends a DENY message as a reply. And if the checking is successful, then network merging is done by notifying other nodes. In this situation, every existing node acts as a new node in the network.

Table 3 Function generate_ip(): for unique address allocation

1.	get my_ip ← a.b.c.d;	16.	**If** d= =0 then
2.	get my_id ← ID;	17.	D=D+1;
3.	cls ← 3ʳᵈ and 4ᵗʰ bit of my_id;	18.	return NEWIP a.b.c.D;
4.	Set static B ← 0, static C ← 0, static D	19.	else call next();
	←0;	20.	D = = AllocIP;
5.	**If** cls = = 01	21.	**If** D ≤ 254 then
6.	if b= = 0 then	22.	return NEWIP a.b.c.D;
7.	B = B+1;	23.	End;
8.	If B≤ 254 then	24.	else
9.	return NEWIP a.B.c.d;	25.	IP is not available;
10.	End;	26.	End;
11.	**If** c= = 0 then	27.	**Else if** cls = = 10 then
12.	C = C+1;	28.	Goto 11;
13.	**If** C ≤ 254 then	29.	**Else if** cls = = 11 then
14.	return NEWIP a.b.C.d;	30.	Goto 16;
15.	End;	31.	End ;

Table 4 Algorithm for network partitioning and merging

Algorithm 3: Algorithm for network partitioning	**Algorithm 4: Algorithm for network merging**
1. if any ALLOC_SUC not received for a long time then	1. If MERGE request received from other partition then
2. set check ← specified value, counter ← 0;	2. check first 2 bits of the ID;
	3. if does not match then
3. send ALIVE message one hop broadcast;	4. send DENY;
	5. End;
4. if ALIVE_REP received then OK;	6. else if the next two bit is same
5. end	7. Check the next 8 bit;
6. else counter ← counter+1;	8. If same then
7. if counter ≤ check goto 3;	9. act as a new node;
8. else call configure ( );	10. notify other nodes;
9. End;	11. Else
	12. End;
	13. Else
	14. End;

4 Conclusion

Our scheme is based on dynamic distributed address allocation where each node is capable of generating IP address for a new node. It is also capable of generating unique IP address. So, DAD is not required. Therefore, the number of message passing is reduced and it reduces the network overhead and requires a less amount of computation; hence, the power consumption is low. We have used a meaningful ID for each node which can distinguish two nodes or two MANETs while situating in one another's radio range. Our algorithm is also capable of handling network merging and partitioning. After a successful allocation of IP address, allocating node broadcasts the IP, ID, and public key of the node which can later help in the routing purpose and the new node is authenticated. As a trusted node has already verified the new node, no further verification is needed. And as corresponding public key is stored, it can be used for later communication with the node. So, it reduces the need for another message asking for the public key of any node.

References

1. Ghosh, U., Datta, R.: A secure dynamic IP configuration scheme for mobile ad hoc networks. Ad Hoc Netw. **9**, 1327–1342 (2011)
2. Droms, R.: Dynamic Host Configuration Protocol. RFC 2131 (1997)
3. Vaidya, N.H.: Weak duplicate address detection in mobile ad hoc networks. In: Proceedings of ACM International Symposium on Mobile Ad Hoc Networking and Computing (MobiHoc02), pp. 206–216 (2002)
4. Weniger, K.: Passive Duplicate Address Detection in Mobile Ad Hoc Networks. WCNC, Florence (2003)
5. Ni, S., Tseng, Y., Chen, Y., Sheu, J.: The broadcast storm problem in a mobile ad hoc network. In: Proceedings of the ACM/IEEE International Conference on Mobile Computing and Networking (MOBICOM), pp. 151–162 (1999)
6. Ghosh, U., Datta, R.: MMIP: A new dynamic IP configuration scheme with MAC address mapping for mobile ad hoc networks. In: NCC (2009)

Dual Security Against Grayhole Attack in MANETs

Ankit D. Patel and Kartik Chawda

Abstract The most critical issue related to the mobile ad hoc networks is the security. Due to various characteristics, MANET is likely to be exposed to the Grayhole attack. This type of attack tends to degrade the network performance by falsifying the route and dropping the packets. The Grayhole attack may take place during the route discovery time as well as during the data transmission time. In this paper, we provide a solution which would mitigate the Grayhole attack during the route discovery phase as well as during the data transmission phase. Our solution detects the attack taking place during any of the two phases.

Keywords MANET · Security · AODV · Grayhole · Promiscuous mode

1 Introduction

Grayhole attack tends to degrade the performance of MANET by sending false replies regarding the route toward the destination [1]. The malicious node behaves as a legitimate node for a specific duration and behaves as a malicious node for remaining time [2]. Thus, Grayhole acts as a slow poison because the probability of the dropping of the packets by the malicious node is highly uncertain [3]. The Grayhole node may behave maliciously during the route discovery time or during the data transmission time. In this paper, we propose a solution that will detect the Grayhole node acting maliciously during the route discovery phase or during the data transmission phase, thus providing dual security against Grayhole attack.

A.D. Patel (✉)
Department of Computer Engineering, Parul Institute of Engineering and Technology, Limda, Vadodara, Gujarat, India
e-mail: ankitpatel.bharuch@gmail.com

K. Chawda
Department of Computer Science and Engineering, Parul Institute of Engineering and Technology, Limda, Vadodara, Gujarat, India
e-mail: rese.paper@gmail.com

© Springer India 2015
L.C. Jain et al. (eds.), *Intelligent Computing, Communication and Devices*,
Advances in Intelligent Systems and Computing 309,
DOI 10.1007/978-81-322-2009-1_4

The rest of the paper is organized as follows: Sect. 2 presents the related work; our proposed approach is explained in detail in Sect. 3; finally, Sect. 4 concludes the paper with the future scope.

2 Related Work

Jhaveri et al. in [1, 4, 5] proposed an approach of peak value calculation. In this approach, the node receiving the RREP packet calculates a peak value based on certain parameters. If the destination sequence number in the RREP packet is greater than the peak value calculated by RREP receiving node, then the node sending that RREP is considered as malicious node. In these schemes, no extra control packets are added; hence, there is no increase in routing overhead. Bindra et al. [6] proposed an approach which makes use of the extended data routing information (EDRI) table. The EDRI table stores the records regarding the number of packets received and the number of packets sent to/from the neighboring node. During the detection of the malicious node, the EDRI entries of two neighbors are matched, and if any mismatch occurs regarding the number of packets, then there is malicious node present. Dhamade and Deshmukh [7] proposed an approach in which the source node stores all the RREPs in a reply request table. Once all the replies are stored, the entries with larger sequence number as compared to the source sequence number are deleted. Then, the entire request reply table is sorted in the descending order, and then, the RREP that is at the top of the table is selected and the route is established through that node. Kariya et al. [8] proposed a course-based detection approach. In this approach, every node monitors the neighboring node in the route. To monitor the neighboring node, every node maintains a buffer named FwdPacketBuffer [8]. Every node before forwarding the packet stores that packet in the buffer and monitors the neighboring node. Once the neighboring node forwards the packet, then that packet is removed from the buffer. In fixed period of time, the monitoring node calculates the overhear ratio, and if that overhear ratio exceeds the threshold value, then the next neighboring node is considered as malicious node.

3 Proposed Methodology

In our proposed approach, we tend to detect the malicious Grayhole node during the route discovery phase as well as during the data transmission phase.

3.1 Route Discovery Phase Detection

In AODV, when the node receives the RREP packet, it checks the destination sequence number in the RREP packet, and if the destination sequence number is greater than the routing table sequence number, then that packet is accepted; otherwise, it is rejected. The malicious node takes the advantage of this fact and replies with the RREP packet having very high destination sequence number. In our approach, the node receiving the RREP packet will compute a threshold value. The threshold value is the highest value that the destination sequence number can take. This threshold value is calculated dynamically on the basis of three parameters: routing table sequence number, number of RREQs sent by the node, and the number of RREPs received by the node. If the destination sequence number in the RREP packet is greater than the threshold value, then that RREP is ignored and is not considered for the route.

The threshold value is calculated by the following formula:

$$\text{Threshold value} = \text{RREQ\_COUNT} + \text{RREP\_COUNT} + \text{RT\_SEQ\_NO}; \quad (1)$$

Figure 1 shows the flowchart for the node receiving RREP packets.

3.2 Data Transmission Phase

It may happen that the malicious node may behave as a genuine node during the route discovery phase. It might not reply with a fake high sequence number RREP. In this case, the malicious node will be involved in the route, and after getting involved in the route, it will start dropping the data packets. In this situation, our route discovery approach would not detect the malicious node.

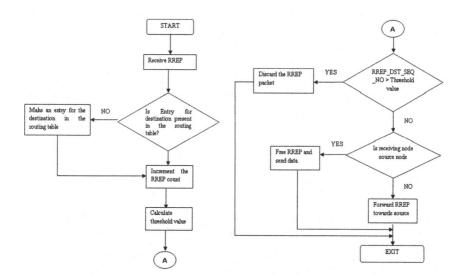

Fig. 1 Flowchart for node receiving RREP

Our proposed approach is that after the route gets established by the above approach, before transmitting the data packets, all the nodes involved in the route will enter in the promiscuous mode. In the promiscuous mode, the node will monitor the next neighboring node in the route. Once the nodes enter in the promiscuous mode, then the data transmission starts. The node forwarding the data packets must maintain the count of data packets sent and should also maintain the count of data packets forwarded by the neighboring node in the route. If the difference between the number of data packets sent by the node and the number of data packets forwarded by the nodes exceeds the particular threshold value, then we can say that the neighboring node of the monitoring node is the malicious node.

In our approach, we maintain the static threshold of 3 % for the detection of the malicious node. We have selected the threshold of 3 % because the normal AODV protocol provides the packet delivery ratio of 97–98 %. If the difference between the number of packets sent and the number of packets forwarded exceeds by 3 %, then the neighboring node that has dropped more than 3 % of the data packets is considered to be malicious node. Once the malicious node gets identified, the monitoring node broadcasts the ALARM packet alerting all the nodes regarding the malicious node.

4 Conclusion and Future Scope

Here, we conclude that the vulnerabilities caused by the Grayhole attack are very serious and they need to be detected as early as possible. The solution presented in this paper provides the dual layer of security. Our approach provides the security during the route discovery phase and the data transmission phase. Thus, the node misbehaving at any of the two phases would get detected in our approach. The theoretical analysis shows that our approach provides high increment of PDR and delay remains as it is in the case of normal AODV protocol. Our future scope involves calculating the dynamic threshold value during the detection of the malicious node in the data transmission phase. Also, we would like to carry out detailed simulations of our approach and compare the results on the basis of PDR, routing overheads, and delays. We would like to optimize our approach in future.

References

1. Jhaveri, R.H., Patel, S.J., Jinwala, D.C.: Improving route discovery for AODV to prevent blackhole and grayhole attacks in MANETs. INFOCOMP. J. Comput. Sci. 11(1), 1–12, 01–02 (2012)
2. Patel, A.D., Chawda, K.: Blackhole and grayhole attacks in MANET. In: International Conference on Information Communication and Embedded Systems (ICICES 2014). IEEE (2014)

3. Patel, M., Sharma, S.: Detection of malicious attacks in MANET a behavioral approach. In: 3rd IEEE International Advance Computing Conference (IACC). IEEE, pp. 388–393 (2013)
4. Jhaveri, R.H., Patel, S.J., Jinwala, D.C.: A novel approach for grayhole and blackhole attack in mobile ad hoc networks. In: Second International Conference on Advanced Computing and Communication Technologies. IEEE, pp. 556–560 (2012)
5. Jhaveri, R.H.: MR-AODV: A solution to mitigate blackhole and grayhole attacks in AODV based MANETs. In: Third International Conference on Advanced Computing and Communication Technologies. IEEE, pp. 254–260 (2013)
6. Bindra, G.S., Kapoor, A., Narang, A., Agrawal, A.: Detection and removal of co-operative blackhole and grayhole attacks in MANETs. In: International Conference on System Engineering and Technology. IEEE, pp. 1–5 (2012)
7. Dhamade, C.S., Deshmukh, H.R.: An efficient way to minimize the impact of the grayhole attack in ad hoc network. Int. J. Emerg. Technol. Adv. Eng. 2(2), 106–110 (2012)
8. Kariya, D.G., Kathole, A.B., Heda, S.R.: Detecting black and gray hole attacks in mobile ad-hoc network using an adaptive method. Int. J. Emerg. Technol. Adv. Eng. 2(1), 37–41 (2012)

I-EDRI Scheme to Mitigate Grayhole Attack in MANETs

Ankit D. Patel, Rutvij H. Jhaveri and Shaishav N. Shah

Abstract Grayhole attack is such an attack which can harm the mobile ad-hoc network (MANET) by misbehaving during data transmission and thus, the network performance is compromised; therefore, it is imperative to detect and prevent this unpredictable attack as early as possible. In this paper, this issue is addressed with a methodology which makes use of the improved extended data routing information (I-EDRI) table to record packet forwarding behavior of neighbors. The proposed method can not only detect a malicious node, but also isolate that node by propagating an alert message to other nodes in the network, and in turn it can improve the network performance significantly.

Keywords MANET · Network layer security · AODV · Grayhole · I-EDRI table

1 Introduction

Grayhole attack is an advancement of the Blackhole attack in which the attacker node tries to fool the sender by sending false information regarding the route to the destination [1]. The malicious node behaves as a legitimate node for a specific duration and behaves as a malicious node for remaining time. In this paper, we propose an improvement over the approach proposed in [2] in which Grayhole

A.D. Patel (✉)
Department of Computer Engineering, Parul Institute of Engineering and Technology,
Vadodara, Gujarath, India
e-mail: ankitpatel.bharuch@gmail.com

R.H. Jhaveri · S.N. Shah
Department of Computer Science and Engineering, SVM Institute of Technology,
Bharuch, Gujarat, India
e-mail: rhj_svmit@yahoo.com

S.N. Shah
e-mail: Shaishavshah1990@yahoo.co.in

© Springer India 2015
L.C. Jain et al. (eds.), *Intelligent Computing, Communication and Devices,*
Advances in Intelligent Systems and Computing 309,
DOI 10.1007/978-81-322-2009-1_5

node is detected by inspecting and recording the packet forwarding behavior of peer nodes using I-EDRI table; the node exceeding the calculated threshold value is isolated by broadcasting ALARM packets in the network. The rest of the paper is organized as follows: Sect. 2 presents the related work; the proposed approach is thoroughly explained in Sect. 3; finally, Sect. 4 concludes the paper with the future scope.

2 Related Work

Jhaveri et al. [1, 3] and Jhaveri [4] proposed schemes in which a node receiving reply packet (RREP) calculates a PEAK value using various parameters; if it finds destination sequence number in the RREP packet greater than the calculated PEAK value, the node is marked as malicious node; the schemes isolate multiple attackers without introducing routing packets to reduce routing overhead. Bindra et al. [2] proposed a method which makes use of extended data routing information (EDRI) table and the refresh packet; every node maintains this table which contains the history of the packets sends and received to/from any neighboring node; the source node compares the EDRI entries of the neighboring nodes and the neighbors of the neighboring nodes and the node with mismatched EDRI entries is marked as a malicious node. Sen et al. [5] proposed a method which makes the use of correspondent node (CN) and the probe packet (PP); an intermediate node (IN) appoints the CN that is found to be most loyal on the basis of the DRI entry; the IN then sends RREQ to its neighbors requesting the route to CN which will receive numerous RREPs; it then sends PP to all neighbors asking them to forward to CN; if the IN receives a negative reply in response to the inquiry about PP, the suspicious value of the node that did not forward the PP gets increased. Sen et al. [6] proposed a method in which the node (IN) generating the RREP has to send the DRI entry of its next hop (NHN). The source node then sends the FREQ request to the NHN. The NHN node replies the FREP with DRI entry of IN. The source node cross checks the entries of IN and NHN and if they match then the node is genuine else IN is malicious.

3 Proposed Approach

We propose an approach that uses of the I-EDRI that is used to record the packet forwarding behavior of neighbors. The Table 1 contains six fields: NODE field represents node IP; FROM field indicates the number of packets that are received from that node; THROUGH field indicates the number of packets that are sent to that node. COUNTER field represents the number of times the node has acted maliciously. STATUS field has binary values in which 0 represents a node is acting as a genuine node for a particular session and 1 indicates a node is detected

Table 1 Improved extended DRI (I-EDRI) table

Node	From	Through	Counter	Status	Timer
B	2	3	2	0	0
C	6	8	1	1	2^2

as malicious for the current session; TIMER field represents the time interval until which the node is considered to be malicious. This value increases with the increase of Counter value.

Table 1 represents an example of I-EDRI table for a node A having neighbors B and C. The first entry in the I-EDRI table states that the node A has received 2 packets form node B and has sent 3 packets to node B. The counter value of 2 represents that node B has been detected as malicious node 2 times. And the Status value stats that node 2 is not behaving maliciously for particular session.

3.1 Proposed Detection Algorithm

1. The source node sends the RREQ packet to the neighboring node.
2. The IN in reply sends the RREP packet and the ID and I-EDRI entry for the next hop node (NHN).
3. The source node now sends further route request (FREQ) to NHN i.e., the NHN on IN via alternate path.
4. The NHN node then sends the FREP packet which consists of the I-EDRI entry of node IN as well as the ID of its NHN.
5. The source node now compares the I-EDRI entries of node IN and node NHN. If the difference between the THROUGH value of IN and the FROM value of NHN exceeds by 5 %, then the IN node is considered to be malicious node.
6. Once a node is detected as malicious, its status value is set to 1, counter value is incremented by 1 and timer value is set according to the counter value.
7. The source node then sends the ALARM packets to all the nodes in the network to alert them regarding the malicious node.
8. Upon receiving the ALARM packet, the node sets the status value for malicious node to 1 and increment the counter value and sets the timer.
9. After the timer time outs, the status value for the malicious node is set to 0 and the malicious node is again considered to be normal genuine node again.
10. This method is carried out by every node. Thus, the node carrying out this procedure becomes source, its next neighbor becomes IN, and the next hop of IN is NHN and the method is repeated.

For example, consider the following network (Fig. 1).

Consider the I-EDRI entries of node IN and node NHN as follows (Table 2 and 3).

Fig. 1 Sample network

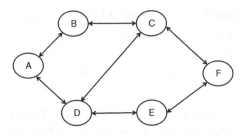

Table 2 I-EDRI entry of node B for node C

Node	From	Through	Counter	Status	Timer
C	56	100	2	0	0

Table 3 I-EDRI entry of node C for node B

Node	From	Through	Counter	Status	Timer
B	90	5	0	0	0

Node A sends the RREQ packet to node B. Node B replies with the RREP packet as well as the I-EDRI entry of the next hop i.e., node C. Node A then sends the FREQ to node C via path A–D–C [2]. Node C then replies with the FREP via same path which contains the I-EDRI entry for node IN [2]. The node A then cross checks the I-EDRI entries of node B and node C. Here the difference between THROUGH and FROM values is $100 - 90 = 10$ and hence the % difference $= 10/100 * 100 = 10\%$; therefore, the node B is considered to be malicious.

Our scheme provides advantages over the scheme explained in [2]. In our approach, we do not need to transmit refresh packet as in [2]. Moreover, as we make use of decimal values in I-EDRI table as compared to binary values in [2], so we can easily detect the Grayhole attack at any time, whereas in [2], only when the suspicious behavior is observed then only the detection procedure starts. Moreover in [2], the values of 0 and 1 in the EDRI table tend to detect the Blackhole attack efficiently, but the Grayhole attack cannot be detected efficiently, whereas our approach the decimal values provides the exact value of the number of packets on the basis of which the Grayhole node can be detected efficiently.

4 Conclusion and Future Work

Security has always been the most talked issue in MANETs. This paper proposes the method which can be used to detect and prevent the Grayhole attack using I-EDRI scheme. The approach can detect the Grayhole nodes even after they have entered in the route between the source and the destination. Our approach detects the malicious node by taking just node other nodes into consideration. In future, we will carry out detailed simulation regarding the work and extend the approach for the detection of the co-operative Grayhole nodes.

References

1. Jhaveri, R.H., Patel, S.J., Jinwala, D.C.: Improving route discovery for AODV to prevent blackhole and grayhole attacks in MANETs. INFOCOMP. J. Comput. Sci. **11**(1), 01–02 (2012)
2. Bindra, G.S., Kapoor, A., Narang, A., Agarwal, A.: Detection and removal of co-operative blackhole and grayhole attacks in MANET's. In: International Conference on System Engineering and Technology. IEEE (2012)
3. Jhaveri, R.H., Patel, S.J., Jinwala, D.C.: A novel approach for grayhole and blackhole attack in mobile ad hoc networks. In: Second International Conference on Advanced Computing and Communication Technologies. IEEE (2012)
4. Jhaveri, R.H.: MR-AODV: a solution to mitigate blackhole and grayhole attacks in AODV based MANETs. In: Third International Conference on Advanced Computing and Communication Technologies. IEEE (2013)
5. Sen, J., Chandra, M.G., Harihara, S.G., Reddy, H., Balamuralidhar, P.: A mechanism for detection of grayhole attack in mobile adhoc networks. IEEE (2007)
6. Sen, J., Koilakonda, S., Ukil, A.: A mechanism for detection of co-operative blackhole attack in mobile ad hoc networks. In: Second International Conference on Intelligent Systems, Modelling and Simulation. IEEE (2011)

Frequency Domain Incremental Strategies Over Distributed Network

Siba Prasad Mishra and Ajit Kumar Sahoo

Abstract This paper presents a frequency-domain adaptive strategy based on incremental techniques. The proposed scheme represents the problem of linear estimation using frequency-domain transformation methods like discrete cosine transform (DCT) and discrete Fourier transform (DFT) in a cooperative manner, where nodes are having the computing ability to find the local estimation in frequency domain and sharing them among the predefined neighbors. This algorithm is distributed and cooperative in nature. In addition to this, it also responds to real-time environments and produces a better result than that of the incremental time-domain adaptive method under colored input. Each node shares information with its immediate neighbors to fully exploit the spatial dimension, thereby lowering the communication burden. Computer simulation result illustrates the performance of the new algorithm.

Keywords Incremental algorithm · Distributed processing · Incremental DCTLMS · Incremental DFTLMS

1 Introduction

Distributed processing is the technique of extracting information from data collected from different nodes spread over a geographical area. In distributed process, nodes collect noisy information, perform local estimation, and then share it with the neighbor node, followed by some defined topology to estimate the parameter of

S.P. Mishra (✉) · A.K. Sahoo
Department of Electronics and Communication Engineering, National Institute
of Technology Rourkela, Rourkela, Odisha 769008, India
e-mail: 212ec6191@nitrkl.ac.in

A.K. Sahoo
e-mail: ajitsahoo@nitrkl.ac.in

© Springer India 2015
L.C. Jain et al. (eds.), *Intelligent Computing, Communication and Devices*,
Advances in Intelligent Systems and Computing 309,
DOI 10.1007/978-81-322-2009-1_6

45

interest. As compared with the centralized solution, distributed solution has better advantage, since it does not require a powerful central processor and extensive amount of communication between node and processor, it only depends upon its local data and the interaction with its immediate neighbors [1]. The distributed processing reduces the communication burden and number of processing [2, 3].

The convergence rate of least mean square (LMS)-type filter is dependent on the autocorrelation matrix of the input data and on the eigenvalue spread of the covariance matrix of the regressor data. The mean square error (MSE) of an adaptive filter using LMS algorithm decreases with time as sum of the exponentials, whose time constants are inversely proportional to the eigenvalue of the autocorrelation matrix of input data [4]. The smaller eigenvalue of autocorrelation matrix of the input results in slower convergence mode, and larger eigenvalues limit the maximum learning rate that can be chosen without encountering stability problem. Best convergence and learning rate results when all the eigenvalues of the input autocorrelation matrix are equal; that is, autocorrelation matrix should be represented in the form of some constant multiplication with the identity matrix [5].

Practically, the input data are colored and the eigenvalues of autocorrelation matrix vary from smallest to the largest. The filter response can be improved by prewhitening the data, but for this, the autocorrelation of the input data should be known. It is difficult to know the autocorrelation of the input data. It can be achieved by using unitary transformation, such as discrete cosine transform (DCT), discrete Fourier transform (DFT), etc. These transformations have decorrelation properties that improve the convergence performance of LMS for correlated input data [5].

Transform domain (which is also called frequency domain) can be applied in two ways: One is blockwise frequency-domain algorithm and other is non-blockwise frequency-domain algorithm. In block wise frequency domain algorithm a block of input data is first transformed then input to the incremental LMS algorithm and in non-block or real-time algorithm the data are continuously transformed by a fixed data-independent transform to de-correlate the input data [5]. DFT-LMS algorithm first introduced by Narayan [6] belongs to a simplest algorithm family because of the exponential nature. But in many practical situations, it was found that DCT-LMS performs better than that of DFT-LMS and other transform domain [7]. In this paper, we interpret the incremental LMS using DCT/DFT algorithm and found that it produces better convergence and performance than previous.

2 Estimation Problem and the Adaptive Distributed Solution

We are interested to estimate the unknown vector w^0 by using the incremental adaptive LMS in frequency-domain method. Let us consider there are N number of nodes having its local zero mean spatial desired data and regressor data d_k and u_k,

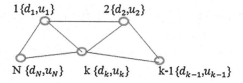

Fig. 1 Distributed network with N nodes accessing space time data

respectively, distributed in a geographical area as shown in the Fig. 1. $k = 1, 2, \ldots, N$, d_k is a scalar, and u_k is regressor vector of size $1 \times M$.

$$\mathbf{U} \triangleq \mathbf{col}\{\mathbf{u_1}, \mathbf{u_2}, \ldots, \mathbf{u_N}\}(\mathbf{N} \times \mathbf{M}) \tag{1}$$

$$d \triangleq \mathrm{col}\{d_1, d_2, \ldots, d_N\}(N \times 1) \tag{2}$$

These quantities collected data across all nodes; the objective is to estimate the $M \times 1$ vector w that solves the distributed solution [1]. The cost function can be decomposed for each node [1] given by

$$J(w) = \sum_{k=1}^{N} J_k(w) \tag{3}$$

Now, let $\phi_k^{(i)}$ be the local estimate of w^0 at node k and time i and let the initial weight assigned at node 1 is $\phi_0^{(i)} \leftarrow w_{i-1}$, and after complete one cycle across node, at the last node, i.e., at node N, it will coincide with w_i, according to steepest descent solution [1]

$$\phi_k^{(i)} = \phi_{k-1}^{(i)} - \mu[\nabla j_k(w_{i-1})]^* \tag{4}$$

Still, it is not purely distributed solution, since in whole updating process, it uses global weight information w_{i-1} in order to find $\nabla J_k(w_{i-1})$; hence, to make it be distributed perfectly, we can use the incremental gradient algorithms which uses the local estimate $\psi_{k-1}^{(i)}$ at each node instead of global information w_{i-1}; that is, the (4) can be written as

$$\phi_k^{(i)} = \phi_{k-1}^{(i)} - \mu\left[\nabla j_k\left(\phi_{k-1}^{(i)}\right)\right]^* \tag{5}$$

The distributed incremental LMS algorithm is summarized as

$$\begin{cases} \phi_0^{(i)} \leftarrow w_{i-1} \\ \phi_k^{(i)} = \phi_{k-1}^{(i)} + \mu_k u_{k,i}^*\left(d_k(i) - u_{k,i}\phi_{k-1}^{(i)}\right) \\ w_i \leftarrow \psi_N^{(i)} \end{cases} \tag{6}$$

3 Frequency-Domain Adaptive Distributed Solution

Transform-domain adaptive filter refers to LMS filter whose inputs are preprocessed with a unitary data-independent transformation. The frequency-domain
transformations are DFT and DCT. This preprocessing improves the eigenvalue
distribution of input autocorrelation matrix of the LMS filter, as a result, its
convergence speed increases. In this paper, we use DCT-LMS and DFT-LMS
frequency-domain approach. In an incremental mode of cooperation, each node
uses its spatial data to estimate the local weight and then share it to the neighboring
node. But the proposed algorithm preprocesses with a unitary process the input
regressor prior to processing, then estimates the weight in frequency domain, and
advances it to the next node for future estimation. It is found that the frequency-
domain approach yields better performance than that of previous algorithm, since
this approach transforms the input data to white form and makes eigen spread
equal to unity results, improving convergence and learning ability [5]. Figure 2
represents the data processing structure of incremental transform-domain adaptive
filter.

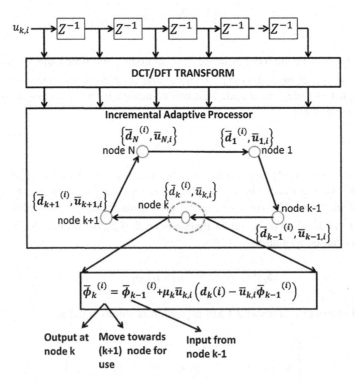

Fig. 2 Data processing structure for transform-domain incremental adaptive LMS algorithm

Table 1 Discrete Fourier transform algorithm

The optimal weight vector w^0 that solves $\min\limits_{w} E\lvert d - uw\rvert^2$ can be approximated iteratively via $w_i = T\bar{w}_i$, T is the unitary DFT matrix $[F_{mk}] = \frac{1}{\sqrt{M}}e^{-\frac{j2\pi mk}{M}}, m, k = 0, 1, \ldots, M-1,$
$S = \text{diag}\{1, e^{\frac{-j2\pi}{M}}, \ldots, e^{\frac{j2\pi(M-1)}{M}}\}, \rho_k(-1) = \epsilon, \bar{w}_{-1} = 0, \bar{u}_{-1} = 0, \text{For } i > 0$
$\bar{u}_i = \bar{u}_{i-1}S + \frac{1}{\sqrt{M}}\{u(i) - u(i - M)\}[1, 1\ldots, 1],$
$\rho_k(i) = \beta\rho_k(i-1) + (1-\beta)\lvert\overline{u_i(k)}\rvert^2, k = 0, 1, \ldots, M-1$
$D_i = \text{diag}(\rho_k(i)), e(i) = d(i) - \bar{u}_i\bar{w}_{i-1}, \bar{w}_i = \bar{w}_{i-1} + \mu D_i^{-1}\bar{u}_i^*, \mu$ is positive step size (usually small) and $0 << \beta < 1$

3.1 DCT-LMS and DFT-LMS Algorithm

The transform-domain algorithms such as DCT-LMS and DFT-LMS [5] are described in (Tables 1 and 2).

4 Simulation

A network consisting of $N = 20$ nodes and each local filter having $M = 10$ taps is considered for simulation study. We take 1,000 iterations and perform 500 independent experiments to get the simulation result. The measurement data $d_k{}^{(i)}$ can be generated by using the data model $d_k{}^{(i)} = u_{k,i}w^0 + v_k{}^{(i)}$ at each node and the vector $w^0 = \text{col}\{1, 1, \ldots, 1\}/\sqrt{M}$, of size $M \times 1$. The background noise is white and Gaussian with $\sigma_v^2 = 10^{-3}$. The excess mean square error (EMSE), MSE, and mean square deviation (MSD) can be plot by using $\lvert u_{k,i}(\bar{\phi}_k{}^{(i)} - \bar{w}^0)\rvert^2$, $\lvert d_k(i) - \bar{u}_{k,i}\bar{\phi}_{k-1}{}^{(i)}\rvert^2$, $\lvert(\bar{\phi}_k{}^{(i)} - \bar{w}^0)\rvert^2$. In this example, the network consists of

Table 2 Discrete cosine transforms algorithm

The optimal weight vector w^0 that solves $\min\limits_{w} E\lvert d - uw\rvert^2$ can be approximated iteratively via $w_i = T\bar{w}_i$, Q is the unitary DCT matrix $[Q_{mk}] = \chi(k)\cos(\frac{k(2M+1)\pi}{2M}), \chi(0) = \frac{1}{\sqrt{M}}, \chi(k) = \frac{\sqrt{2}}{\sqrt{M}}$
$S = \text{diag}\{1, e^{\frac{-j2\pi}{M}}, \ldots, e^{\frac{j2\pi(M-1)}{M}}\}, \rho_k(-1) = \epsilon(\text{a small value}), \bar{w}_{-1} = 0, \bar{u}_{-1} = 0, \text{and repeat for } i \geq 0$
$\delta(k) = [u(i) - u(i-1)]\cos(\frac{k\pi}{2M}), c(k) = (-1)^k[u(i - M) - u(i - M - 1)]\cos(\frac{k\pi}{2M}), \gamma(k) = \alpha(k)[\delta(k) - c(k)], \bar{u}_i = \bar{u}_{i-1}Q - \bar{u}_{i-2}, +, \gamma(0)\gamma(1) [\ldots\gamma(M-1)],$
$\rho_k(i) = \beta\rho_k(i-1) + (1-\beta)\lvert\overline{u_i(k)}\rvert^2 k = 0, 1\ldots m-1$
$D_i = \text{diag}(\rho_k(i)), e(i) = d(i) - \bar{u}_i\bar{w}_{i-1}, \bar{w}_i = \bar{w}_{i-1} + \mu D_i^{-1}\bar{u}_i^*, \mu$ is step size (usually small) and $0 << \beta < 1$

$N = 20$ nodes, with each regressor of size $(1 \times M)$ collecting data by observing a time-correlated sequence $\{u_k^{(i)}\}$, generated as

$$u_k^{(i)} = \alpha_k u_k^{(i-1)} + \beta_k z_k^{(i)}, \ i > -\infty$$

Here, $\alpha_k \in [0, 1)$ is the correlation index and $z_k^{(i)}$ is a spatially Gaussian-independent process with unit variance and $\beta_k = \sqrt{\sigma_{u,k}^2(1 - \alpha_k^2)}$ [1]. The resulting regressor has Toeplitz covariance matrices $R_{u,k}$, with correlation sequence $r_k(i) = \sigma_{u,k}^2(\alpha_k)^{|i|} i = 0, \ldots, M - 1$. The input regressor power profile $\sigma_{u,k}^2 \in (0, 1]$, the correlation index $\alpha_k \in (0, 1]$, and the Gaussian noise variance $\sigma_{v,k}^2 \in (0, 0.1]$ are chosen at random. The algorithms such as DCT-LMS and DFT-LMS, which are described in Sect. 3, are used to update the tap weights at each node. The step size used for all simulation is 0.03. The values of node power profile and correlation index used in this simulation are given by

$$\sigma_{u,k}^2 = [0.2 \ 0.5 \ 0.8 \ 0.1 \ 0.5 \ 0.8 \ 0.7 \ 0.4 \ 0.9 \ 0.9 \ 0.3 \ 0 \ 0.2 \ 0.6 \ 0.2 \ 0.7 \ 0.9 \ 0.4 \ 0.5 \ 0.7]$$
$$\alpha_k = [0.8 \ 0 \ 0.7 \ 0.4 \ 0.8 \ 0.5 \ 0.7 \ 0.4 \ 0.3 \ 0.2 \ 0.2 \ 0.6 \ 0.3 \ 0.5 \ 0.2 \ 0.6 \ 0.4 \ 0.9 \ 0.9 \ 0.6]$$

The convergence rate and performance of MSE, EMSE, and MSD simulation results are shown in Figs. 3, 4, and 5. The simulation result clearly shows that the convergence rate and performance of DCT-LMS using incremental strategies is better than that of the rest.

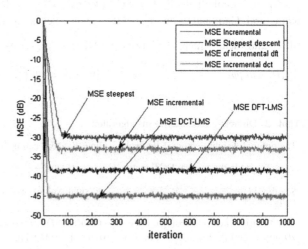

Fig. 3 Transient MSE performance at node 1

Fig. 4 Transient EMSE
performance at node 1

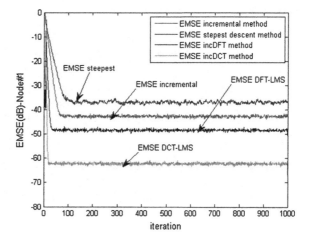

Fig. 5 Transient MSD
performance at node 1

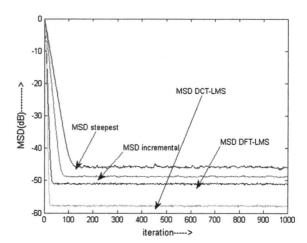

5 Conclusion

The simulation results of the proposed transform-domain incremental adaptive
algorithm not only gives better steady-state performance but also improves the
convergence rate under colored data. These algorithms are very useful for better
convergence of adaptive filter, since it is really impossible to construct a prewhitening filter to produce unity eigen spread.

References

1. Lopes, C.G., Sayed, A.H.: Incremental adaptive strategies over distributed networks. IEEE Trans. Sign. Proces. **55**(8), 4057–4064 (2007)
2. Estrin, D., Pottie, G., Srivastava, M.: Instrumenting the world with wireless sensor networks. In: Proceedings IEEE International Conference Acoustics, Speech, Signal Processing (ICASSP), pp. 2033–2036. Salt Lake City, UT (2001)
3. Wax, M., Kailath, T.: Decentralized processing in sensor arrays. IEEE Trans. Acoust. speech signal process. **Assp-33**(4), 1123–1129 (1985)
4. Widrow, B., Stearns, S.D.: Adaptive Signal Process. Englewood Cliffs, NJ (1985)
5. Sayed, A.H.: Fundamentals of Adaptive Filtering. Wiley, New York (2003)
6. Narayan, S.S., Peterson, A.M., Narsimha, M.J.: Transform Domain LMS algorithm. IEEE Trans. Acoust. Speech Signal Process. **ASSP-31**, 609–615 (1983)
7. Francoise, B.: Transform domain adaptive filters. IEEE Trans. Signal process. **43**(2), 422–431 (1995)

Modeling and Analysis of Cross-Layer Uplink Scheduler in IEEE 802.16e Standard

M. Vijayalaxmi, Swaroop R. Puranik and Linganagouda Kulkarni

Abstract Wireless interoperable microwave access, abbreviated as WiMAX is a standard of IEEE 802.16, a wireless communication technology that has a main goal to achieve the quality of service (QoS) in a secured environment. The Wi-MAX MAC layer is designed to support different types of applications and services having different QoS requirements. Apart from providing high throughput and less delay, a scheduling algorithm at MAC Layer should also take into account the QoS classes and service requirements. We propose an uplink scheduling algorithm for IEEE 802.16e, WiMAX multimedia standard at MAC Layer of the base station, which supports four types of QoS traffic classes (UGS, rtPS, nrtPS, and BE) and focuses mainly developing a mathematical model based on theoretical concepts of statistics. We use NS3 for simulation of the algorithm. Proposed algorithm provides interesting results and thus enhances throughput and minimizes delay and jitter.

Keywords IEEE 802.16e · MAC layer · QoS · Service time · Arrival time

1 Introduction

IEEE 802.16e also called wireless MAN covers a metropolitan area of several kilometers. A WiMAX base station can provide a maximum data rate of up to 70 Mbps, 5–15 km for mobile stations, and wireless access range up to 50 km for

M. Vijayalaxmi (✉) · S.R. Puranik · L. Kulkarni
Department of Computer Science, BVB College of Engineering and Technology, Hubli 580031, Karnataka, India
e-mail: vijum11@rediffmail.com

S.R. Puranik
e-mail: rp.shishu@gmail.com

L. Kulkarni
e-mail: lgk@bvb.edu

© Springer India 2015
L.C. Jain et al. (eds.), *Intelligent Computing, Communication and Devices*,
Advances in Intelligent Systems and Computing 309,
DOI 10.1007/978-81-322-2009-1_7

53

fixed stations. The main advantage of WiMAX over the wired technology is the rapid delivery of services in the remote areas. Thus, in this way, WiMAX networks accelerate the broadband wireless technology by increasing performance and reliability of services offered by ISPs. Some of the main features of WiMAX are incorporated in 4G networks; hence, the proposed algorithm that is described in this paper can also be used in the 4G networks.

1.1 MAC Layer Architecture

MAC layer is divided in to three layers i.e., Convergence Sub layer receives the data from higher layer and forwards to CPS (Common Part Sub-layer) and also sorts the incoming MACSDUs (Service Data Unit) by the connections to which they belong. In MAC CPS, protocol data units (PDUs) are constructed, connections are established and bandwidth is managed. The Security sub layer addresses the authentication, establishment of keys and encryption [3].

1.2 QoS Classes

IEEE 802.16e standard specifies the four quality of service (QoS) traffic classes as mentioned in Table 1. To meet QoS requirements especially for voice and video transmission, with the delay and jitter constraints, the key issue is how to allocate resources among the users.

1.3 Scheduler at MAC Layer

The general scheduler at MAC layer consists of call admission control (CAC) that is a mechanism which determines whether there is sufficient network bandwidth to establish a real-time session of acceptable quality without violating the QoS of already accepted calls. An uplink algorithm at the BS has to coordinate its decision with all the SSs. Mapping refers to transmitting of packets to physical layer.

Table 1 QoS class in IEEE 802.16

Service class	Font size and style
Unsolicited grant service (UGS)	Supports CBR services such as T1/E1 emulation and VoIP without silence suppression
Real-time polling service (rtPS)	Supports real-time services with variable size data on a periodic basis, such as MPEG and VoIP with silence suppression
Non-real-time polling service (nrtPS)	Supports non-real-time services that require variable size data grant bursts on a regular basis, such as FTP
Best effort (BE)	For application that does not require QoS, such as Web-surfing

The reminder to this paper is organized into three sections: Sect. 1, we provided the basic information regarding WiMAX and its QoS. In Sect. 2, we will go through some related works with respect to this paper. In Sect. 3, we will give the architecture and complete design of the proposed scheduler. In Sect. 4, we will develop a theoretical mathematical model based on this proposed scheduler. We will discuss results and analyze it with NS3 simulator. Finally, the conclusions are presented in Sect. 5.

2 Literature Survey

Pujolle and Achir [1] presented case study in order to analyze and discuss several solutions developed to guarantee QoS management of a mobile WiMAX system. The limitation is that it was supported for UGS class even though it was developed for ErTPS, UGS, and rtPS. Guesmi and Maaloul [2] studied the performance of different homogenous algorithms and WRR algorithm guarantees the QoS of each service class in WiMAX networks. The opportunistic algorithms focused on exploiting the variability in channel conditions in WiMAX and maintaining the fairness between the SSs [3]. So, the problem arises while optimizing the scheduling algorithm to uplink the traffic load for the rtPS class. A hybrid scheduler is very efficient instead of a homogenous scheduler. WRR is good for better throughput, and WFQ is best for end-to-end delay and real-time classes [4].

3 Methodology

To provide the solution to the problems defined for uplink traffic, let us follow the three step solution. Scheduler design is very simple as it uses two homogeneous schedulers (WFQ + WRR) to form a hybrid scheduler. Also, we define queues that are implemented in this scheduler. Each user has incoming queues and outgoing queues for each above defined traffic classes. Since we are concentrating on uplink traffic, we bind with outgoing queue.

3.1 Design of Scheduler

In the first stage, each user's traffic class queues are associated to a queue. The scheduler arranges the packets using WFQ algorithm for each traffic class associated with each user. In the second stage, output of the stage 1 WFQ schedulers is en-queued in three queues F1, F2, and F3 where packets of F1 belong to UGS traffic class, F2 belong to rtPS traffic class, and F3 belong to nrtPS traffic class. For

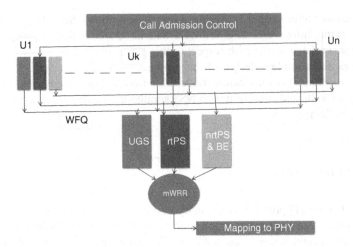

Fig. 1 Proposed scheduler architecture

a particular time slot k given to each traffic class, we will fetch the packets as shown in the Fig. 1. For this type of queuing mechanism of the scheduler, we will develop a model that will be dealt in the next section.

4 Mathematical Model

For the proposed algorithm, the queuing system is shown in Fig. 2. We use the queuing model theories as described in [6]. The queuing mechanism for proposed scheduler has three user queues called stage-1 queues and traffic class queues called stage 2. As mentioned earlier, the packets from stage 1 queues will be sent to stage 2 queues using WFQ algorithm. Then to transmit packets from stage 2 queues, it uses modified weighted round robin (mWRR) algorithm.

Considering the Fig. 2, we will define few terms relating to the queuing theory model.

Let λ be the arrival rate, μ_n be the service rate, $S \rightarrow$ switch over time.

Let $P_n \rightarrow$ offered time $\rightarrow \rho = \lambda/\mu$, λ_{ni} be arrival rate of user queue, μ_{ni} be the service rate of user queue, where n denotes the number of users and $i = 1, 2,$ or 3.

Let λ_1, λ_2, and λ_3 be the average arrival time and μ_1, μ_2, and μ_3 be service time at stage 2 queues.

Now consider the Fig. 3. Let λ_{11} and μ_{11} be the arrival and service time for one user with respect to one service class queue with ρ_0 as traffic offered. Let λ_1 and μ_1 be the arrival rate and service rate of one traffic class queue wit traffic offer ρ_1.

From the definition of ρ,

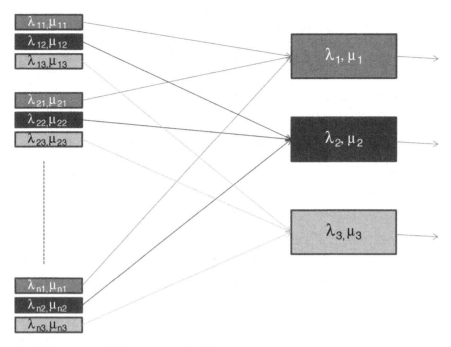

Fig. 2 Network model for proposed scheduler

Fig. 3 Network model for one set of queue

$$\rho_0 = \lambda_1 / \mu_1 \tag{1}$$

$$\rho_0 = \sum_{i=1}^{n} \frac{\lambda_{i1}}{\mu_{i1}} \tag{2}$$

$$\rho_1 = \sum_{i=1}^{n} \frac{\lambda_{i2}}{\mu_{i2}} \tag{3}$$

Since arrival time λ_1 is average service time of all n users of respective user queues,

$$\lambda_1 = \frac{\mu_{11} + \mu_{21} + \mu_{31} + \mu_{41} + \cdots + \mu_{n1}}{n} \tag{4}$$

And in general, we can define as:

$$\lambda_n = \frac{\mu_{n1} + \mu_{n2} + \mu_{n3} + \mu_{n4} + \cdots + \mu_{nm}}{n} \tag{5}$$

where $m = 1, 2$ and 3.

Therefore, we can derive from (5) arrival time for stage 2 queues as follows:

$$\lambda_1 = \sum_{i=1}^{n} \frac{\mu_{i1}}{n} \tag{6}$$

$$\lambda_2 = \sum_{i=1}^{n} \frac{\mu_{i2}}{n} \tag{7}$$

$$\lambda_3 = \sum_{i=1}^{n} \frac{\mu_{i3}}{n} \tag{8}$$

4.1 Probability of Packets Being Served

From Fig. 3, we can model a stochastic process whose queues are specified by pairs (k_0, k_1) where $k_i \geq 0$ and $i = 0, 1$ is the number of packets being served. The change of queue occurs when completion of service at one of the stage queues or upon external arrival. The proof of the following is described in [6]. So, it can be easily shown by direct substitution that following equation is the solution for probability of packets being served.

$$p(k_0, k_1) = (1 - \rho_0)\rho_0^{k0} * (1 - \rho_1)\rho_1^{k1} \tag{9}$$

But before defining in general the probability being served, we must define ρ_0 and ρ_1.

Since ρ_0 is the traffic offered at stage 1 queue,

$$\rho_0 = \lambda_{11}/\mu_{11} \tag{10}$$

Let k_{in} be the number of packets in the queue where n is number of users $n = 1$, $2,\ldots, n$ and $i = 1, 2, 3$. At the stage 2 queue, the traffic offered ρ_1 is given by,

$$\rho_1 = (\lambda_1/\mu_1) * k_1 \tag{11}$$

where k_1 is given by equation

$$k_i = (X_i * T_i)/Q_i \tag{12}$$

X_i is the time slot allocated and T_i is throughput for respective traffic class and $i = 1, 2, 3$.

λ_1 is given by Eq. (6); hence, Eq. (12) can be written as

$$\rho_1 = \left(\left(\sum_{i=1}^{n} \frac{\mu_{i1}}{n} \right) / \mu_1 \right) * k_1 \tag{13}$$

Hence, if we combine both the Eqs. (11) and (13), we get probability of packets being served for Fig. 2, i.e., for one set of queue can be defined as,

$$p(k_0, k_1) = (1 - \rho_0)\rho_0^{k0} * (1 - \rho_1)\rho_1^{k1} \tag{14}$$

$$p(k_0, k_1) = \left(1 - \frac{\lambda}{\mu_0} \right)\left(\frac{\lambda}{\mu_0} \right)^{k0} * \left(1 - \sum_{i=1}^{n} \frac{\mu_{in}}{n} / \mu_1 \right)\left(\sum_{i=1}^{n} \frac{\mu_{in}}{n} / \mu_1 \right)^{k1} \tag{15}$$

Therefore, in general, for the proposed scheduler and for the network as shown in the Fig. 3, the probability of packets being served is given by,

$$\sum_{i=1}^{n} p(k_0, k_1) = \sum_{j=1}^{3} \left\{ \left(1 - \frac{\lambda_i}{\mu_i} \right)\left(\frac{\lambda_i}{\mu_i} \right)^{k_{0j}} * \left(1 - \sum_{i=1}^{n} \frac{\mu_{1ij}}{n} / \mu_{ij} \right) * \left(\sum_{i=1}^{n} \frac{\mu_{1ij}}{n} / \mu_{ij} \right)^{k_{1j}} \right\} \tag{16}$$

From this Eq. (16), we can directly find the number of packets dropped. The probability of packets being dropped is given by,

$$p[\text{Packets Dropped}] = \left\{ 1 - \sum_{i=1}^{n} p(k_0, k_1) \right\} \tag{17}$$

Hence, number of packets dropped is

$$\text{No of Packets} = \left\{ 1 - \sum_{i=1}^{n} p(k_0, k_1) \right\} * TR$$

where TR is number of packets received.

Now if we consider the switch over time S_i that is sum of time required to switch from stage 1 to stage 2 queues and switch over time in stage 1 and stage 2 queues. We must consider this switch over time because to calculate the overall delay in the network. The switch over time for stage 1 queues is,

$$S_1 = \sum_{i=1}^{n} (S_i)$$

The switch over time for stage 2 queues is,

$$S_2 = \sum_{i=1}^{n} (S_i)$$

The switch over time between stage 1 and stage 2 queues is S_{12}, Hence, the overall switch over time is given by,

$$S = S_1 + S_2 + S_{12} \tag{18}$$

4.2 Average Waiting Time

Here, we consider the average waiting time that each packet spends in the queue because we need to find the total delay in the network caused by the scheduler. Hence, this parameter will directly affect on the total delay parameter of QoS in the network. This parameter will also tell that how the network behavior at peak traffic to support multimedia data. The waiting for a packet is given by,

$$W_q = \lambda/(\mu * (\mu - \lambda)) \tag{19}$$

Hence, the waiting time W_q for one traffic class (say UGS) is derived as follows:

$$W_{qu} = W_{s1} + W_{s2} \tag{20}$$

where W_{s1} is waiting time at stage 1 queue and W_{s2} is waiting time at stage 2 queue.

$$W_{qu} = \left\{ \sum_{i=1}^{n} \lambda_i/(\mu_i * (\mu_i - \lambda_i)) \right\}/n + \lambda_u/(\mu_u * (\mu_u - \lambda_u)) \tag{21}$$

Therefore, for other two queues, i.e., for rtPS and (nrtPS + BE), it can be defined as:

$$W_{qr} = \left\{ \sum_{j=1}^{n} \lambda_j/(\mu_j * (\mu_j - \lambda_j)) \right\}/n + \{\lambda_r/(\mu_r * (\mu_r - \lambda_r))\} \tag{22}$$

$$W_{qnb} = \left\{ \sum_{k=1}^{n} \lambda_k/(\mu_k * (\mu_k - \lambda_k)) \right\}/n + \{\lambda_{nb}/(\mu_{nb} * (\mu_{nb} - \lambda_{nb}))\} \tag{23}$$

Hence, for overall network, the average waiting time is given by

$$W_q = \left(W_{qu} + W_{qr} + W_{qnb}\right)/3 + S \tag{24}$$

where S is switch over time as defined in the Eq. (13).

4.3 Average Time Spent by a Packet in Queue W_s

This parameter will help us to analyze the jitter parameter of QoS. We can do this by taking the difference of the successive packets since jitter is defined as the variation of delay. Thus, we can define this by the following equation,

$$W_s = W_{\text{previous}} * \rho_{\text{previous}} - W_{\text{current}} * \rho_{\text{current}} + \left(S_{\text{previous}} - S_{\text{current}}\right) \tag{25}$$

where W_{previous} is the average time spent in the queue by previous packet and W_{current} is the average time spent in the queue by current packet, and S_{previous} and S_{current} average time spent is given by,

$$W = 1/(\mu - \lambda) \tag{26}$$

Hence, from Eq. (21), W_s is given by,

$$\begin{aligned} W_s = \rho_{\text{previous}}/\left(\mu_{\text{previous}} - \lambda_{\text{previous}}\right) - \rho_{\text{current}}/\left(\mu_{\text{current}} - \lambda_{\text{current}}\right) \\ + \left(S_{\text{previous}} - S_{\text{current}}\right) \end{aligned} \tag{27}$$

5 Result Analysis

For the implementation, we used NS-3.18 with bandwidth -10 MHz, AODV as routing protocol, simulation time is 100 s, modulation type is OFDM_QPSK, and varying SSs (0–24).

As we see in Fig. 4, throughput for UGS and rtPS class of proposed scheduler is almost the same throughout the simulation showing the constant nature which ensures that it can have a control over the network at peak traffic. As the number of SS of nrtPS and BE class exceeds 12, the throughput decreases at a faster rate, thereby giving preference to real-time (UGS and rtPS) classes. Variation in delay shows the result for the proposed scheduler in Fig. 5 and as we increase the number of SSs and thereby introducing the heavy traffic for real-time classes.

Variation in jitter shows almost same results as it has shown for delay. As from the Fig. 6, jitter performance is best for UGS class as we see a constant variation of jitter throughout the simulation. Similarly, there is constant variation of jitter in relative comparison of nrtPS and BE classes.

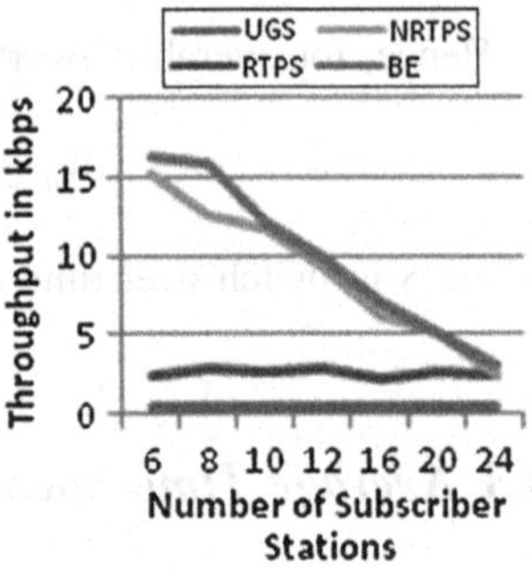

Fig. 4 Throughput variation
for proposed scheduler

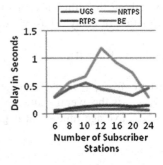

Fig. 5 Delay variation for
proposed scheduler

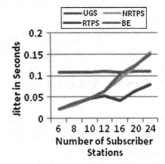

Fig. 6 Jitter variation for
proposed scheduler

6 Conclusion

In this paper, we proposed an uplink scheduler for IEEE 802.16e standard that supports multimedia. For this uplink scheduler, we build a mathematical model that uses theoretical concepts of queuing theory. We derive the equations for probability of packets being served that helps in calculating number of packets dropped, average waiting time for a packets that spent time in waiting in the queue that will define the delay parameter of QoS and lastly average time that a packet spent in the queue that describes the jitter in the network. From the result analysis section, we can conclude that we achieved the objectives of QoS required for IEEE 802.16e standard. Thus, this model describes the QoS parameters that will support real-time traffic classes.

References

1. Pujolle, S.K.G., Achir, K.B.N.: A cross-layer radio resource management in WiMAX systems. In: Hincapie, R. (ed.) ISBN 978-953-307-956-1. InTech Publisher, Published online 03, Feb 2012, Published in print edition February, 2012
2. Guesmi, H., Maaloul, S.: A cross-layer Qos based scheduling algorithm WRR design in Wimax base stations. Am. J. Electr. Electron. Eng. 1(1), 1–9 (2013). doi:10.12691/ajeee-1-1-1. Available online at Science and Education Publishing, http://pubs.sciepub.com/ajeee/1/1/1
3. IEEE 802.16e-2005.: IEEE standard for local and metropolitan area networks: part 16—air interface for fixed and mobile broadband wireless access systems amendment 2: physical and medium access control layers for combined fixed and mobile operation in licensed bands, Feb 2006
4. Puranik, S., Vijayalakshmi, M., Kulkarni, L.: A survey and analysis on scheduling algorithms in IEEE 802.16e (WiMAX) standard. 79(12), ISBN: 973-93-80878-12-4, Publisher IJCA, Oct 2013
5. Ali, D.M., Noordin, K.A., Dimyati, K., Idris, A.: Modeling and performance analysis of uplink scheduling algorithm in mobile WiMAX systems. Int. J. Phys. Sci. 6(14), 3487–3501 (2011)
6. Trivedi, K.: Probability and statistics with reliability, queuing, and computer science applications. ISBN-81-203-0508-6, Aug 1999

A Probabilistic Packet Filtering-Based Approach for Distributed Denial of Service Attack in Wireless Sensor Network

Sonali Swetapadma Sahu and Manjusha Pandey

Abstract Wireless sensor networks (WSNs) are widely used networks that have lured attention of varied research fields due to their numerous ranges of applications. They have limited energy and power consumption, memory, communication, and computation capabilities. They are also distributed and randomly deployed. Due to the above-listed features, they are prone to various security threats and attacks. Distributed denial of service (DDoS) attack is one among them. These attacks aim at flooding the victim with abundant packets so as to exhaust its resources and cripple its capacity to receive desired packets and give its response accordingly. The network becomes congested and the victim becomes either unresponsive leading to denial of service or its response gets delayed. In this paper, we propose a mitigation mechanism that will curb the attempts of the attackers aiming to flood the WSN so as to cause denial of service with multitude of packets within a time span.

Keywords Wireless sensor network · Distributed denial of service · Flooding · Probabilistic packet filtering system · Mitigation

1 Introduction

A wireless sensor network (WSN) is a collection of a number of nodes that has sensing, processing, and communication abilities to monitor the real-world surroundings [1]. It has diverse range of applications such as military surveillance,

S.S. Sahu (✉) · M. Pandey
School of Computer Engineering, Kalinga Institute of Industrial Technology,
Bhubaneswar, India
e-mail: sonali90sahu@gmail.com

M. Pandey
e-mail: manjushapandey82@gmail.com

© Springer India 2015
L.C. Jain et al. (eds.), *Intelligent Computing, Communication and Devices*,
Advances in Intelligent Systems and Computing 309,
DOI 10.1007/978-81-322-2009-1_8

earth/environment monitoring, healthcare monitoring, agriculture, etc. As sensors are cheaper, they are densely distributed across the area to monitor specific events. The WSNs are mostly deployed in public and uncontrolled places; therefore, security issues come to the forefront. Moreover, they have dearth of resources because of limited processing power, memory, and energy. So, they need to be dealt with security mechanisms to prevent them from being attacked by intruders.

A distributed denial of service (DDoS) attack is a combined effort of malicious users to victimize a system by preventing it from functioning as desired by bombarding the victim with lot of packet requests within a stipulated time slot so as to overwhelm the victim, thereby causing denial of service to the legitimate users. The flooding-based DDoS attack has become most dreaded form of attack where the victim network elements are assaulted with huge volumes of attacking traffic.

HCF (Hof Count Filter) [2] is a filtering technique that deals with detection of DDoS attack and dropping of deceptive packets. It finds out the genuineness of a packet by analyzing the number of hops that the packet goes through till it reaches its destination. It creates an IP to hop count mapping table to analyze and then accordingly to identify spoofed packets in the learning state and discard them in the filtering state. Although HCF requires moderate amount of storage, updating IP2HC mapping table after a fix time slot can be hectic [3]. Zhang et al. [4] proposed a model based on marking packets. The attack paths are traced and reconstructed using the information from the packets that are marked. Then, according to the information obtained, attacked packets are filtered. Though the model has reduced cost, high credibility, and low false alarm rate, overhead is incurred to mark every packet coming into the router so as to be able to traceback to the source [5, 6]. DAT [7] is a defense system that monitors users' behavior to determine it is malicious or not. The model provides varied services according to the differently behaving users. Filter, rate limiter, and scheduler are also incorporated to restrict services to malicious users. System's throughput improves significantly and response time and detection accuracy are outperformed. Chen et al. [8] present a Heimdall architecture that defends DDoS attack. It comprises of a puzzle generator that generates puzzle once the attack is detected, victim is identified, and a change aggregation tree is built, puzzle verifier that provides clients with the puzzle along with ID and then verifies the puzzle solution submitted by the clients is correct or not, and a puzzle resolver that finds solution to the puzzle. Although the mechanism safeguards established connections and opens valid channels between the users and the protected server, there is much possibility that the edge routers might become the targets of a coordinated DDoS attack.

2 Proposed Architecture and Methodology

The proposed system is deployed on the sensor nodes of the WSN as shown in Fig. 1. There are attacking nodes whose aim is to flood the network as well as the base station so as to cause denial of service to the legitimate nodes. The proposed

Fig. 1 Probabilistic packet
filtering system

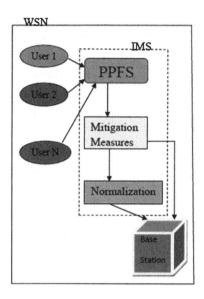

framework will find the malicious nodes and mitigate their efforts in bombarding the network with huge number of packets. PPFS is probabilistic packet filtering system. We assume there are n users including suspicious and malicious users. When users request for certain service from the base station, then those requests are fetched to integrated mitigation system (IMS). IMS's task will be to mitigate DDoS attacks by adhering to certain mitigation measures. According to the packet flow, packet rate is computed. PPFS probabilistically mitigates the incoming traffic flow by following the mitigation methodology. When packets arrive in huge number within a stipulated time, packet rate is reduced depending on the flow. When packet rate falls far below a certain threshold indicating that the attackers have reduced their sending rate, normalization is done to allow them to send packets in a normal fashion. Under normalization, packet rate is increased. Packets from the attackers whose flow rate has been reduced, packets from the previous attacker/s (assuming they have started behaving normally) whose rate has been normalized, and the packets from the genuine nodes are fetched to the base station. Base station then analyzes the requested service and accordingly provides service.

In Fig. 2, Initialization, Packet Sending, and Packet Receiving are functions defined in our mechanism. In Initialization, counters are initialized. Packet Sending function sends packet after a certain timer expires. Timer x is used by intermediate nodes to send packets downstream in a controlled manner after an attacker is identified. Packet Receiving function receives packet from the upper layer nodes. A vector is created for each node to store its related information. When an arrived packet's node id does not match to the node id present in the vector, then that node's record is added to the vector. If the time interval of sending packets is greater than certain threshold t1 (indicating that the packets are not sent by a node in a flooded manner), then it is considered as a normal node.

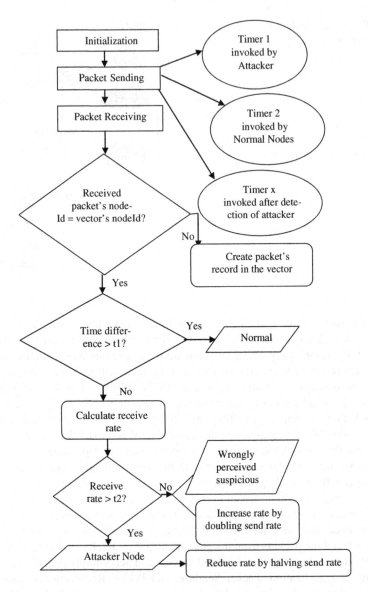

Fig. 2 Flowchart of proposed methodology

Otherwise, receiving rate from that node is calculated. If it exceeds certain threshold t2, then it is considered as an attacker node and the node that receives packets from the attacker reduces its sending rate by halving it (sending rate of normal nodes is taken to be 1). Another case may be that a node might have appeared malicious by observing the rate at which another node receives packets from it. But if gradually that assumed malicious node reduces the rate of

Fig. 3 Packets received and
packets forwarded by
intermediate nodes

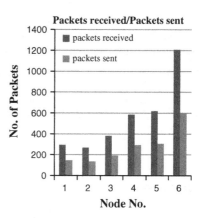

transmitting packets, the node that receives from it increases its sending rate
proportionately so that packets are forwarded in a normal way. If the above case
does not hold true, then it will be inferred that the nodes have been wrongly treated
as suspicious.

3 Result Analysis

In Fig. 3, the graph depicts average number of packets received by nodes from the
attackers and average number of packets it then forwards to downstream nodes. A
total of 20 nodes are taken into consideration and assumed that there are two
attacker nodes that do not receive packets rather only send. Aggregated values are
taken for each 3 nodes. Using our PPFS mechanism, the number of packets further
sent by a node is nearly half of the number of packets it receives from the
attackers. This indicates that the packet flow rate is reduced which gradually
reduces the flooding of packets from the attacker side.

4 Conclusion

DDoS attack has become predominant nowadays. In the WSN, it is imposing a
major threat. Therefore, greater focus is required to overcome the attack. Due to
the influence of the attackers, each node was flooded with tremendous amount of
packets and this in turn flooded the base station. Our work is the first attempt to
mitigate the attack in the WSN. It is observed that the packet receiving rate of the
nodes present in the downstream path of the network reduces considerably with the
reduction in the sending rate. The sending rate has been slashed down

probabilistically by the intermediate nodes. This ensures that the high value of receiving rate from the attackers has been controlled. This way flooding has been minimized, and therefore, the base station can uninterruptedly provide expected service.

In the future work, we would make our system energy efficient [9]. A buffer mechanism can be used that would discard packets after a certain limit from the attackers. This would control congestion and minimize bandwidth consumption in the WSN.

References

1. Pandey, A., Tripathi, R.C.: A survey on wireless sensor networks security. Int. J. Comput. Appl. 3(2), 975–8887 (2010)
2. Mopari, I.B., Pukale, S.G., Dhore, M.L.: Detection and defense against DDoS attack with IP spoofing. In: International Conference on Computing, Communication and Networking, IEEE, 2008
3. Velauthapillai, T., Harwood, A., Karunasekera, S.: Global detection of flooding-based DDoS attacks using a cooperative overlay network. In: Fourth International Conference on Network and System Security, IEEE, 2010
4. Zhang, Y., Wan, Z., Wu, M.: An active DDoS defense model based on packet marking. In: Second International Workshop on Computer Science and Engineering, IEEE, 2009
5. Zhou, Z., Qian, Z., Tian, X., Xie, D.: Fast traceback against large-scale DDoS attack in high-speed internet. In: IEEE, 2009
6. Liang, F., Zhao, X., Yau, D.: Real time IP traceback with adaptive probabilistic packet marking. J. Softw. 14(5), 1005–1010 (2003)
7. Liu, H.I., Chang, K.C.: Defending systems against tilt DDoS attacks. In: The 6th International Conference on Telecommunication Systems, Services and Applications, IEEE, 2011
8. Chen, Y., Ku, W.S., Sakai, K., DeCruze, C.: A novel DDoS attack defending framework with minimized bilateral damages. In: IEEE CCNC, 2010
9. Pandey, M., Verma, S.: Energy consumption patterns for different mobility conditions in WSN. Scientific Research, 2011

Location-Based Coordinated Routing Protocol for Wireless Sensor and Actor Networks

Biswa Mohan Acharya and S.V. Rao

Abstract Wireless sensor and actor networks (WSAN) are composed of low-powered, low-resourced, densely deployed sensor nodes and high-powered, resource-rich, sparsely deployed actor nodes to perform distributed sensing and acting tasks, respectively. Sensor nodes sense the physical phenomenon and report to actor nodes at earliest possible time for appropriate action by the actor nodes. The major objective of WSAN is to have the desired action correspondent to the reported event with higher precession. Due to the coexistence of sensor and actor nodes, one of the major challenges in such networks is communication and coordination mechanisms among the sensor and actor nodes. We have considered the problem of communication and coordination and proposed an efficient model based on geometric structure called Voronoi diagram. Our protocol is based on clustering (virtual grid) and Voronoi region concept. Simulation results demonstrate that the proposed protocol outperforms in terms of throughput, packet delivery ratio, average delay, and normalized routing overhead.

Keywords Wireless sensor and actor networks · Voronoi region · Coordination protocol

B.M. Acharya (✉)
Department of Computer Applications, Institute of Technical Education and Research,
Siksha 'O' Anusandhan University, Khandagiri Square, Bhubaneswar, India
e-mail: biswaacharya@soauniversity.ac.in

S.V. Rao
Department of CSE, IIT Guwahati, Guwahati, India
e-mail: svrao@iitg.ac.in

© Springer India 2015
L.C. Jain et al. (eds.), *Intelligent Computing, Communication and Devices*,
Advances in Intelligent Systems and Computing 309,
DOI 10.1007/978-81-322-2009-1_9

1 Introduction

Wireless sensor and actor networks (WSAN) are capable of observing the physical environment, processing the data, making decisions based on the observations, and performing appropriate actions. Sensor nodes gather information about the physical environment by automatically collaborating to sense, collect, and process data and transmit to some specific actor nodes. Actor nodes are physically equipped with actuation units (mechanical device) in order to take action in the event area. The objective of WSAN is to get the event information from the physical environment, process the sensed data, take decision based on those data, and perform the required action to deal with the detected event effectively. Due to dense deployment of sensor nodes, an event is detected by many sensor nodes and all of them communicate to actor nodes, which may lead to more number of transmissions, congestion in the network, and redundant action by actor nodes. Hence, WSAN need an efficient coordination among sensor and actor nodes.

We address this challenge by proposing a model for WSAN that organizes the WSAN in Voronoi regions such that each region contains only one actor node and it is the nearest to all sensor nodes inside that region. We also reduce the number of communication among the sensor nodes by assuming grid architecture and following sleep/wake schedule among sensor nodes inside a grid that can increase the network life.

2 Literature Review

A comprehensive survey on various research issues of WSAN and the future aspects has been pointed out by the authors in [1]. The authors have discussed the architecture of WSAN and the different issues like sensor–sensor, sensor–actor, and actor–actor coordination along with the research challenges. The unique characteristics of WSAN like real-time requirements and coordination issues that is our main focus is also discussed at length.

A coordination protocol framework for WSAN is addressed with a proposed sensor–actor coordination model based on event-driven clustering paradigm where cluster formation is triggered by an event so that clusters are created on the fly to optimally react to the event and provide reliability with minimum energy consumption [2].

In Ref. [3], coordination and communication problems with mobile actors are studied and a hybrid location management scheme is introduced to handle mobility of actors with minimum energy consumption. Actor nodes broadcast location updates based on Voronoi region as their scope of communication.

Authors in [4] focus on three aspects of coordination namely sensor–sensor coordination, sensor–actor coordination, and actor–actor coordination. As the numbers of nodes in WSAN are quite large, researchers show that sensor network protocols and algorithms should be scalable and localized.

An energy efficient layered routing scheme is described in [5] for semi-automated architecture, where the network field is divided into different sized (overlapped) actor fields, which covers all the sensor nodes. The communication is transformed into two layers. In each actor field, actor is the center of the network and sensor nodes transmit information to the actor node. Actor nodes communicate with other actor nodes and sink directly but sensor nodes communicate with other sensor and actor nodes hop by hop.

A sensor–sensor coordination protocol for WSAN based on clustering and Voronoi region concept is proposed in [6]. This protocol creates clusters consisting of sensors detecting the same event and forwards to the nearest actor. However, the event-driven clustering approach maintains many paths between cluster members and cluster heads, which is possible because of more number of communication. Also, here, there is periodic transmission of information that maintains the path among the sensors.

3 The Procedure

Our procedure consists of three phases, and they are as follows.

3.1 Initialization Phase

Sensor and actor nodes are deployed uniformly throughout the network area under consideration. Actor nodes create Voronoi region around themselves consisting of all the sensor nodes that are closer to itself than any other actor nodes as shown in Fig. 1.

The predefined route from sensor node to actor node reduces the cost of route establishment during occurrence of events or forwarding of event information.

Every sensor node gathers the information regarding its neighboring sensor nodes inside the same grid and associated with the same actor node for node scheduling, which is a major requirement for sensor–sensor coordination.

3.2 Detection and Reporting Phase

When an event occurs, it is detected by the active sensor node inside the grid(s) in which it occurs and transmitted along the optimal path toward the nearest actor node as shown in Fig. 2. This phase refers to sensor–actor coordination in wireless sensor–actor networks. The active sensor nodes in other grids that are one hop away from actor node directly transmit without transmitting to the active sensor node inside actor node's own grid.

Fig. 1 Network area divided
into grids and Voronoi
regions

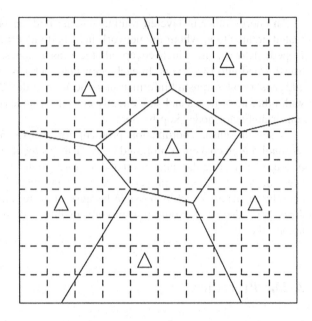

Fig. 2 Greedy forwarding
from sensor to actor

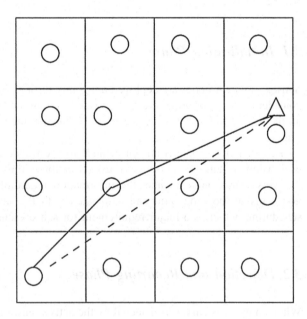

3.3 Action Phase

Actor node aggregates and processes to take the required action. It is the responsibility of actor node to ensure that all the locations are covered. When new event information is received from within its Voronoi region by the actor node after initiating action, it adjusts its power level to take care of the new locations. The objective of our protocol is to send the sensed information at the earliest to the actor node with reduced redundancy and for that we propose the data aggregation approach.

4 Implementation and Result Analysis

To measure our success in meeting the design goals, we simulated our approach with NS-2.33 and analyzed the performance in terms of throughput, packet delivery ratio, and average delay.

We observed the above-mentioned performance metrics with respect to the variation of traffic load (CBR packets) in the network for three different number of actor nodes.

Figure 3 shows the relationship between CBR packet interval (seconds) and average delay (seconds) for three cases of number of actors. With increase in the number of actor nodes, the average delay decreases because with more number of actor nodes, the actor nodes become closer to the event area and hence delay decreases.

Figure 4 describes the relationship between traffic load and throughput (bps) of the network. It clearly shows that more the number of packets sent, more the throughput is. With change in number of actors, there is not that much difference

Fig. 3 CBR versus average delay

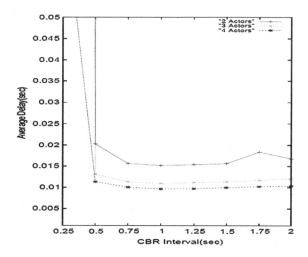

Fig. 4 CBR versus
throughput

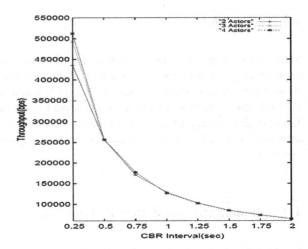

Fig. 5 CBR interval versus
packet delivery ratio

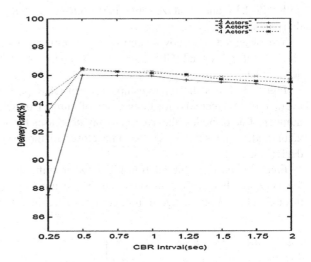

for CBR interval of up to 0.5 from 2.0. But when the CBR interval is 0.25 or more, there is a remarkable change that depicts our limitation of load.

Figure 5 shows the relationship between CBR packet interval (seconds) and the packet delivery ratio (percentage). From the figure, it is quite clear that our assumption of keeping one sensor node active per grid keeps the system connected and there is very less packet drops in the network. Hence, it can be concluded that there is less data loss in our protocol up to the limitation of traffic load. However, when the traffic load is more (CBR interval of 0.25 in this case) than our limitation, there is a sudden drop in delivery ratio. Also, the figure claims that the more the number of actors, more is the delivery ratio because the actors become closer to the event area and hence packet drop rate is reduced.

5 Conclusion

Our approach tries to improve the throughput and packet delivery ratio, while minimizing routing overhead and delay, by choosing a stable path for packet forwarding through static routes.

The protocol emphasizes on coordination among sensor and actor nodes for WSAN based on Voronoi region and virtual grid and reduces congestion and redundancy in the network. The advantage is that as for each sensor node, there is only one closest actor node based on the Voronoi region, the problem of deciding the nearest actor node is removed. Sensor–sensor coordination issue is taken care by dividing the network area into grids and scheduling nodes within each grid to keep only one sensor node active at any time for event detection and reporting. Sensor–actor coordination is achieved by data aggregation among sensors and following simple greedy forwarding approach.

References

1. Akyildiz, I.F., Kasimoglu, I.H.: Wireless sensor and actor networks: research challenges. Ad Hoc Netw. **2**, 351–367 (2004)
2. Melodia, T., Pompili, D., Gungor, V.C., Akyildiz, I.F.: A distributed coordination framework for wireless sensor and actor networks. In: Proceedings of 6th ACM International Symposium on Mobile Ad Hoc Networking and Computing, May 2005
3. Melodia, T., Pompili, D., Akyildiz, I.F.: A communication architecture for mobile wireless sensor and actor networks. In: Proceedings of IEEE SECON, 2006
4. Yuan, H., Ma, H., Liao, H.: Coordination mechanism in wireless sensor and actor networks. In: Proceedings of the First International Multi-Symposiums on Computational Sciences (IMSCCS'06), 2006
5. Peng, H., Huafeng, W., Dilin, M., Chuanshan, G.: Elrs: an energy efficient layered routing scheme for wireless sensor and actor networks. In: Proceedings of the 20th IEEE International Conference on Advanced Information Networking and Applications (AINA'06), 2006
6. Bouhafs, F., Merabti, M., Mokhtar, H.: A coordination protocol for wireless sensor and actor networks. PGP Net (2006)

Energy Saving Performance Analysis of Hierarchical Data Aggregation Protocols Used in Wireless Sensor Network

Bhagwat Prasad Chaudhury and Ajit Kumar Nayak

Abstract Since the inception of wireless sensor networks (WSN), many protocols have been designed and developed to address various issues. Energy efficient data aggregation has been a challenge for researchers working in this field. In this work, we have investigated the performance of some state-of-the-art hierarchical data aggregation protocols used in WSN from the energy conservation perspective when information is gathered at the sink node. The primary objective of data aggregation is to gather and aggregate data in a way that the network lifetime is enhanced as well as power is conserved. Two well-accepted aggregation protocols namely low energy adaptive clustering hierarchy (LEACH) and power efficient gathering in sensor information systems (PEGASIS) are considered for evaluation on a common network scenario. Analysis of these protocols has been done through extensive simulation and is found that the simulation results give an insight into theoretical properties only without providing enough information regarding its practical performance. We conclude with possible future research directions.

Keywords Data aggregation · Routing protocols · Wireless sensor networks · Energy efficiency

B.P. Chaudhury (✉)
Department of Information Technology, Silicon Institute of Technology, Bhubaneswar, India
e-mail: bhagwat@silicon.ac.in

A.K. Nayak
SOA University, Bhubaneswar, India
e-mail: ajitnayak@soauniversity.ac.in

© Springer India 2015
L.C. Jain et al. (eds.), *Intelligent Computing, Communication and Devices*,
Advances in Intelligent Systems and Computing 309,
DOI 10.1007/978-81-322-2009-1_10

79

1 Introduction

With the advancement in technology, sensor networks that are composed of small and cost-effective sensing devices are equipped with wireless radio transceiver for environment monitoring and other applications have become feasible. The key advantage of using these small devices is that they do not require any infra-structure such as electric mains for power supply and wired lines for Internet connections to collect data, nor do they need human interaction while deployment. These sensor nodes can sense the environment by collecting information from their surroundings, and work cooperatively to send the data to a base station, or sink, for further processing and analysis.

The main goal of data aggregation algorithms is to gather and aggregate data in an energy efficient manner so that network lifetime is enhanced. Wireless sensor networks (WSN) provide an increasingly attractive method of data gathering in distributed system architectures and dynamic access through wireless connectivity.

In this paper, we investigate and compare routing algorithms based on module connectivity and data transmission method, energy consumption, and target investigation. In Sect. 2, WSN routing protocols are illustrated; in Sect. 3, we consider the performance evaluation of low energy adaptive clustering hierarchy (LEACH) and power efficient gathering in sensor information systems (PEGASIS) protocols; and finally, in Sect. 4, conclusion and references are provided.

2 Wireless Sensor Network Routing Algorithms

Routing protocols, based on network structure, may be categorized into three classes. Flat routing algorithms use symmetric nodes that have similar function-ality in data gathering, transmission, and power consumption; whereas hierarchical routing algorithms configure nodes into several clusters and in each cluster, the node with higher energy level is selected as the cluster head (CH). Further in location-based routing algorithms, sensors use geographical information to send data to specified regions.

2.1 Flat Routing Protocols

- *SPIN* Sensor protocol for information via negotiation is an adaptive protocol [1] that uses negotiation and resource adaptation to recover flooding method's deficiencies. In this protocol, each node sends a metadata that present attributes of main information and negotiate them with other nodes. By using this method, no additive information will be transferred in network.

- SPIN functionality is divided into three steps: advertise new data, request for data, and sending actual data. When a sensor receives new data from environment, it sends an ADV message to its neighbors. Afterward, if the neighbors require these data, they will send REQ message and finally actual data will be sent to them.

- *Directed Diffusion*: Directed diffusion protocol [2] is used for higher performance and functionality. The sink that is interested in receiving messages floods the interest message in network. Each node that receives the interest message from the neighbor node keeps it in an interest cache table. Each message has a gradient. In the next step, interests with their gradients are sent to other neighbors. By investigating the interest cache, the source node that is the information producer sends the required information to interested neighbor nodes. Finally, the requisite data are received by sink. The sink node sends a positive reinforcement message on the path on which data have been received to amplify it. In this manner, a path is created between source and destination.

- *Rumor Routing*: This protocol is derived from directed diffusion protocol and is a candidate for applications with impossible geographic routing [3]. In this protocol, requests are sent to the nodes that have sensed a specific event, instead of flooding them in the whole network.

2.2 Hierarchical Routing Algorithms

- *LEACH* Low energy adaptive clustering hierarchy is the first propound protocol for clustering in WSN [4]. It uses the idea of rotational clustering method. In this protocol, a network with N nodes and K CHs, with probability of $P_i(t)$, each node will introduce itself as CH in each round. After selecting the CHs, they send advertisement messages with CSMA protocol. These messages are extended in the manner that each node in network receives at least one advertisement message. By receiving the strongest advertisement message, other nodes will join to its cluster.

- In the next step, CHs prepare a TDMA scheduling program to manage data transfer from cluster member nodes. This will prevent from data collision and will also reduce energy consumption. Finally, by receiving TDMA scheduling program in cluster nodes, steady state phase will start. In this phase, nodes send their specific data to CH and CHs receive, aggregate, and finally send them to the destination.

- *PEGASIS*: The power efficient gathering in sensor information systems protocol is proposed as an improvement to LEACH protocol [5]. In this protocol, only one node has direct connection to the sink and the other nodes are connected to the most nearest node to receive required data. The node aggregates the received information with its own data and extracts a packet that is sent to the nearest node on the path. The path selection is done using greedy algorithm and it starts from sink. Generally, data fusion reduces the size of transmitted data from source to destination. When 1–100 % of nodes are dead, this protocol has 100–300 % improvement as compared to LEACH.

- *TEEN* Threshold sensitive energy efficient network protocol is designed to prevent unexpected alterations in environmental parameters [6]. This capability is very important in time-sensitive applications, especially in reaction operation networks. TEEN follows data-centered model and after clusters creation, CHs send threshold levels, one soft and one hard, to their member nodes that are used to receive data. These threshold levels are used to activate nodes in different conditions and will change node status to transfer state.

- *EADAT* An energy aware distributed heuristic called energy aware data aggregation tree has been proposed to construct and maintain a data aggregation tree in sensor networks [7]. The algorithm is initiated by the sink that broadcasts a control message. The sink assumes the role of the root node in the aggregation tree. The control message has five fields: *ID, parent, power, status,* and *hopcnt* indicating the sensor *ID*, its parent, its residual power, the status (leaf, non-leaf node or undefined state), and the number of hops from the sink. After receiving the control message for the first time, a sensor *V* sets up its timer to T_V. T_V counts down when the channel is idle. During this process, the sensor *V* chooses the node with the higher residual power and shorter path to the sink as its parent. This information is known to node *V* through the control message. When the timer times out, the node *V* increases its hop count by one and broadcasts the control message. If a node *U* receives a message indicating that its parent node is node *V*, then *U* marks itself as a non-leaf node. Otherwise, the node marks itself as a leaf node.

The process continues until each node broadcasts once, and the result is an aggregation tree rooted at the sink. The main advantage of this algorithm is that sensors with higher residual power have a higher chance to become a non-leaf node. To maintain the tree, a residual power threshold P_{th} is associated with each sensor. When the residual power of the sensor falls below P_{th}, it periodically broadcasts help messages for T_d time units and shuts down its radio. A child node upon receiving a help message switches to a new parent. Otherwise, it enters into a danger state. If a danger node receives a hello message from a neighboring node *V* with shorter distance to the sink, it invites *V* to join the tree.

2.3 Location-Based Protocols

- *GAF* Geographic adaptive fidelity protocol divides network into virtual grids and, in each grid, nodes run different rules cooperatively [8]. Grid nodes activate a node for a specific period of time after which the node goes into sleep mode. This node is responsible to monitor and report network activities to the sink. Each node by using GPS presents its position in virtual grid. So, GAF saves energy and increases network lifetime by turning off the unnecessary

Table 1 Comparison between LEACH and PEGASIS

	Network lifetime	Organization type	Characteristics
LEACH	High	Cluster	Randomized cluster head rotation, non-uniform energy drainage across different sensors
PEGASIS	Very high	Chain	Global knowledge of the network is required. Considerable energy savings compared to LEACH

nodes in network. The main point in GAF is to select cell dimension in grid in a manner that nodes can coordinate with their neighbor nodes.

- *GEAR* Geographic and energy aware routing protocol uses geographic information to send requests to required regions [9]. The main idea is to send interest messages of directed diffusion to specific regions instead of whole network. By reducing the number of messages, it saves energy as compared to directed diffusion protocol.

We were motivated toward performance analysis of hierarchical routing protocols because, in general, sensor nodes create a local network hierarchy of one or more levels under certain criteria that are busy in aggregating and sending data to the base station. Moreover, hierarchical routing protocols have proved to have considerable savings in total energy consumption. The less the energy consumption, the more the network lifetime is. It has also been demonstrated from practical perspectives that hierarchical routing is an appealing routing approach for sensor networks, and above all, a hierarchical routing infrastructure can be autonomously bootstrapped and maintained by the nodes. In particular, we have chosen LEACH and PEGASIS for comparison. The overview of these protocols is given in Table 1.

2.4 Related Work

From the time WSN are on the floor, researchers are constantly working on routing issues, routing protocols, their performance, and possible modifications for betterment. Most researchers have focused on performance comparison between LEACH and PEGASIS protocols only from the category of hierarchical routing protocols, and some have suggested modifications to the above-stated protocols. Some of the recent developments and comparisons are listed below.

Sittalatchoumy et al. [10] proposed a modified version of PEGASIS to reduce the energy consumption in WSN. They modified the standard PEGASIS protocol and observed that the resulting routing protocol required minimum energy as compared to normal PEGASIS.

Pu-Tai Yang and Seokcheon Lee proposed a hierarchical routing algorithm named spanning tree of residual energy (STORE) for wireless multimedia sensor networks (WMSNs) [11]. The two key factors that they have considered in designing routing algorithms for these networks are as follows: the energy cost for data aggregation and the other being the complex data aggregation behavior that results in various compression ratios. They considered four different types of data aggregation models, and the simulation results indicated that their algorithm achieves a better lifetime in comparison with other traditional routing algorithms such as PEGASIS, LEACH, and many others.

Madheswaran and Shanmugasundaram [12] performed an extensive review on enhancements of LEACH algorithm. Each version of LEACH solved a couple of limitations of its predecessor. Their conclusion states that any future research on LEACH algorithm will aim to use multi-hop communication pattern.

Shankar et al. [13] carried out a performance evaluation of LEACH protocol and showed that energy dissipation in LEACH can be reduced by a factor of 8 and also concluded that LEACH dissipates energy evenly among the sensors.

3 Simulation Setup

A common network scenario has been considered for the evaluation and performance comparison of LEACH and PEGASIS protocol. All simulations are performed using MATLAB in an ideal environment where we assume no connectivity loss, no data loss, no packet loss, etc. These extensive and exhaustive simulations not only give insight into theoretical properties of protocols but also provide enough information for predicting the behavior and practical performances of these protocols.

In our work, we assume a simple model for radio hardware energy dissipation where the transmitter dissipates energy to run the radio electronics and the power amplifier, and the receiver dissipates energy to run the radio electronics. For the experiments described here, both the free space (d2 power loss) and the multi-path fading (d4 power loss) channel models are used, depending on the distance between the transmitter and receiver. If the distance is less than the threshold, the free space (fs) model is used; otherwise, the multi-path (mp) model is used.

In this model, radio dissipates Eelec = 50 nJ/bit to run the transmitter or receiver circuitry and Eamp = 10 pJ/bit/m^2 for the transmit amplifier. From the analysis of these parameter values, it is found that receiving a message is not a low-cost operation and hence the protocols should have to minimize not only the transmit distances but also the number of transmit and receive operations for each message.

The simulation parameters of our model are shown in Table 2.

Some of the important terminologies are described as under.

Table 2 Simulation parameters

Network parameters	Value
Network size	100 m × 100 m
Number of sensor nodes	100–250
Initial energy of sensor nodes	0.25 J
Packet size	2,000 bits
Data aggregation energy	5 nJ/bit
Communication range	100 m
Amplification energy(d > d0)	$Efs = 10$ pJ/bit/m^2
Amplification energy(d ≤ d0)	$Emp = 0.0013$ pJ/bit/m^2

Number of Nodes Alive: It is the total number of sensor nodes having residual energy greater than 10 % of its initial energy.

Instantaneous Network Energy: It is the total energy of the network at time t. It can be computed as,

$$Et = \sum_{i=1}^{n} E_i, \text{ Where } E_i \text{ is the residual energy of node } i$$

Thus, the energy consumed at time t (Ec) can be calculated as the difference between total initial network energy (E) and the instantaneous network energy (Et).

$$Ec = E - Et$$

Round: A composite phase of cluster/chain setup, data aggregation, and data transmission from sensor nodes to sink node.

Our simulations are performed under two different network scenarios. In scenario 1, the network density is fixed (100 nodes) but the round varies. In this scenario, we have calculated the number of nodes alive after each round as well as the network Ec in each round.

In the alternate scenario, the number of rounds is kept fixed (1,000) but the network density is varied from 100 to 250.

3.1 Result and Analysis

LEACH is one of the fundamental and powerful routing protocols that are designed for hierarchical networks, but LEACH weak points led to the design of other protocols such as PEGASIS. In LEACH, data collection is done by forming

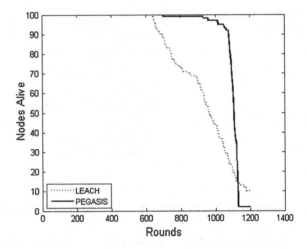

Fig. 1 Nodes alive after each round

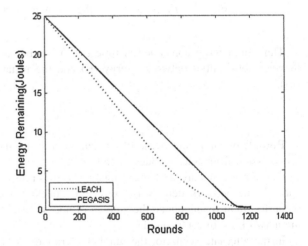

Fig. 2 Network energy consumed after each round. Radio model

clusters. Each cluster member forwards its own-sensed data to the CH. The CH has to aggregate and forward the resultant data to sink. Forming cluster in each round consumes major percentage of network energy. PEGASIS increases network lifetime and decreases overhead on CHs, but it needs to keep the information of neighbor nodes and it leads to network overhead.

Even though we have carried out simulations for 1,200 rounds, we focus our attention till 1,000 rounds because it is found that beyond this value, the network is not performing any useful work. Figure 1 shows the number of nodes that are alive after each round. From the graph, it is evident that PEGASIS keeps the network alive for a longer period of time in comparison with LEACH because it eliminates the

Fig. 3 Nodes alive at
different densities

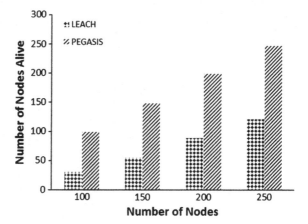

Fig. 4 Energy consumed at
different densities

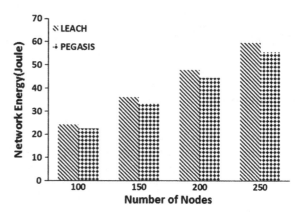

overhead of dynamic cluster formation by limiting the number of transmissions and
receives among all nodes, and using only one transmission to the BS per round.

Figure 2 shows the energy consumption in each round. PEGASIS has low
energy consumption because only one node, the leader, is responsible to transmit
the aggregated data to the sink. The network overhead is thus less in comparison
with LEACH.

The routing protocols have been simulated with different node densities to
know about their nature and effect of density on network overhead. The network
density is increased by increasing the number of nodes in a fixed area as shown in
above figures. Figure 3 shows the number of nodes alive after 1,000 rounds of
iteration for different network densities. It is seen that the number of nodes alive is
always more in PEGASIS than LEACH irrespective of network density. Figure 4
depicts the network Ec at different network sizes for LEACH and PEGASIS. From
the bar graph, it is concluded that the performance of PEGASIS is better and is
independent of network density. Figure 5 indicates the network overhead of the

Fig. 5 Network overhead at different densities

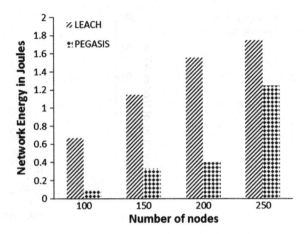

simulated protocol. Undoubtedly, PEGASIS outperforms LEACH but from the graph, it is evident that overhead increases with increase in density.

4 Conclusion

The flexibility, fault tolerance, high sensing fidelity, low cost, and rapid deployment characteristics of WSN create many new application areas for remote sensing that would make sensor networks an integral part of our lives in recent future. This paper investigates and compares routing algorithms in WSN. WSN routing protocols are divided into flat, hierarchical, and location-based protocols. We have presented a comparative study of data aggregation algorithms in WSN. All of them focus on optimizing important performance measures such as network lifetime, data latency, data accuracy, and energy consumption. Efficient organization, routing, and data aggregation tree construction are the three main focus areas of data aggregation algorithms. We have described the main features, advantages, and disadvantages of each of the data aggregation algorithms.

PEGASIS outperforms LEACH by eliminating the overhead of dynamic cluster formation, minimizing the distance that non-leader nodes must transmit, limiting the number of transmissions and receives among all nodes, and using only one transmission to the BS per round. Nodes take turns to transmit the fused data to the BS to balance the energy depletion in the network and preserve robustness of the sensor Web as nodes die at random locations. Distributing the energy load among the nodes increases the lifetime and quality of the network. Our simulations show that PEGASIS performs better than LEACH irrespective of network size. PEG-ASIS shows an even further improvement as the size of the network increases.

References

1. Kulik, J., Heinzelman, W.R., Balakrishnan, H.: Negotiation-based protocols for disseminating information in wireless sensor networks. Wireless Netw. **8**, 169–185 (2002)
2. Intanagonwiwat, J., Govindan, R., Estrin, D.: Directed diffusion: a scalable and robust communication paradigm for sensor networks. In the Proceedings of the 6th Annual ACM/IEEE International Conference on Mobile Networking (MobiCom'00), Boston, MA, Aug (2000)
3. Braginsky, D., Estrin, D.: Rumour routing algorithm for sensor networks. In the Proceedings of the First Workshop on Sensor Networks and Applications (WSNA), Atlanta, GA, Oct (2002)
4. Heinzelman, W., Chandrakasan, A., Balakrishnan, H.: Energy-efficient communication protocol for wireless microsensor networks. In Proceedings of the Hawaii Conference on System Sciences, Jan (2000)
5. Du, K., Wu, J., Zhou, D.: Chain-based protocols for data broadcasting and gathering in sensor networks. In: International Parallel and Distributed Processing Symposium, Apr (2003)
6. Manjeshwar, A., Agrawal, D.P.: TEEN: a routing protocol for enhanced efficiency in wireless sensor networks. In: Proceeding of IPDPS 2001 Workshops, (2001)
7. Ding, M., Cheng, X., Xue, G.: Aggregation tree construction in sensor networks. In: 2003 IEEE 58th Vehicular Technology Conference, vol.4, no.4, pp. 2168–2172, Oct 2003
8. Xu, Y.; Heidemann, J.: Geography-informed energy conservation for ad hoc routing. In: Proceedings of the 7th Annual International Conference on Mobile Computing and Networking (MobiCom '01), Rome, Italy, 16–21 Jul (2001)
9. Yu, Y., Govindan, R., Estrin, D.: Geographical and energy aware routing: a recursive data dissemination protocol for wireless sensor networks. In: Technical Report UCLA/CSD-TR; Citeseer: University Park, PA, USA, (2001)
10. Sittalatchoumy, R., Sivakumar, L.: Performance enhancement of energy efficient protocols for application in WSN. J. Theoret. Appl. Info. Technol., vol. 60 no.1, 10th Feb (2014)
11. Yang, P.T., Lee, S.: Spanning tree of residual energy based on data aggregation for maximizing the lifetime of wireless multimedia sensor networks. In: Proceedings of the 2014 International Conference on Industrial Engineering and Operations Management Bali, Indonesia, Jan 7 9, 2014
12. Madheswaran, M., Shanmugasundaram, R.N.: Enhancements of LEACH algorithm for wireless networks: a review. ICTACT J. Commun. Technol. **4**(4) Dec (2013)
13. Shankar, M., Sridar, M., Rajani, M.: Performance evaluation of LEACH protocol in wireless network. Int. J. Sci. Eng. Res., **3**(1), Jan-(2012)

Performance Analysis of IEEE 802.11 DCF and IEEE 802.11e EDCF Under Same Type of Traffic

Rameswari Biswal and D. Seth

Abstract Mobile ad hoc network is adaptive computing and self-configuring network that automatically form their own infrastructure without support of any base station. The IEEE standard 802.11 protocol is generally used for wireless LAN. This standard 802.11 specifies DCF mode of Mac protocol that does not support quality of service (QoS) as all stations have same priority. To develop the basic performance in real-time application such as video and audio, QOS is essential. So, the standard 802.11e specifies EDCF mode of Mac protocol that enhances the DCF to provide prioritized QOS. Most of the performance of this protocol have evaluated for ideal environment. In this paper, the performance of Mac protocol is evaluated through simulation by using QUALNET software without assigning any priority for non-ideal environment by using performance matrix such as throughput, delay, and jitter.

Keywords MANET · IEEE 802.11 · IEEE 802.11e · MAC protocol · QoS

1 Introduction

In MANET, nodes communicate directly with each other with the help of inter-mediate nodes through single hop or multi-hop to enable data transfer. So, each node acts as host and router. Mobile ad hoc networks are widely used in military and battlefield as they can access anytime in anyplace without fixed infrastructure. The ad hoc networks are advantageous for router free and conserved energy. The

R. Biswal (✉) · D. Seth
School of Electronics Engineering, Kalinga Institute of Industrial Technology University,
Bhubaneswar, Odisha, India
e-mail: rameswaribiswal@gmail.com

D. Seth
e-mail: dsethfet@kiit.ac.in

© Springer India 2015
L.C. Jain et al. (eds.), *Intelligent Computing, Communication and Devices*,
Advances in Intelligent Systems and Computing 309,
DOI 10.1007/978-81-322-2009-1_11

91

issues for designing ad hoc networks are medium access scheme, routing, and quality of service (QOS) provisioning. Mac protocol plays an important role for transmission of packets in shared channel. The provision of QOS is used to determine better network performance and to deliver the information in batter way. QOS can be quantified by maximum throughput, delay, jitter, and minimum bandwidth. Routing protocol maintains path from one node to other node [1]. The IEEE 802.11 wireless local area network is widely used due to simplicity, flexibility, and cost effective.

1.1 IEEE 802.11 Mechanism

This standardized network specifies a fully distributed Mac scheme known as DCF that operates in carrier-sense multiple access collision avoidance (CSMA/CA). CSMA/CA determines the status of the channel, whether the channel is busy or idle. During transmission of frame if the channel is busy, the station defers for an extra time interval DIFS with a back off counter. The frame is transmitted when the counter becomes zero [2, 3]. The receiving station sends ACK frame after a short inter-frame space (SIFS), which is shorter than DIFS indicating successful transmission (Fig. 1).

1.2 IEEE 802.11e Mechanism

DCF provides equal access probabilities among channels with same priorities, which is the main drawback for multiservice network. So, EDCF enhances the DCF to provide different priorities by assigning four access categories (ACs) to each frame. In EDCF, before transmission, an AC senses the medium whether the medium is busy or free. It starts back off process if the medium remains idle for an interval equal to AIFS [AC]. AC freezes back off counter when the channel senses busy. ACK is received by the sender from receiver indicating successful transmission. An AC with smaller values of contention window has higher priority and the AC with large AIFS has lower priority [4, 5].

2 Simulation Model and Performance Analysis

The performance of Mac protocol will be measured through throughput, delay, and jitter by using QUALNET simulator.

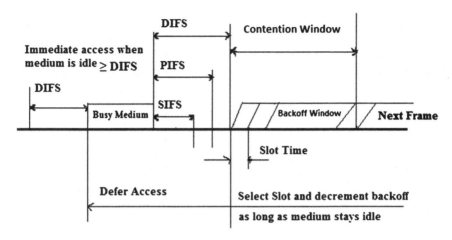

Fig. 1 DCF channel access mechanism [3]

2.1 Simulation Model

In this performance comparison, QUALNET 5.2 [6] network simulator with AODV as routing protocol is used. The performance matrices like throughput, delay, and jitter are calculated. The network scenario is designed using different nodes and mobility, set up with terrain size (1,500 × 1,500 m). The node placement strategy used is uniform. The nodes are mobile and are interconnected through fixed application traffic CBR to compare the performance of MAC protocol. The experiment is continued for different cases and, in all these cases, the value of throughput, delay, and jitter is noted (Fig. 2).

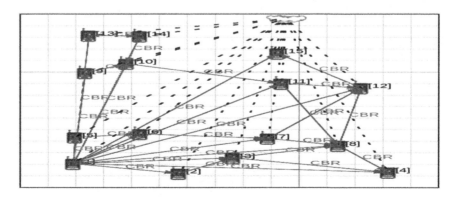

Fig. 2 Run time simulator (10 node, 15 CBR)

Table 1 Simulation parameter

PHY	WIRELESS PHY
Mac protocol	802.11, 802.11e
Propagation model	Two-ray ground reflection
No. of node	5,6,10,12,20,25
Traffic	CBR (15)
Mobility	Low, medium, high
Mobility model	Random waypoint
Maximum speed	4 m/s
Pause time	3 s
Shadowing model	Lognormal with constant (shadowing mean—4.0,10.0,15.0)
Routing	AODV

2.2 Simulation Parameter

The simulation parameters used to set up scenario are listed in the table above (Table 1).

2.3 Results

By using QUALNET 5.2 simulator, the system model is simulated. The values of throughput, delay, and jitter are noted for DCF and EDCF by using different node densities, traffic densities, mobility, and shadowing and then by using these values graphs are plotted using Origin 6.1 software. For performance analysis, we analyzed the throughput, delay, and jitter with respect to number of CBR connection (Figs. 3, 4, 5, 6, 7, 8, 9 and 10).

Fig. 3 Throughput versus CBR

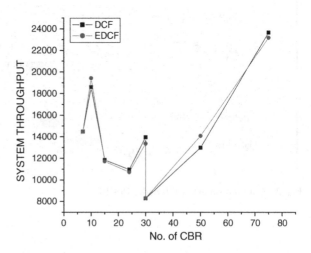

Fig. 4 Delay versus CBR

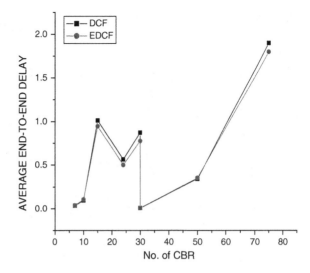

Fig. 5 LM throughput versus CBR

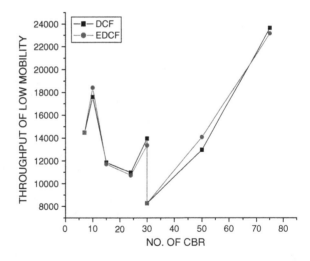

Fig. 6 MM delay versus
CBR

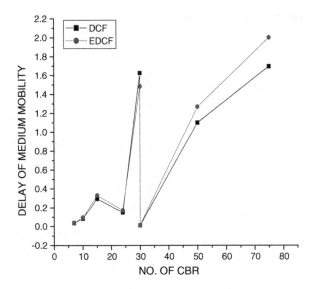

Fig. 7 HM jitter versus CBR

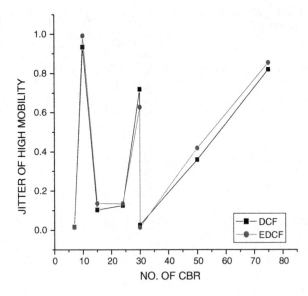

Fig. 8 Throughput versus CBR

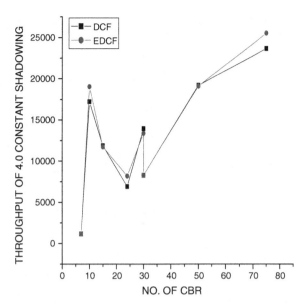

Fig. 9 Delay versus CBR

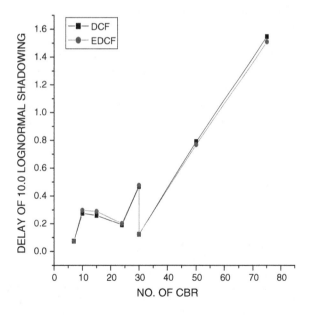

Fig. 10 Throughput versus
CBR

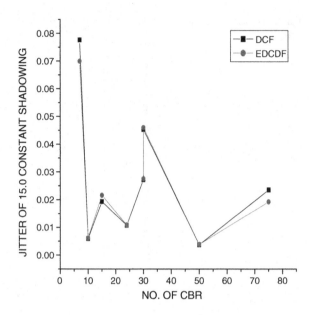

3 Conclusion

Based on the objective of this paper, the details of the literature for DCF and
EDCF were investigated. It was found that DCF does not support QOS where as
EDCF support QOS. The performances of both the assured protocol were evaluated by using QUALNET software. The parameters used to evaluate routing
protocols are node density, traffic density, mobility, and shadowing.

In this investigation, it was found that the performance of DCF and EDCF is
found to be almost same when same type of traffic used without assigning any
priority.

References

1. Manoj, B.S., Siva Ram Murty, C.: Ad hoc wireless networks. In: PEARSON education (2007)
   ISBN 978-81-317-0688-6
2. Kumar, S., Raghavan, V.S., Deng, J.: Medium access control protocols for ad hoc wireless
   networks: a survey. J. Ad Hoc Netw. 4(3), 326–358 (2006)
3. Bianchi, G., Choi, S., Tinnirello, I.: Performance study of IEEE 802.11 DCF and IEEE
   802.11e EDCA. In: University of Roma Tor Vergata, Italy (2007) 63–103
4. Dridi, K., Javaid, N., Djouani, K., Daachi, B.: Performance study of IEEE 802.11e QoS In
   EDCF-contention-based static and dynamic scenarios. In: IEEE International conference on
   Electronics, Circuits, and Systems (ICECS) (2009) 840–843
5. Tamer, N.C., Doudane, Y.G., El Hassan, B.: A complete and accurate analytical model for
   802.11e EDCA under saturation conditions. In: IEEE/ACS International Conference on
   Computer Systems and Applications, AICCSA (2009) 800–807
6. Qualnet Network Simulator–Scalable wireless networks http://www.scalable-networks.com/

A Distributed Prime Node-ID Connected Dominating Set for Target Coverage Using Adjustable Sensing Range

Anita Chandra, Nachiketa Tarasia, Amulya Ratan Swain
and Manish Kumar

Abstract The wireless sensor network (WSN) consisting of a large number of autonomous sensors with limited battery. It is a challenging aim to design an energy efficient routing protocol along with original coverage which can save the energy and thereby extend the lifetime of the network. However, in the context of WSN, connected dominating set (CDS) principle has emerged as the most popular approach for energy efficient routing mechanism in WSNs. In this paper, the target coverage is achieved with adjustable sensing range to the proposed distributed CDS based on prime node-ID (P-CDS) modeled in unit disk graph (UDG) (Wan et al. in Distributed construction of CDS in wireless ad hoc networks, 2002) [1]. P-CDS has time complexity $O(n)$ and message complexity of $O(n)$. Theoretical analysis and simulation results are also presented to verify efficiency of our approach.

Keywords Connecting dominating set (CDS) · Maximal independent set (MIS) · Prime node ID · Target coverage · Adjustable sensing range

1 Introduction

Wireless ad hoc network are characterized by dynamic topology, multi-hop communication and has limited resources. Since there is no fixed infrastructure or centralized management in wireless sensor networks (WSN) and has very less

A. Chandra (✉) · N. Tarasia · A.R. Swain
School of Computer Science, KIIT University, Bhubaneswar, Odisha, India
e-mail: chandraanita1@gmail.com

N. Tarasia
e-mail: ntarasiafcs@kiit.ac.in

A.R. Swain
e-mail: swainamulya@gmail.com

M. Kumar
School of Computer Science, BPUT, Bhubaneswar, Odisha, India
e-mail: manishislampur@yahoo.co.in

© Springer India 2015   99
L.C. Jain et al. (eds.), *Intelligent Computing, Communication and Devices*,
Advances in Intelligent Systems and Computing 309,
DOI 10.1007/978-81-322-2009-1_12

(a)   **(b)**

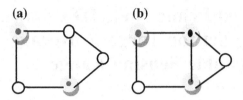

Fig. 1 **a** Dominating set. **b** Connected dominating set

energy a connected dominating set (CDS) has been proposed as virtual backbone through which efficient routing, transmission, activity scheduling, and connectivity can be maintained in network. Nodes in CDS [2] is called as dominator and other nodes are called as dominate. The CDS problem usually has been modeled in unit disk graph (UDG), in which each node has the same transmission range.

In Fig. 1a and b, red nodes represent dominating set (DS) of network and black node is connector to dominators set, collectively it form a CDS. A set is dominating if every node in the network is either in the set or a neighbor of a node in the set. When a DS is connected, it is called a CDS. Since the nodes in a CDS may have heavy load working as the central management agents, minimizing the size of the CDS can greatly help with reducing transmission interference and the number of control messages. This characteristic of wireless networks induces us to take into account another factor of a CDS, the diameter, which is the longest shortest path between any pair of nodes in the CDS. It is desirable to get a CDS with diameter and size as small as possible. To reduce the traffic during communication and prolong network lifetime, it is desirable to construct a minimum CDS (MCDS).

In this paper, we provide a distributed and multiple initiator algorithm for finding MCDS in single phase rather than two different phase in UDG having time complexity of $O(n)$ and message complexity of $O(n)$.

The structure of this paper is organized as follows: Sect. 2 provides related research works on CDS. Section 3 contains proposed distributed P-CDS algorithm. Section 4 deals with the performance analysis. Section 5 will conclude the paper and the future work.

2 Related Work on Connected Dominating Set

In this section, we investigate few algorithmic research works on CDS in WSN. Constructing a MCDS is NP-complete problem [3]. Following are some existing algorithms for construction of CDS. Marathe et al., Guha Khuller, Ruan et al.

proposes centralized algorithm based on graph coloring and spanning tree with performance ratio as 10, $2(1 + H(\Delta))$, $3 + \ln(\Delta)$. Wan et al. [4] proposes a distributed algorithm based on quasi-global information with approximation ratio of 8 and $O(n)$ as time complexity and $O(n \log n)$ as the message complexity, where 'n' is the number of nodes. The other type is to find an initial CDS and then prune some redundant nodes to attain MCDS. Wu et al. [5] proposes a distributed algorithm with message complexity as $O(m)$ and time complexity of $O(\Delta^3)$ and the approximation ratio at most $O(n)$ where, Δ and m are the maximum degree in graph and number of edges, respectively.

3 Proposed Algorithm

3.1 Network Model

In this section, we state our assumption regarding the development of the ad hoc wireless network model.

1. The nodes do not have any topological information in advance. They do not even have the knowledge of location of their neighbors.
2. Each node has a unique node ID along with remaining energy and degree.
3. Node exchange packet to identify their single hop neighbors and ascertain their degree, energy, and node ID.
4. All the nodes are deployed in a 2-D plane, static, uniform, and transmission range is same as modeled in UDG.

3.2 Proposed Approach

Our distributed P-CDS multiple initiator scheme is based on prime node ID, energy, and degree are constructed in single phases. Sink flood INFO_MSG to all nodes to set flag as true if node is prime else false. In this, multiple initiators initiate to construct CDS. Initiators mark itself as BLACK and add themselves in DS and prepare neighbors list with their remaining energy and degree. Initiators broadcast BLACK MSG to its neighbors marking themselves as GRAY node. Among GRAY nodes, the highest remaining energy nodes selected as connector, whereas in case of tie, tie-breaker factor be degree of neighbor nodes, set color to GREEN from GRAY. Now, connector also maintains information about covered and uncovered neighbor list with their remaining energy. GREEN nodes as powerful connectors find maximal independent set in prime node ID of network. There are three different cases while finding MIS in prime nodes-ID of network discussed in Sect. 3.2.1. This process is repeated until every GRAY node have empty uncovered list.

3.2.1 Algorithm: Determination of P-CDS of a Graph

Input: $G(V, E)$—A connected undirected graph where each node has unique ID, energy, and degree. Initially all nodes are white.
Output: DS—Set of all BLACK dominator nodes; C—Set of all GREEN connectors; D—Set of all GRAY nodes are dormant nodes.

1. Initialize P-CDS, DS, and C to $\emptyset$.
2. Multiple initiators initiate in network $G(V, E)$ marked as BLACK and add themselves in DS. DS—DS U $\{u, v, \}$.
3. /*selecting u, v as dominator */
4. $N(u) = \{i, j, k \in V \mid d(u, i), d(u, j), d(u, k) < r\}$ i.e., r-sensing range, N-neighbor nodes. Neighbor nodes of initiators mark as GRAY node.
5. GRAY nodes having highest energy (E_h) marked as connectors, mark as GREEN node. C—C U $\{i\}$.
   /*selecting i as connectors by dominators */
6. Find MIS as DS of graph in prime nodes-ID by powerful connector nodes.
7. There are three possibilities of choosing MIS or dominators:-

   a. If one prime node ID in neighbor of connectors, then add it to DS.
   b. If there are more than one prime nodes-ID in its neighbors, neighbor with the highest remaining energy are marked as BLACK node.
   c. If no prime node-ID in neighbors of connectors then select one non-prime node among all neighbors with the largest remaining energy.

8. Repeat steps 1–6 until Gray node end with empty uncovered neighbors list and form P-CDS with BLACK nodes, GREEN node, and GRAY nodes.
   /*Dominators & Connectors collectively form the P-CDS*/

We know that after forming CDS, many nodes are in sleep mode leaving sensing holes [6] in network. So, active nodes in CDS implement concept of adjustable sensing range [7] for discrete target coverage. In which actives nodes of P-CDS have different sensing ranges $(r_1, r_2, r_3 \ldots r_n)$. As after forming P-CDS in network, it will adjust its required sensing range to cover all targets in network and prevent from coverage hole. Suppose there is redundant energy to any target with two minimum sensing range of two nodes, turn off of the nodes with minimum ranges and can cover it with its only one node with required sensing range. Like this, we can achieve target coverage and connectivity along with extending lifetime in the network.

3.2.2 Diagrammatical Illustration of Algorithm

P-CDS construction is illustrated using graph coloring technique. Initially, all nodes in graph of network are marked as white. BLACK color for dominators, GREEN color for connectors, and GRAY color for sleeping nodes (Fig. 2).

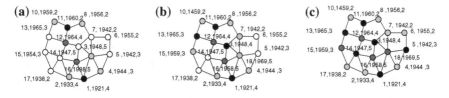

Fig. 2 **a** Node ID 12, 16 are connectors of dominators 1, 11. **b** Node ID 3, 13 dominators of connectors 12, 16. **c** P-CDS constructed

4 Performance Analysis

The performance of the P-CDS is evaluated through simulation on Castalia [8]. The IEEE 802.11 protocol is used at MAC layer. The metrics used for performance evaluation are the total number of connectors in CDS, the lifetime of network after formulation of MCDS automatically increases (Fig. 3).

As expected there is a reduction in connectors' node as in P-CDS there is only one connector selected between pairs of dominators during construction of CDS. In other algorithm after finding MIS, we find out connectors to it using spanning tree or Steiner tree. Like many of the other problems that arise in CDS construction, Steiner tree problem is also NP-Hard. In Steiner tree sometimes we have to select more than one connectors to connect MIS between pairs of dominators.

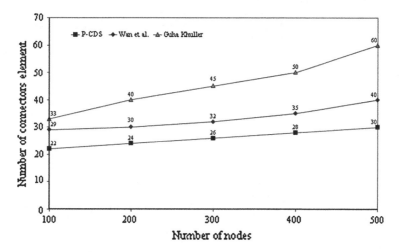

Fig. 3 Shows the number of connectors in backbone structure in P-CDS

5 Conclusion

In this paper, we presented a distributed CDS formed in single phase modeled in UDG called P-CDS. P-CDS has time complexity $O(n)$ and message complexity $O(n)$ and also preserve coverage using adjustable sensing range. The main goal of our algorithm is to construct a maximal independent set in prime node-ID along with powerful connectors in same phase. There is no need to use Steiner tree or spanning tree to interconnect the MIS. Through simulation results, we have seen that selecting only one connector per pair of dominators help to reduce the size of CDS. In addition, choosing a node with largest adjusting sensing range as a dominator can further reduce CDS size. Our future work will be toward CDS construction with disk graphs with bidirectional links (DGB).

References

1. Wan, P-J., Alzoubi, K.M., Frieder, O.: Distributed construction of connected dominating set in wireless ad hoc networks. In: 21st Annual Joint Conference of the IEEE New York, USA, Jun 23–27 (2002)
2. Karbasi, A.H., Atani, R.E.: Application of dominating sets in wireless sensor networks. Int. J. Sec. Appl. **7**(4), Jul (2013)
3. Garey, R.M., Johnson, S.D.: Computers and Intractability: A guide to the Theory of NP-Completeness. Freeman, San Francisco (1978)
4. Cheng, X., Ding, M., Du, D.H.: The construction of connected dominating set in ad hoc wireless networks. In: Wireless Communications and Mobile Computing, (2004)
5. Wu, J., Li, H.: On calculating connected dominating set for efficient routing in Ad hoc wireless networks. In: Proceedings of (DIALM'99), Aug. (1999)
6. Zhao, E., Yao, J., Wang, H., Lv, Y.: A coverage hole detection method and improvement scheme in WSNs. In: IEEE (2011)
7. Cardei, M., Wu, J., Lu, M., Pervaiz, MO.: Maximum network lifetime in wireless sensor networks with adjustable sensing ranges In: IEEE (2005)
8. Castalia A simulator for Wireless Sensor Networks and Body Area Networks Version 3.2 User's Manual

Token Based Group Local Mutual Exclusion Algorithm in MANETs

Ashish Khanna, Awadhesh Kumar Singh and Abhishek Swaroop

Abstract In this paper, a generalization of the group mutual exclusion (GME) problem based upon the concept of neighborhood has been proposed and named as group local mutual exclusion (GLME). The neighborhood is defined based upon the location of shared resources, and no synchronization is required between nodes in two different neighborhoods. A token-based solution of the GLME has also been proposed. The algorithm satisfies safety, starvation freedom, and concurrent occupancy properties. To the best of our knowledge, it is the first token-based algorithm to solve GLME problem in MANETs.

Keywords Ad hoc network · Local mutual exclusion · Neighborhood · Resource allocation

1 Introduction

The efficient and effective use of resources is a highly desirable property in MANETs. Due to dynamic topology, limited bandwidth, limited battery power, and low processing capability, algorithms designed for synchronization in static networks cannot be directly applied in MANETs. Therefore, several algorithms for

A. Khanna (✉)
MAIT, Delhi 110085, India
e-mail: ashishk746@yahoo.com

A.K. Singh
NIT, Kurukshetra, Haryana 136119, India
e-mail: aksinreck@rediffmail.com

A. Swaroop
Galgotias University, Greater Noida, UP, India
e-mail: asa1971@gmail.com

© Springer India 2015   105
L.C. Jain et al. (eds.), *Intelligent Computing, Communication and Devices*,
Advances in Intelligent Systems and Computing 309,
DOI 10.1007/978-81-322-2009-1_13

various resource allocation problems, e.g., mutual exclusion [1], k-mutual exclusion [2], and group mutual exclusion (GME) [3] have been proposed in the literature.

The GME problem [4] deals with two fundamentally opposite issues, namely mutual exclusion and concurrency. In GME, each request is associated with a type or group. The processes requesting the same group are allowed to be in their CS simultaneously. However, the processes requesting different groups must execute their CS in mutually exclusive way. GME problem has been modeled as congenial talking philosopher (CTP) problem by Joung [5].

Attiya et al. [6] introduced the concept of local mutual exclusion for MANETs. In local mutual exclusion, no two neighboring nodes can enter CS simultaneously; however, the nodes which are not neighbors can be in their CS simultaneously. On the other hand, in classical or global mutual exclusion, no two nodes (how far apart) can be in CS simultaneously. Kogan [7] claims that in MANETs, the local mutual exclusion problem has more potential applications in comparison with global mutual exclusion problem. Khanna et al. [8] proposed k-local mutual exclusion problem (KLME) in MANETs and solved it using a token-based approach. Wu et al. [9] defined a new problem named LGME as a variant of GME problem specially suited for VANETs and proposed a coterie-based solution for LGME. However, their concept of conflicting nodes is different from the neighborhood concept used in [8].

In KLME, the neighborhood has been defined as the independent smaller unit of a larger geographical area. The larger area is divided in smaller neighborhoods based upon the location of shared resources. Each set of shared resources placed at one location can be accessed in a particular neighborhood only. In this paper, the group local mutual exclusion (GLME) problem with the concept of neighborhood (similar to [8]) has been handled using a token-based approach. An interesting application of the problem may be the situation in which multiple replicas of a read only data base are available at different locations. At each location, due to limited amount of buffer, only a subset of database may be in the buffer and the processes (in the range of the current replica) requesting the same portion of database may be in CS simultaneously; however, the processes requesting the different portion have to wait. Moreover, the processes in the range of another replica may be in CS using a different portion of the data base. The system model and assumptions similar to used in [8] have been considered.

Following properties must be satisfied for GLME:

Mutual exclusion: No two processes in the same neighborhood requesting different groups (resources) can be in their CS simultaneously.
Bounded delay: A process attempting to enter CS will eventually succeed.
Concurrent entering: If some processes in the neighborhood are interested in entering CS using a group and no process in the neighborhood is interested in a different group, then the processes can concurrently enter CS

2 Related Works

The dining philosopher's problem [2] is an interesting variant of mutual exclusion problem in static networks. Although the solution to dining philosopher's problem works as a solution to mutual exclusion problem, it has a special property that the failure of a node does not affect the entire system. Recently, Sharma et al. [10] presented a detailed survey of the mutual exclusion in MANETs. The dining philosopher's problem has been extended by Attiya et al. [6]. Kogan [7] explained that unlike conventional mutual exclusion, local mutual exclusion algorithms have better failure locality. Khanna et al. [8] presented a KLME using token-based approach in MANETs. In KLME, the concept of neighborhood is based upon the location of sets of resources scattered in the entire area; all the nodes in the range of one set of resources comprise one neighborhood. Wu et al. [9] defined a new problem named LGME as a variant of GME problem specially suited for VANETs and proposed a coterie-based solution for LGME. However, they took the example of automatic vehicles and considered the traffic lanes as resources. The vehicles with crossing paths are considered conflicting. On the other hand, the vehicles with non-crossing paths are non-conflicting and no synchronization is required among these. The concept of local versus non-local used in LGME is different from the concept of neighborhood based upon resource location used in KLME. In the present paper, a variant of GME problem using the concept of neighborhood [8] has been proposed and solved using a token-based approach. Recently, also lot of research is going on mutual exclusion problem and its variants [11–13] in MA-NETs; however, the problem of local mutual exclusion and its variants are still unexplored.

3 GLME Algorithm

3.1 Working

In this section, brief working of our algorithm is discussed in MANETs. To the best of our knowledge, this is the first token-based solution for GLME problem in MANETs. In this algorithm, n numbers of nodes are considered in local neighborhood. After initialization, node 1 is elected as token holder, and token information is broadcasted, so that neighboring nodes may transfer their request to token holder.

When a node i receives a non-stale token information, if the node i is in waiting state with the same group of request and if node i is part of token's follower list, it enters CS. Otherwise, it will send the request for CS to the sender. In case, node i is in remainder section and is part of token's follower list it will send leaving CS message.

When any node requests for CS, there are following possibilities. (a) Node is idle token holder node. It enters CS. (b) Token holder but not in CS and its group type match the current group. The node enters CS. (c) Token holder, not in CS, and its group type do not match the current group. The request is added in token queue. (d) The node does not have token information. It will wait for token information. (e) The node has token information. It will forward request to token holder.

When the token holder receives a non-stale request, request acknowledgment is sent to the token holder and further actions are performed on the receiving node.

When node i receives token, if it is in waiting state with matching group of request, it will enter CS and will send permission to enter CS to all the requesting processes in token queue with matching group. The node i will take further actions shown in pseudocode. A node receiving permission to enter CS as follower will enter CS.

In case a token holder node i exits from CS, if number of followers are zero, and token queue is empty, its state is changed to holding idle token. Otherwise, if token queue is not empty, the token is transferred to the node at front of token queue and other nodes having requests for matching group in token queue are also permitted to enter CS. However, if followers are still in CS, token holder state is changed accordingly. In case a non-token holder node comes out of CS, it sends release CS message to token holder.

When token holder receives a release message from any follower node, follower node is removed from token follower's list. In case token holder is not in CS and it is holding token with some followers, appropriate actions are taken as shown in pseudocode. In case a non-token holder receives release CS from any follower, this information is stored in the local data structure of receiving node for future action.

When a node is about to leave the neighborhood, various actions are performed depending upon its state. In case leaving node is in CS, its state is changed to waiting state which does not happen in GME algorithm and is a special feature of GLME algorithm. Similarly, handling of joining and leaving of any new node is presented in pseudocode.

3.2 Data Structure

Data structure at node i $(1 \leq i, j \leq n)$

$state_i$: The current state of $node_i$: REM: remainder, W: waiting, HI: holding idle token, HF: holding token with followers but not in CS, CF: critical section as follower, and CT: critical section as a token holder.

$n\_foll\_leave_i$: stores information of foll_leaves_CS when node is non-token holder.

$token\_id_i$: stores the node id of token holder or $\emptyset$ at node i.

RN_i: request no. of node i which is incremented when $node_i$ request.

$g\_node\_type_i$: type of group for which node has requested.

node_req_list_i: stores the request number and its corresponding group number of neighboring nodes.

node_leader_no_i: contains the value that how many times new token holder has been elected as token holder according to node_i

req_ack_i: Boolean variable indicating request acknowledgment is received.

n: total no. of nodes in neighborhood.

Data Structure at token holder

T_Q: A FIFO queue which stores nodes requests with group and request number.

gp_t_type: stores the group being accessed by token holder in current session.

no_t_follower: number of nodes currently in CS as follower.

t_foll_list: node ids to which per_grant has been sent.

LN: an array to store the sequence number of latest served request of very node.

token_leader_no: number of times the leader has changed.

3.3 Messages

1. *token_info_i* (*j, t_foll_list, token_leader_no, gp_t_type*): Node *i* broadcasts token information along with other useful information.
2. *req_CS_i* (*i, g_i,RN_i, Token_id_i*): Node *i* sends the request message to the token holder.
3. *token(j, T_Q, gp_t_type, no_t_followers, t_foll_list, LN)*: Node receiving token becomes token holder.
4. *req_ack(j, LN_i)*: Token holder sends this message to the node from which it has received request for CS as an acknowledgment of request.
5. *per_grant* (*g_x, j*): Token holder sends this to *j* to allow *j* to enter CS as follower.
6. *foll_release_CS(j, RN_i)*: sent by follower to token holder after coming out of CS.
7. *I_am_leaving(i, n_foll_leave_i, node_req_list_i)*: broadcasted by leaving node.
8. *I_have_join(i)*: broadcasted by a node entering neighborhood to all nodes.

3.4 Pseudocode of Algorithm at Node i

```
a) Initialization
g_node_type_i = Ø; node_leader_no=0; req_ack=false
for(i=1 to n) state_i = REM; token_id_i=Ø; RN_i=0; G_i=Ø;
n_foll_leave_i= Ø; g_node_type_i=Ø; node_leader_no_i=0
for{i=1 to n}
   for{j= 1 to n} node_req_list_i[j] = {0, Ø}
TokenGenerate()
 Let node 1 is initially elected as token Holder
 token_id = 1; T_Q=Ø; token_leader_no=token_leader_no + 1
 node_leader_no_i = token_leader_no; gp_t_type = Ø
 no_t_followers = 0; t_foll_list = Ø
 for(i=1 to n) LN[i] = 0
token_id_1 = 1; Broadcast token_info
b) Node i receives token information from node j
  if (token_leader_no > node_leader_no_i)
   token_id_i = j; node_leader_no_i = token_leader_no
   if (state_i = W && g_node_type_i = gp_t_type)
      if ( i ε t_foll_list) enter CS;call exit_CS;rec_ack=true
      else  send req_CS          // rec_ack_i = false
      if (state_i = REM && i ε t_foll_list)send leave_CS to j
   else reject the message
c) Node_i request for CS with group g_i
   state_i  = W; req_ack_i = 0 ; RN_i = RN_i +1;g_node_type = g_i
   add request in node_req_list_i
   if(token_id_i = i)
      if(state_i=HI) enter CS; call exit_CS; state_i = CT
      if(state_i = HF && gp_t_type=g_i)
         state_i = CT; enter CS; Call exit_CS
   if(token_id = Ø) wait for token info
   else send req_CS to token_id_i .
d) node i receives req_CS from node j
   if (node_req_list_i[i]<RN_j )
    if(token_id_i=i) send req_ack to node j; node_req_list[i]=RN_j
       if(j ε t_foll_list) remove i from t_foll_list
       if(state_i=CT)
         if(gp_t_type = g_i) LN[j] = RN_j; no_t_followers ++
            send per_grant to j
       if (state_i = HF)
          if(gp_t_type = g_i &&  T_Q = Ø )
             LN[j] = RN_j; send per_grant to j; no_t_followers++
       if(state_i = HI) token_leader_no +=1;
        node_leader_no =token_leader_no; token_id_i = j
        Send token to node j; Broadcaste token_info
      else append the request in T_Q
     else store the request in node_req_list_i
   else reject the message
e) node i receives token
   Remove nodes in n_foll_list from t_foll_list; Decrement
   number of follower correspondingly; update T_Q with
   node_req_list_i; token_id_i = i
   if( i ε T_Q ) remove i from T_Q
```

```
   if (state_i = W)
     if(gp_t_type = g_i) state_i = CT; enter CS; call exit_CS( );
        for(k=1 to n )
           if(k ε T_Q && g_k = gp_t_type)
             send per_grant to node k; no_t_followers ++
             add k in t_foll_list
           else
             if (no_t_foll = 0) state_i = CT; gp_t_type = g_i
              for(k=1 to n )
               if(k ε T_Q && g_k = gp_t_type)
                Send per_grant to node k; no_t_followers ++
                add k in t_foll_list
     if(state_i = CF) state_i = CT; no_t_follower --; remove i
        from t_foll_list
     if(state_i = REM)
        if(T_Q = ∅) state_i = HI
        else send token to node(X) at front at T_Q
           token_id_i = X; gp_t_type = g_x;Broadcast token_info
           for(k=1 to n )
            if(k ε T_Q && g_k = gp_t_type)
             Send per_grant to node k; no_t_follower ++
f) node i receives req_ack(j,LN_i)
   if(RN_i[i] = LN_i ) req_ack_i = true
g) node_i receives per_grant(g_x,j) from node j
   state_i = CF; enter CS; call exit_CS
h) node_i exit from CS
   if(token_id_i = i)
     if(t_foll_list = 0)
       if(T_Q = ∅) state_i = HI
       else token_leader_no +=1
         node_leader_no =token_leader_no; state_i =REM
       if(x ε  front of T_Q) gp_t_type = g_x; remove x from T_Q
          for(k=1 to n )
             if(k ε T_Q && g_k = g_x )
                Send per_grant to k; no_t_followers ++; add K to
                t_foll_list; Send token to x; Broadcast token_info()
     else state_i = HF
   else state_i = REM; send foll_release_CS to token_id_i
i) node i receives release of CS message from follower.
   if(token_id_i = i)
     if(LN_i = RN_i) remove j from t_foll_list; no_t_followers--
     if(state_i = CT) do nothing
     if(state_i = HF)
        if (no_t_followers = 0)
           if(T_Q = ∅) state_i = HI
           else  let x is node at front of T_Q; gp_t_type = g_x
           remove x from T_Q
           for(k=1 to n)
            if(k ε T_Q && g_k = g_x)
              send per_grant to k; no_t_followers ++; add k in
              t_foll_list; Send token to X; broadcast token_info
           else do nothing       //no_t_follower ≠ 0
     else reject the message
```

```
   else Store foll_release_CS info n_foll_leave list
j) node i is about to leave the neighborhood.
   if(token_idᵢ=i)
     if(stateᵢ = CT or stateᵢ=HF)
        if(stateᵢ=CT)       stateᵢ = W
        else     stateᵢ = REM
        if(no_t_foll ≠ 0 )
           token_leader_no += 1; node_leader_no = token_leader_no
           token_idᵢ = X (lowest id node of t_foll_list)
           broadcast token_info; Send token to X
        else
           if(T_Q≠∅) Send token to node at front of T_Q; store
                similar nodes having similar groups Of request into
                t_foll_list and send per_grant to respective nodes;
                broadcast token_info
           else send token to lowest id in neighbourhood
     if(stateᵢ = HI) send token to lowest id in neighbourhoo
       else
           if(stateᵢ = CF) send foll_leave_CS to token_id; stateᵢ = W
           if(stateᵢ = W ) stateᵢ = W
     else do nothing
     broadcast I_am_leaving message
k) node i receives I_am_leaving from j.
   remove leaving node ( j ) from all lists
   update n_foll_leaveⱼ node_req_listⱼ received
   if(token_idᵢ = i ) Delete j from token's data structure
l) new node i joins neighborhood
   send I_have_join to all node in neighborhood.
   wait for token_info; apply initialization to node i
m) node i receives I_have_join.
   add entry of node j into local data structure
   if(token_idᵢ = i)   send token_info to node j
```

4 Performance Analysis

In the best case, the requesting token holder node can directly enter CS without any message(s) exchanged in the best case. In average case, where $k + 1$ processes enter in CS (K as follower and one as token holder), the number of messages exchanged for these $k + 1$ CS executions will be n (broad cast new leader) + $(k + 1)$ (request) + $(k + 1)$ (request acknowledgement) + k (permission grant) + k (follower_leave_CS). Hence, no. of messages/CS = $\{(n) + (k + 1) + (k + 1) + k + k\}/(k + 1)\}$. The worst case will occur when there is no follower; that is, $k = 0$, and in this case, the messages/CS will be $n + 2$ $(O(n))$.

The synchronization delay is considered at heavy load where as the waiting time is considered at light load. The heavy load synchronization delay of our algorithm is T (where T is the maximum message propagation delay). The light load waiting time of the algorithm is $2T$.

5 Conclusions and Future Scope

The present paper proposes a token-based GLME algorithm for MANETs. To the best of our knowledge, the proposed algorithm is the first token-based GLME algorithm for MANETs. Algorithm satisfies the safety condition, starvation freedom, and concurrent occupancy. Due to the lack of space, the proof of correctness and dynamic analysis could not be presented and will be presented in the full version of the paper. The algorithm also handles link breakage and related dynamic changes in MANETs. To develop a fault-tolerant version of the proposed algorithm can be a possible future research direction.

References

1. Swaroop, A.: Efficient Group Mutual Exclusion Protocols for Message Passing Distributed Computing System doctoral diss. NIT, Kurukshetra (2009)
2. Dijkstra, E.W.: Hierarchical ordering of sequential processes. Acta Informatica 1(2), 115–138 (1971)
3. Kshemkalyani, A.D., Singhal, M.: Distributed Computing Principles, Algorithms, Systems. Cambridge University Press, Cambridge (2009)
4. Joung, Y.J.: Asynchronous group mutual exclusion. Distrib. Comput. 13(4), 189–206 (2000)
5. Joung, Y.J.: The congenial talking philosopher's problem in computer networks. Distrib. Comput. 15(3), 155–175 (2002)
6. Attiya, H., Kogan, A., Welch, J.L.: Efficient and robust local mutual exclusion in mobile ad hoc networks. IEEE Trans. Mob. Comput. 9(3), 361–375 (2010)
7. Kogan, A.: Efficient and Robust Local Mutual Exclusion in Mobile Adhoc Networks. In: Masters Research thesis, Technion-Israel Institute of Technology, Israel (2008)
8. Khanna, A., Singh, A.K., Swaroop, A.: A leader-based k-local mutual exclusion algorithm using token for MANETs. (Accepted) J. Inf. Eng. (JISE) Taiwan (2014)
9. Wu, A.W., Cao, J., Raynal, M.: A Generalized Mutual Exclusion Problem and Its Algorithm. In: AoxueLuo, Weigang Wu, International Conference on Parallel Processing (42), pp. 300–309 (2013)
10. Sharma, B., Bhatia, R.S., Singh. A.K.: DMX in MANETs: Major Research Trends Since 2004. In: Proceedings of the International Conference on Advances in Computing and Artificial Intelligence ACAI'11, pp. 50–55 (2011)
11. Hosseinabadi, G., Vaidya, N.H.: Exploiting Opportunistic Overhearing to Improve Performance of Mutual exclusion in Wireless Ad Hoc Networks. In: Proceeding of International Conference on Wired/Wireless Internet Communications (WWIC), pp. 162–173 (2012)
12. Parameswaran M., Hota, C.: Arbitration-based reliable distributed mutual exclusion for mobile ad hoc networks. In: Proceedings of Modelling and Optimization in Mobile, Ad Hoc & Wireless Networks, vol. 11, pp. 380–387 (2013)
13. Wu, W., Cao, J., Yang, J.: Fault tolerant mutual exclusion algorithm for mobile ad hoc networks. Pervasive Mobile Comput. 4, 139–160 (2008)

A Distributed Secured Localization Scheme for Wireless Sensor Networks

Lakshmana Phaneendra Maguluri, Shaik Mahaboob Basha, S. Ramesh and Md Amanatulla

Abstract Nowadays, location-aware security policies play a key role in wireless sensor networks. In this paper, we propose to develop a distributed secured localization scheme that validates the reliability of location information associated with event reports. We make the anchor nodes are deployed such that they form connected sensor coverage and their routing information to these anchor nodes, thereby avoiding the sink overload problem. Since this scheme involves distributed anchor nodes for verification, it has less overhead and delay. By simulation results, we show that our proposed scheme attains good delivery ratio with reduced delay and overhead.

Keywords Security issues on localization · Anchor nodes · Coverage algorithm

1 Introduction

Minimizing the energy consumption for communication and information processes are advisable [1]. Academic and industrial areas use wireless sensor networks with great concern. They give way for several applications such as military, industrial, scientific,

L.P. Maguluri (✉)
Department of Computer Science and Engineering,
Gudlavalleru Engineering College, Krishna District, India
e-mail: narayanacsegec@gmail.com

S.M. Basha · S. Ramesh
Department of Computer Science and Engineering,
Sree Vahini Institute of Science & Technology, Krishna District, India
e-mail: email2mahaboob@gmail.com

S. Ramesh
e-mail: samineni.ramesh@rediffmail.com

M. Amanatulla
Department of Computer Science and Engineering,
Nimra Institute of Science & Technology, Krishna District, India
e-mail: amanatulla@gmail.com

© Springer India 2015
L.C. Jain et al. (eds.), *Intelligent Computing, Communication and Devices*,
Advances in Intelligent Systems and Computing 309,
DOI 10.1007/978-81-322-2009-1_14

civilian, and commercial. Few applications which cannot be effective, unpleasing, and dangerous and which cost much for the human observations and traditional sensors can be implemented successfully by using wireless sensor networks [2].

Therefore suitable mechanism may be implemented for protecting localization techniques from the new form of attacks [3]. In addition to location calculation, tracking, and deployment, the basic problem in sensor networks is the coverage and connectivity. Coverage and connectivity are two key factors to a successful WSN. In our previous work, we design a protocol that validates the reliability of location information associated with event reports. The main drawbacks of this protocol are sink overload and confidentiality.

To overcome the above drawbacks, in this paper, we express the following in Sect. 2. Presents Coverage Algorithm for anchor node deployment Sect. 3. Presents the encryption and decryption mechanisms Sect. 4. Presents the route discovery process Sect. 5. Presents the performance evaluation overhead delays and network delays Sect. 6. Conclusion.

2 Coverage Algorithm for Anchor Node Deployment

For connecting a sensor cover of near-optimal size, a greedy algorithm is designed. At each stage, a communication path is selected which connects the existing sensors with the incompletely covered sensors. At this stage, the path which has been selected is added to the selected sensors. Consider a sensor network consisting [4] of the set of sensors $\{S_1, S_2 \ldots S_n\}$. Each sensor S_i has a sensing region SRi associated with it. Also consider a query Q over a region R in the network.

1. Let S be the set of sensors whose sensing region intersects with R. Let C denote the set of sensors selected by the algorithm at a given stage. Let R_C be the region covered by C.

   $C = \{S_i\}$, for some $S_i \in S$.

2. While $(R \not\subseteq R_C)$

   2.1. Let CS be the set of candidate sensors in Q, i.e., the set of sensors in S–C whose sensing region intersects with some sensor in C.
   Max Gain = 0;

   2.2. For each $c_i \in CS$

      2.2.1. Find the max gain candidate path CP_i for the candidate sensor c_i, i.e., a candidate path CP_i with maximum gain such that $CP_i = < S_{i0}, S_{i1}, S_{i2} \ldots S_{il} >$ for some l, where $S_{il} \in C$, $S_{i0} = c_i$, and S_{ij} can communicate directly to $S_i(j-1)$.
      2.2.2. Gain = (No. of valid subset covered by the region $((SR_{i0} \cup SR_{i1} \cup \ldots SR_{il}) - RM))/l$;
      2.2.3. If (Gain > Max Gain)

Max Gain = Gain;
$CP = CP_i$;
End if;
End for;
$C = C \cup CP$;
End while;
RETURN C;

3 Encryption and Decryption

After the anchor node selection, each sensor nodes communicate with these anchor nodes a_i, using a symmetric key $K_{a,\,i}$. The sensor nodes send the encrypted RREQ packet using this key to the anchor node.

$$S_i \xrightarrow[{[E(\mathrm{RREQ})]}]{K_{a,i}} a_i$$

Each anchor node decrypts the RREQ packets and finally all the RREQ packets are encrypted and transmitted to the sink.

$$a_i \xrightarrow[{[E(\mathrm{RREQ})]}]{K_{a,s}} \mathrm{Sink}$$

4 Route Discovery Process

In the proposed protocol, once a node S want to send a packet to anchor node A, it initiates the route discovery process by constructing a route request RREQ packet. It contains the source and anchor node ids and location of source and a MAC computed over the accumulated path with a key shared by each node.

From the Fig. 1, we can see that there are two paths

$$R : S \rightarrow A \text{ and}$$
$$\bar{R} : \bar{S} \rightarrow A$$

for the anchor node A.

When RREQs of both R and $\bar{R}$ reaches the anchor node A, it decrypts the RREQs and from the received MAC values, it calculates

$$D(Ni, \bar{N}i) = \sum_{i=1}^{4} \|N_i - \bar{N}_i\|^2 \quad \text{by} \tag{1}$$

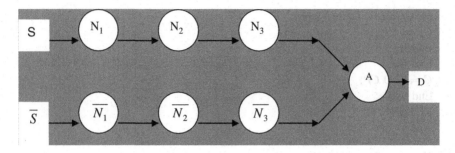

Fig. 1 Route request process

Then it checks the value of V based on which the trust values are incremented or decremented for the corresponding nodes.

If $V < th_1$ then

$$CCN_i = CCN_i + \delta,$$

Else

$$CCN_i = CCN_i - \delta,$$

End if.

where th_1 is the minimum threshold value for V and δ is the scale factor for increment or decrement. The process is repeated for various time intervals and finally the value of credit counter is checked.

If $CCN_i > th_2$ then

RREP is sent

Else

The source is considered malicious,
RREQ is discarded

End if.

Where th_2 is the minimum threshold value for CCN.

5 Performance Evaluation

We evaluate our distributed secured localization scheme (DSLS) through NS2 [5] simulation. We use a bounded region of $1,000 \times 1,000$ m^2, in which we place nodes using a uniform distribution. We assign the power levels of the nodes such that the transmission range and the sensing range of the nodes are all 250 m [6]. In our simulation, the channel capacity of mobile hosts is set to the same value: 2 Mbps. We use the distributed coordination function (DCF) of IEEE 802.11 for

Fig. 2 Nodes versus delay

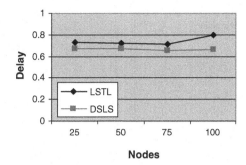

Fig. 3 Nodes versus overhead

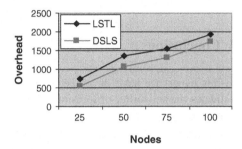

Fig. 4 Attackers versus delay

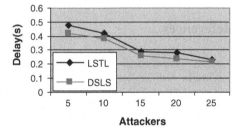

wireless LANs as the MAC layer protocol. The simulated traffic is Constant Bit Rate (CBR). In order to test the scalability, the number of nodes is varied as 25, 50, 75, and 100.

Figure 2 shows the end-to-end delay occurred for both DSLS and LSTL. As we can see from the figure, the delay is less for DSLS, when compared to LSTL.

Figure 3 shows the overhead for both DSLS and LSTL. As we can see from the figure, the overhead is less for DSLS, when compared to LSTL. The number of attacker nodes is varied as 5, 10, 15, 20, and 25 in a 100 nodes scenario.

Figure 4 shows the end-to-end delay occurred for both DSLS and LSTL. As we can see from the figure, the delay is less for DSLS, when compared to LSTL.

Fig. 5 Attackers versus
overhead

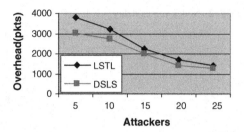

Figure 5 shows the overhead for both DSLS and LSTL. As we can see from the
figure, the overhead is less for DSLS, when compared to LSTL.

6 Conclusions

In this paper, we have developed a distributed secured localization scheme that
validates the reliability of location information associated with event reports. We
make the network distributed by including a set of anchor nodes across each route
in the network. Also we have included authentication and encryption mechanisms
to make the network more secured. Since this scheme involves distributed anchor
nodes for verification, it has less overhead and delay.

References

1. Alfaro, J.G., Barbeau, M., Kranakis, E.: Secure localization of nodes in wireless sensor
   networks with limited number of truth tellers. In: Proceedings of the Seventh Annual
   Communication Networks and Services Research Conference (2009)
2. Chen, H., et al.: A secure localization approach against wormhole attacks using distance
   consistency. EURASIP J. Wireless Commun. Networking **2010**, 8 (2010)
3. Soosahabi, R., Naraghi-Pour, M., Perkins, D., Bayoumi, M.A.: Optimal probabilistic
   encryption for secure detection in wireless sensor networks information forensics and
   security. IEEE Transactions on Digital Object Identifier, vol. 9, PP. 375–385 (2014). doi: 10.
   1109/TIFS.2014.2298813
4. Shih, K.P., Deng, D.J., Chang, R.S., Chen, H.C.: On connected target coverage for wireless
   heterogeneous sensor networks with multiple sensing units. Sensors **2009**(9), 5173–5200
   (2009). doi:10.3390/s90705173
5. Elbasi, E., Ozdemir, S.: Secure data aggregation in wireless multimedia sensor networks via
   watermarking. In: Application of Information and Communication Technologies (AICT), 2012
   6th International Conference on Digital Object Identifier, PP. 1–6 (2012). doi: 10.1109/
   ICAICT.2012.6398501
6. Ben Othman, S., Alzaid, H., Trad, A., Youssef, H.: An efficient secure data aggregation
   scheme for wireless sensor networks. In: Information, Intelligence, Systems and Applications
   (IISA), 2013 4th International Conference on Digital Object Identifier, PP. 1–4 (2013). doi: 10.
   1109/IISA.2013.6623701

Characteristic Function for Variable-Bit-Rate Multimedia QoS

Neela Rakesh and Hemanta Kumar Pati

Abstract Multimedia users are numerously increasing with the increase in Internet users. To be able to provide QoS to such users, it is important to understand how VBR multimedia QoS varies with its bandwidth characteristics. In this paper, we proposed a generic mathematical model for variable-bit-rate multimedia QoS which is applicable to various wireless network systems. This paper comprises of characteristic function for variable-bit-rate multimedia QoS using bandwidth terminology, such as minimum bandwidth of the connection and maximum bandwidth acceptable condition. We have provided numerical results required for verification and validation of the proposed model. This model is very useful for understanding how the wireless system behaves in terms of QoS while allocating bandwidth to variable-bit-rate multimedia traffic.

Keywords Soft real time · Variable-bit-rate traffic · Multimedia · QoS · Geometric distribution

1 Introduction

In recent personal communication systems, there is enormous demand for multimedia applications. Multimedia is defined as information that combines and interacts with multiple forms of media data (e.g., text, speech, audio, image, video graphics, animation, and possibly various formats of documents) [1]. From the networking perspective, multimedia types can be classified as either real-time

N. Rakesh · H.K. Pati (✉)
Department of Computer Science and Engineering, IIIT Bhubaneswar, Bhubaneswar 751003, Odisha, India
e-mail: hemanta@iiit-bh.ac.in; h_pati_hindol@yahoo.com

N. Rakesh
e-mail: a112010@iiit-bh.ac.in; theidea.rakesh@gmail.com

© Springer India 2015
L.C. Jain et al. (eds.), *Intelligent Computing, Communication and Devices*,
Advances in Intelligent Systems and Computing 309,
DOI 10.1007/978-81-322-2009-1_15

media type (RT) and non-real-time media type (NRT). Real-time media types require either hard or soft bounds on end packet delivery/jitter, while NRT media types such as text and image files do not have any strict constraints, but may have rigid constraints on error. Further, RT media types are classified as discrete media (DM) or continuous media (CM), depending on whether the data are transmitted in discrete quantum as a file or message, or continuously as a stream of messages with message interdependency. The RT continuous type of media can be further classified as delay tolerant or delay intolerant. The term 'delay tolerant' signifies that such media type can tolerate higher amounts of delay than their intolerant counterparts, without significant performance degradation [2]. We have identified and analyzed the requirements that a distributed multimedia application may enforce on communication network. Broadly, these requirements are divided into two categories: traffic requirements and functional requirements. Multimedia traffic requirements include limits on real-time parameters, such as delay, jitter, bandwidth, and reliability, and functional requirements include support for multimedia services such as multicasting, security, mobility, and session management. The traffic requirements can be met only by introducing newer protocols over the TCP/IP networking stack. The functional requirements are not an absolute necessity.

Wireless systems (such as mobile cellular networks and satellite communications) have a tremendous demand for multimedia traffic such as video on demand and video conferencing. Multimedia traffic puts a heavy bandwidth demand for these wireless networks. In ensuring the effective dissemination of compressed multimedia data over personal communication systems, the main challenge occurs while integrating previous and next-generation systems. Provisioning of QoS to such system has been difficult due to change in bandwidth allocation, adaptability, and its mechanisms. Multimedia traffic is used to indicate the transmitted information which comes from various source characteristics with different qualities. A multimedia QoS model for each multimedia application is more valuable to identify the QoS patterns.

The rest of this paper is organized as follows. In Sect. 2, we present the related work. In Sect. 3, we present our characteristic function for variable-bit-rate multimedia QoS. Section 4 presents simulation results. Finally, in Sect. 5, we conclude this paper.

2 Related Works

The most critical components in personal communication systems are data compression, quality of service, communication protocols, and effective digital management. Among the most critical components, quality of service plays the most important aspect. Quality of service is defined as well-controllable behavior of a system according to quantitatively measurable parameters (defined as per ISO) [3].

In [4, 5], it is found that modeling multimedia traffic and generating desired results are expensive, and often, it is difficult to generate reasonable results. Statistical and mathematical traffic models are mostly used to provide a better understanding for various traffic characteristics. Using such models, different realizations that represent actual data can be obtained by varying model parameters. A good traffic model captures the characteristics of multimedia applications, which include video streaming, VOIP, and online gaming, which often demand seamless real-time delivery. Recent efforts to reduce bandwidth requirements while maintaining the quality of the multimedia services have led to data compression schemes. In these compression schemes, we have found two stream types, namely constant-bit-rate (CBR) stream and variable-bit-rate (VBR) stream, and the selection of stream depends on the numerical variable. Autocorrelation function (ACF) used by different stochastic processes captures the long-range and short-rage dependencies between frame sizes. VBR exhibits both long-range-dependent (LRD) and short-range-dependent (SRD) properties.

The work reported in [6] discusses about the mechanism of QoS mapping between multiuser session and personal communication system. It also discusses about adaptation control techniques. The work reported in [7] discusses about QoS mapping between video applications and networks. It also discusses about QoS mapping control in the network. The work reported in [8] discusses about architecture components and its functionality from user and system perspectives. In this paper, we proposed a model to map bandwidth allocation to QoS of variable-bit-rate multimedia. This mapping can be used to analyze system behavior. In the following section, we present our proposed model.

3 Proposed Model for Mapping QoS of Variable-Bit-Rate Multimedia

The model proposed in this paper is for variable-bit-rate multimedia QoS. This model helps to map system's QoS according to the amount of bandwidth allocated. We assume that when a user requests a new connection or roams into a new cell, it provides the following information such as (1) type of traffic (CBR and VBR); (2) desired amount of bandwidth of the connection; (3) minimum bandwidth acceptable condition (only for the VBR); and (4) maximum bandwidth acceptable condition (only for VBR traffic) [9]. Here, QoS of VBR multimedia traffic is characterized by geometric distribution, which is based on the above-described bandwidth values. Analysis of VBR multimedia QoS in personal communication systems is essential because it facilitates a better understanding of such systems providing divergent multimedia services. We describe our analysis using characteristic function.

Characteristic Function: Characteristic function defines completely about its probability distribution. It serves as an important tool for analyzing the random variable and its properties.

Generalized Power Series Function Definition [10, 11]: Generalizing formal power series plays an important role in mathematics. Let us consider coefficients of real numbers $\{a_i : i = 0, 1, 2, 3, \ldots\}$. The generalized power series function can be expressed in closed form without any expansion as given in the following expression:

$$G(s) = \sum_{i=0}^{\infty} a_i s^i \tag{1}$$

The power series variable is s for which the sum converges. There exists a radius of convergence denoted by $R \ (\geq 0)$ such that (i) sum converges absolutely if $|s| < R$ and (ii) sum diverges if $|s| > R$.

a. *Probability Generating Function Definition*: Let a random variable X selected from the sequence. It is written as follows:

$$p_k = P(X = k), \quad k = 0, 1, 2, 3\ldots \tag{2}$$

The probability generating function (PGF) of X can be written as given in the following expression:

$$G_X(s) = \sum_{k=0}^{\infty} p_k s^k = E(s^X) \tag{3}$$

If PGF, $G_X(1) = 1$, and it indicates that the series converges absolutely for $|s| \leq 1$. Also, $G_X(0) = p$.

b. *Geometric Random Variable Definition*: If $p_k = pq^{k-1}$, $k = 1, 2, 3\ldots$ and $q = 1 - p$, then

$$G_X(s) = \frac{ps}{1 - qs} \text{ if } |s| < q^{-1} \tag{4}$$

c. *Uniqueness Theorem*: If X and Y are random variables having generating functions G_X and G_Y, respectively, then $G_X(s) = G_Y(s)$ for all s iff $P(X = k) = P(Y = k)$ for $(k = 0, 1, \ldots)$, i.e., if and only if X and Y have the same probability distribution.

Proof Let $G_X(s) = G_Y(s)$ be denoted by (*a*) and $P(X = k) = P(Y = k)$ for ($k = 0$, 1,…) be denoted by (*b*). Here, the condition to prove is (*a*) implies (*b*). Let radii of convergence of G_X and G_Y be ≥ 1, having unique power series expansion:

$$G_X(s) = \sum_{k=0}^{\infty} s^k P(X = k) \tag{5}$$

$$G_Y(s) = \sum_{k=0}^{\infty} s^k P(Y = k) \tag{6}$$

If $G_X = G_Y$, these two power series have identical coefficients.

d. *Proposed Characteristic Function for Bandwidth Allocation*: Characteristic function using geometric distribution is given, based on the bandwidth values such as minimum bandwidth and average bandwidth values.

Function Definition: Let us consider a random variable *Breq* (i.e., *bandwidth requested* is a discrete random variable taking nonnegative values). Other parameters to be used in the function definition are described in the following.

p: Value generated from pseudo-random numbers from the system,
$q = 1 - p$,
B_{min}: Minimum bandwidth required for connection,
B_{max}: Maximum bandwidth that can contribute to the QoS of the connection, and
B_{alloc}: Bandwidth allocated by the system

The characteristic function definition is given in the following.

Case 1: If $B_{req} < B_{min}$, then $B_{alloc} = 0$ and $G_{QoS}(B_{alloc}) = 0$
Case 2: If $B_{min} \leq B_{req} \leq B_{max}$ and if $|B_{req}| < q^{-1}$, then $B_{alloc} = B_{req}$ and $G_{QoS}(B_{alloc}) = \frac{pB_{alloc}}{1-qB_{alloc}}$
Case 3: If $B_{max} < B_{req} < \infty$ and if $|B_{req}| < q^{-1}$, then $B_{alloc} = B_{req}$ and $G_{QoS}(B_{alloc}) = \frac{pB_{max}}{1-qB_{max}}$

4 Numerical Results

Bandwidth allocation is one among the most fundamental and important aspects in personal communication systems. Multimedia traffic characteristics in personal communication systems are, however, not known well because (i) the introduction of multimedia services is rather up to date and (ii) the necessary bandwidth from the operational network is rarely available. In the presence of such limitations, our

Table 1 Simulation parameters

Parameter	Values	Description
p	0–1	Pseudo-random value generated by the system
q	0–1 ($q = 1 - p$)	Pseudo-random value generated by the system
B_{min}	1 Mbps	Minimum bandwidth which is required for the connection
B_{max}	4 Mbps	Maximum bandwidth that can contribute to the QoS of the connection

findings on the bandwidth characteristics for variable-bit-rate multimedia will be helpful. The parameters and their respective values used in the simulation study are summarized in Table 1.

a. *QoS of VBR Multimedia with offered bandwidth*: If we examine the bandwidth allocation of a wireless network system from Fig. 1. It shows how the acceptance rate of bandwidth is observed throughout the system. Further, it describes the following. Firstly, if the requested bandwidth is less compared to minimum bandwidth required for the connection, then the connection is not established. Secondly, if the requested bandwidth is greater compared to the maximum bandwidth acceptable condition, it shows that allocated bandwidth is higher compared to the maximum acceptable bandwidth value and is represented by a uniform constant QoS same as to that of the maximum acceptable bandwidth condition. Therefore, the system should allocate the maximum bandwidth accepting value but not higher values, while the connection's requested bandwidth is more than the maximum bandwidth acceptable condition. Thirdly, if the requested bandwidth is between the minimum bandwidth required for the

Fig. 1 QoS for VBR multimedia traffic versus offered bandwidth

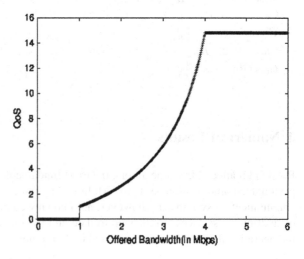

Fig. 2 QoS of different systems for similar bandwidth allocation

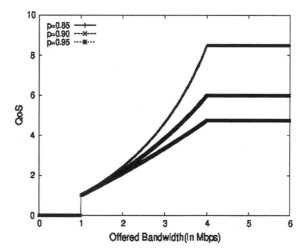

connection and maximum acceptable bandwidth value, then the allocation is following the geometrical increasing rate. Hence, this is the typical behavior of the wireless system using bandwidth acceptance values.

b. *Comparison of various systems offering VBR multimedia for similar type of bandwidth allocation*: QoS for variable-bit-rate multimedia over different systems allocated with same bandwidth characteristic is examined. These results are useful while comparing the various system behaviors at a particular requested bandwidth value. Figure 2 illustrates how the typical mathematical function is applicable to various wireless network systems to examine QoS for similar bandwidth allocation.

c. *Comparison of VBR multimedia QoS in a single system with different bandwidth characteristics*: Fig. 3 illustrates that QoS provided by the system is affected by differing the bandwidth characteristics.

Fig. 3 QoS of a single system with different bandwidth characteristics

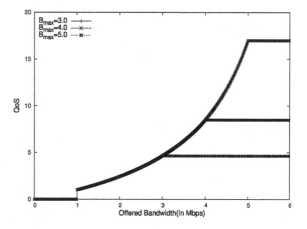

Fig. 4 $G_{QoS}(B_{alloc})$ versus
frequency ($G_{QoS}(B_{alloc})$)

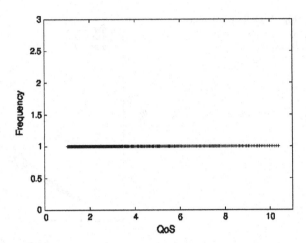

d. *Verification and Validation*: To verify whether allocated bandwidth is found
confirming the proposed model, we tried to prove it by plotting $G_{QoS}(B_{alloc})$
versus frequency of $G_{QoS}(B_{alloc})$ in Fig. 4. It is observed from Fig. 4 that each
value of the $G_{QoS}(B_{alloc})$ is unique and its repetition is not allowed. By
uniqueness theorem that is defined in Sect. 3, we can say that $G_{QoS}(B_{alloc})$ is
strictly following the geometrically increasing function. Hence, the proposed
work is found to be true by verification.

5 Conclusion

Multimedia users are numerously increasing. To be able to provide QoS to such
users, it is important to understand how VBR multimedia QoS varies with its
bandwidth characteristics. To this end, we proposed a model which is found to be
suitable to model QoS of VBR multimedia with its bandwidth characteristics. This
model can be useful for designing protocols for QoS provisioning and call
admission control schemes for serving VBR multimedia applications over any
wireless network such as mobile system and satellite systems.

References

1. Hwang, J.N.: Multimedia networking: from theory to practice. Cambridge University Edition,
   Cambridge (2009)
2. Khanvilkar, S., Bashir, F., Schonfeld, D., Khokhar, A.: Multimedia networks and
   communication. http://multimedia.ece.uic.edu/04-8.pdf. (2004)

3. Sahinoglu, Z., Tekinay, S.: On multimedia networks: self-similarity traffic and network performance. IEEE Commun. Mag. **37**, 48–52 (1999)
4. Tanwir, S., Perros, H.: A survey of VBR video traffic models. IEEE Commun. Surv. Tutorials **15**(4), 1778–1802 (2013)
5. Akyildiz, I.F., Melodia, T., Chowdury, K.R.: A wireless multimedia sensor networks: a survey. IEEE Wirel. Commun. **14**(6), 32–39 (2007)
6. Cerqueira, E., Veloso, L., Curado, M., Monteiro, E., Mendes, P.: QoS mapping and adaptation control for multi-user sessions over heterogeneous wireless networks. MobiMedia (2007)
7. Shin, J., Kim, J.G., Kim, J.W., Lee, D.C., Jay Kuo, C.C.: Aggregated QoS mapping framework for relative service differentiation-aware video streaming. Packet Video Workshop (2001)
8. Cerqueira, E., Veloso, l., Neto, A., Curado, M.: A unifying architecture for publish-subscribe services in the next generation IP networks. IEEE Global Telecommunications Conference, pp. 1–5 (2006)
9. Oliveira, C., Kim, J.B., Suda, T.: An adaptive bandwidth reservation scheme for high-speed multimedia wireless networks. IEEE JSAC. **16**(6), 858–874 (1998)
10. http://www.am.qub.ac.uk/users/g.gribakin/sor/Chap3.pdf
11. Papoulis, A., Pillai, S.U.: Probability, Random Variables and Stochastic Process, 4th edn. Tata Mc-Graw hill, New Delhi (2009)

A Fully Simulatable Oblivious Transfer Scheme Using Vector Decomposition

Manoj Kumar and I. Praveen

Abstract Oblivious transfer is one of the most basic and widely used protocol primitives in cryptography. It can be described as a two-party protocol used for interaction between a sender and a receiver. A 1-out-of-2 oblivious transfer is the interaction between a sender and a receiver in which a sender has two strings m_0 and m_1. At the end of the interaction, receiver learns exactly one of the strings m_0 and m_1, while the sender learns nothing. Lindell showed how to achieve efficient and fully simulatable non-adaptive oblivious transfer under decisional Diffie–Hellman (DDH) problem, Nth residuosity and quadratic residuosity assumptions, as well as the assumption that homomorphic encryption exists. We propose a scheme based on this protocol under the assumption namely vector decomposition problem. Our scheme is non-adaptive and fully simulatable.

Keywords Vector decomposition problem · Oblivious transfer

1 Introduction

Oblivious transfer (OT) was introduced by Rabin [11] and generalized by Even et al. [3]. It is the two-party protocol where one party, the sender (master authority), has N-bit message $M_1,...,M_N$ and the other party, the receiver (user),

M. Kumar (✉)
TIFAC Core in Cyber Security, Amrita Vishwa Vidyapeetham (University),
Coimbatore 641 112, Tamil Nadu, India
e-mail: manojkumar0688@gmail.com

I. Praveen
Department of Mathematics, Amrita Vishwa Vidyapeetham (University),
Coimbatore 641 112, Tamil Nadu, India
e-mail: praveenvchandran@gmail.com

© Springer India 2015
L.C. Jain et al. (eds.), *Intelligent Computing, Communication and Devices*,
Advances in Intelligent Systems and Computing 309,
DOI 10.1007/978-81-322-2009-1_16

wants to learn some information corresponding to the indices $\sigma_1,\ldots,\sigma_k \in [1, N]$. At the end of the interaction, receiver obtains $M_{\sigma_1},\ldots,M_{\sigma_k}$ without getting the other messages and the sender will not be able to know the indices [5, 6]. A 1-out-of-2 variant of the OT includes a sender with two bit strings m_0 and m_1. At the end of the interaction, receiver learns exactly one of the strings m_0 or m_1 and the sender does not learn anything. Efficiency and security of the oblivious transfer protocol is vital. Security of OT protocol is defined via simulation. Lindell [7] showed how to achieve efficient and fully simulatable non-adaptive oblivious transfer. Lindell [7] used decisional Diffie–Hellman (DDH), Nth residuosity and quadratic residuosity assumptions, as well as the assumption that homomorphic encryption exists. We propose a scheme based on this protocol. Security of our protocol is based on vector decomposition problem (VDP).

Vector decomposition problem (VDP) was introduced as an alternative for discrete logarithm problem (DLP) or computational Diffie–Hellman problem (CDHP). VDP was initially proposed by Yoshida [12]. This was further analyzed by Duursma and Kiyavash [2] and Gal-braith and Verheul [4]. Only super-singular elliptic curves of genus 1 are appropriate for the applications based on VDP. Due to the existence of MOV reduction and FR reduction, these curves are vulnerable to attack. Hence, higher-genus curves are preferred for the use of VDP. Such curves were given in [4]. Our scheme is also a homomorphic scheme as in Lindell.

The remaining of this paper is organized as follows: Sect. 2 provides basic definitions and preliminaries; in Sect. 3, we provide our proposed oblivious transfer scheme; and Sect. 4 concludes the paper.

2 Preliminaries

A bilinear map is a function $e{:}G_1 \times G_2 \rightarrow G_3$, where G_1, G_2, and G_3 are groups of large prime order r. The function $e(.,\,.)$ is such that for all $u \in G_1$, $v \in G_2$, $e(u^a,\,v^b) = e(u,\,v)^{ab}$, where a and b are integers [9]. Due to the property that they associate pairs of elements from G_1 and G_2 with an element in G_3, bilinear maps are termed as pairings.

Definition 2.1 A bilinear pairing on (G, G') is an efficiently computable map $\varphi{:}G \times G{:} \rightarrow G'$, which is bilinear and non-degenerate, i.e., for all $a, b, c \in G$,

   (i) $\varphi(a + b,\,c) = \varphi(a,c)\varphi(b,c)$

   (ii) $\varphi(a,b + c) = \varphi(a,b)\varphi(a,c)$

   (iii) $\varphi(P,P) \neq 1$.

2.1 Vector Decomposition Problem

The VDP is a computationally hard problem for certain curves [1].

Definition 2.2 *Vector Decomposition Problem*: Let V be a vector space over the field F_p and $\{P_1, P_2\}$ is a basis for V. Let $Q \in V$. Compute the element $R \in V$ such that $R \in \langle P_1 \rangle$ and $Q - R \in \langle P_2 \rangle$. For a fixed base $\{P_1, P_2\}$, VDP is defined as follows: given $Q \in V$, find R as above.

m-torsion point on an elliptic curve E is defined as the collection of points of order m denoted by $E[m]$. $E[m] = \{P \in E/mP = P_\infty\}$. Here, P_∞ is identity of the group of all points on the elliptic curve. The collection of m-torsion points on an elliptic curve E is a vector space under elliptic curve addition and scalar multiplication. If the CDHP on a one-dimensional subspace is hard, then VDP is also hard.

Theorem [12] *The VD Problem on V is at least as hard as CDH Problem on $V' \subset V$ if for any $e \in V'$ there are isomorphism f_e, $\phi_e \colon V \to V$ which satisfy the following conditions*

- *For any $v \in V$, $\phi_e(v)$ and $f_e(v)$ are effectively defined and can be computed in polynomial time.*
- *$\{e, \phi_e(e)\}$ is an F-basis for V*
- *There are $\alpha_1, \alpha_2, \alpha_3 \in F$ with*

$$f_e(e) = \alpha_1 e$$
$$f_e(\phi_e(e)) = \alpha_2 e + \alpha_3 \phi_e(e),$$
$$\alpha_1, \alpha_2, \alpha_3 \neq 0$$

- *The elements $\alpha_1, \alpha_2, \alpha_3$ and their inverses can be computed in polynomial time.*

All super-singular curves satisfy these conditions. For genus 2 curves it can be shown that the curves of the form $y^2 = x^6 - ax^3 + 1$ and $y^2 = x^6 - ax^3 - 3$ satisfy these conditions [2].

2.2 Trapdoor VDP

To use VDP in cryptography, setting up a trapdoor is required [4, 8].

Definition 2.3 Let G be a group of exponent r and order r^2. Let $f\colon G \to G$ be a group isomorphism computable in polynomial time. A pair of elements $S, T \in G$ is an eigenvector base with respect to f if

1. $G = \langle S, T \rangle$
2. $f(S) = \lambda_1 S$ and $f(T) = \lambda_2 T$ for some distinct nonzero $\lambda_1, \lambda_2 \in Z/rZ$

Definition 2.4 An eigenvector base $\{S, T\}$ is said to be a distortion eigenvector base if there is group homomorphism $\phi_1: \langle S \rangle \rightarrow \langle T \rangle$ and $\phi_2: \langle T \rangle \rightarrow \langle S \rangle$ computable in polynomial time and if an integer d not congruent to 0 (mod r) is given such that $\phi_2(\phi_1(S)) = dS$.

Proposition Let $\{S, T\}$ be a distortion Eigen vector base for V normalized such that $T = \phi(S)$. Let $u_{11}, u_{12}, u_{21}, u_{22} \in Z/rZ$ be such that $u_{11}u_{22} - u_{12}u_{21} \neq (\text{mod } r)$. Let $P_1 = u_{11}S + u_{21}T$ and $P_2 = u_{12}S + u_{22}T$. Given $Q \in V$, if one knows the u_{ij}, then one can solve VDP of Q to the base $\{P_1, P_2\}$ [4, 10].

Trapdoor $V_{\text{Deco}}(c, \langle P1 \rangle, X, B)$
The function trapdoor can be used for vector decomposition. Let $E = F_p$ be a hyper elliptic curve and $F: E \rightarrow E$ be an endomorphism. Let $A = \{S, T\}$ be a basis of $E[m]$, the set of m-torsion points on E and S, $T \rightarrow E/F_p$ is such that $F(S) = \lambda_1 S$ and $F(T) = \lambda_2 T$. $X = (u_{ij})$ is a matrix where $u_{ij} \in F_p$. P_1 and P_2 are two points generated by X such that $P_1 = u_{11}S + u_{12}T$ and $P_2 = u_{21}S + u_{22}T$ and $u_{22}u_{11} - u_{12}u_{21} \neq 0$. Let $B = \{P_1, P_2\}$. Hence, B also forms a basis.

Consider the projection operator $P_j(c) = \frac{(F - \lambda_i)(c)}{\lambda_j - \lambda_i}$ and a distortion map $\phi_{ij}(b_j) = b_i$ where $b_1 = S$, $b_2 = T$ forms a distortion eigenvector base along with the distortion map. Also $(t_{ij}) = X^{-1}$.

$$V_{\text{Deco}}(c, \langle P_1 \rangle, X, B) = t_{11}u_{11}\phi_{11}(\text{Pr}_1(c)) + t_{11}u_{12}\phi_{21}(\text{Pr}_1(c))$$
$$+ t_{21}u_{11}\phi_{12}(\text{Pr}_2(c)) + t_{21}u_{12}\phi_{22}(\text{Pr}_2(c)).$$

Theorem Let $\{S, T\}$ be a distortion Eigen vector base for V normalized such that $T = \phi_1(S)$. Let $u_{11}, u_{12}, u_{21}, u_{22} \in Z/rZ$ be such that $u_{11}u_{22} - u_{12}u_{21} \neq 0(\text{mod } r)$. Let $P_1 = u_{11}S + u_{21}T$ and $P_2 = u_{12}S + u_{22}T$. Let $c = mP_1 + rP_2$. Then, VDeco$(c, \langle P_1 \rangle, X, (P_1, P_2)) = mP_1$ [4, 8].

3 Proposed Oblivious Transfer Using Vector Decomposition

Our scheme includes five steps executed by either the sender P_1 or the receiver P_2.

Step 1 *Key Generation*: This step is executed by receiver P_2.

- P_2 runs setup algorithm to generate public key P_k and private key S_k. Here, $P_k = (R_1, R_2)$ and $S_k = X$
- P_2 chooses random bit $x_i \in \{0, 1\}$ and defines $k^{x_i} = r^{x_i}R_2$ and $k^{1-x_i} = R_1 + r^{1-x_i}R_2$.

Step 2 *Initialization*: This step is executed by both sender and receiver.

- P_1 chooses a random number $s \in F_p$ and sends $Com_h(s)$ to P_2.
- P_2 chooses a random number $r \in F_p$ and sends $Com_b(r)$ to P_1.
- P_1 and P_2 then send decommitments to $Com_h(s)$ and $Com_b(r)$, respectively, set $x = rs$.

Step 3 *Request*: This step is executed by the receiver P_2. P_2 sends (k^0, k^1) pair with the random numbers r^0 and r^1 used in the encryption process. P_2 also sends a bit y_j so that

- If $\sigma = 0$, then $y_j = x_j$ and
- If $\sigma = 1$, then $y_j = 1 - x_j$

Step 4 *Respond*: This step is executed by the sender P_1. Sender verifies y_j and performs the following.

- If $y_j = 0$, then set $k = k^0$ and $k^* = k^1$
- If $y_j = 1$, then set $k = k^1$ and $k^* = k^0$

Sender performs the homomorphic encryption as follows

- $C_0 = uk^* + m_0R_1 + aR_2 + xR_1$
- $C_1 = vk^* + m_1R_1 + bR_2 + xR_1$, where u, v, a, b are random numbers.

Step 5 *Complete*: This step is executed by the receiver P_2. After getting (C_0, C_1), P_2 performs $VDeco(C_\sigma, R_1, X, (R_1, R_2)) = t_\sigma$ and $t_\sigma - xR_1 = m_\sigma R_1$. $m_\sigma = D\log_{R_1}(m_\sigma R_1)$.

3.1 Correctness of Our OT Scheme

Case 1: when $\sigma = 0$

(a) If $x_j = 0$, then $y_j = 0$. Take $k = k^0 = r^0R_2$ and $k^* = k^1 = R_1 + r^1R_2$.
Hence, $C_0 = ur^0R_2 + m_0R_1 + aR_2 + xR_1 = (m_0 + x)R_1 + (ur^0 + a)R_2$

$C_1 = v(R_1 + r^1R_2) + m_1R_1 + bR_2 + xR_1 = (v + m_1 + x)R_1 + (vr^1 + b)R_2 VDeco(C_0, R_1, X, (R_1, R_2))$
$= (m_0 + x)R_1(= t_0)$ and $VDeco(C_1, R_1, X, (R_1, R_2)) = (v + m_1 + x)R_1$

Hence, the receiver P_2 gets only m_0.

(b) If $x_j = 1$, then $y_j = 1$. Take $k = k^1 = r^1R_2$ and $k^* = k^0 = R_1 + r^0R_2$. Hence,

$C_0 = ur^1R_2 + m_0R_1 + aR_2 + xR_1 = (m_0 + x)R_1 + (ur^1 + a)R_2$
$C_1 = v(R_1 + r^0R_2) + m_1R_1 + bR_2 + xR_1 = (v + m_1 + x)R_1 + (vr^0 + b)R_2$
$VDeco(C_0, R_1, X, (R_1, R_2)) = (m_0 + x)R_1, VDeco(C_1, R_1, X, (R_1, R_2)) = (v + m_1 + x)R_1$

Hence, the receiver P_2 gets only m_0.

Case 2: when $\sigma = 1$

(a) If $x_i = 0$, then $y_j = 1 - x_i = 1$. Then, $k = k^1 = R_1 + r^1 R_2$ and
     $k^* = k^0 = r^0 R_2$. Hence,

$$C_0 = u(R_1 + r^1 R_2) + m_0 R_1 + a R_2 + x R_1 = (u + m_0 + x)R_1 + (ur^1 + a)R_2$$
$$C_1 = vr^0 R_2 + m_1 R_1 + b R_2 + x R_1 = (m_1 + x)R_1 + (vr^0 + b)R_2$$
$$VDeco(C_0, R_1, X, (R_1, R_2)) = (u + m_0 + x)R_1, VDeco(C_1, R_1, X, (R_1, R_2)) = (m_1 + x)R_1$$

Here, the receiver gets only m_1.

(b) If $x_j = 1$, then $y_j = 1 - x_i = 0$. Then, $k = k^0 = R_1 + r^0 R_2$ and
    $k^* = k^1 = r^1 R_2$ Then,

$$C_0 = u(R_1 + r^0 R_2) + m_0 R_1 + a R_2 + x R_1 = (u + m_0 + x)R_1 + (ur^1 + a)R_2$$
$$C_1 = vr^1 R_2 + m_1 R_1 + b R_2 + x R_1 = (m_1 + x)R_1 + (vr^1 + b)R_2$$
$$VDeco(C_0, R_1, X, (R_1, R_2)) = (u + m_0 + x)R_1, VDeco(C_1, R_1, X, (R_1, R_2)) = (m_1 + x)R_1$$

Here, the receiver gets only m_1.

4 Conclusion and Future Work

Oblivious transfer is traditionally described as a two-party protocol between sender and receiver. Here, maintaining user privacy is the important concern. In this paper, we provide OT 1-out-of-2 oblivious transfer scheme using vector decomposition, which insures user privacy. Our scheme can be extended to a *K*-out-of-*N* oblivious transfer. The proposed scheme is a two-party scheme and can be extended further to multiparty protocol where multiple users use a single database.

References

1. Balasubramanian, R., Koblitz, N.: The improbability that an elliptic curve has sub exponential discrete log problem under the Menezes-Okamoto-Vanstone algorithm. J. Cryptology. **11**(2), 141–145 (1998)
2. Duursma, I., Kiyavash, N.: The vector decomposition problem for elliptic and hyperelliptic curves. J. Ramanujan Math. Soc. **20**(1), 5976 (2005)
3. Even, S., Goldreich, O., Lempel, A.: A randomized protocol for signing contracts. In: CRYPTO 1982, pp. 205210 (1982)
4. Galbraith, S.D., Verheul, E.: An analysis of the vector decomposition problem. In: Cramer, R. (ed.) PKC 2008. LNCS, vol. 4939, pp. 308327. Springer, Heidelberg (2008)
5. Green, M., Hohenberger, S: Blind identity-based encryption and simulatable oblivious transfer. In ASIACRYPT '07, vol. 4833 of LNCS, pp. 265–282 (2007)

6. Green, M., Hohenberger, S: Universally composable adaptive oblivious transfer. In ASIACRYPT, pp. 179–197 (2008)
7. Lindell, Y.: Efficient fully-simulatable oblivious transfer. In: Malkin, T.G. (ed.) CT-RSA 2008. LNCS, vol. 4964, pp. 5270. Springer, Heidelberg (2008)
8. Okamoto, T., Takashima, K.: Homomorphic encryption and signatures from vector decomposition. In Pairing, pp. 57–74 (2008)
9. Praveen, I., Sethumadhavan, M.: An efficient pairing computation, 1st international conference on security of internet of things (SecurIT 2012), pp. 145–149, 2012. ISBN: 978–1–4503–1822–88
10. Praveen, I., Sethumadhavan, M.: An application of vector decomposition problem in public key cryptography using homomorphic encryption, international conference on emerging research in computing, information, communication and applications-ERCICA (2013)
11. Rabin, M.O.: How to exchange secrets by oblivious transfer, technical report TR-81, Aiken Computation Laboratory, Harvard University (1981)
12. Yoshida, M.: Inseparable multiplex transmission using the pairing on elliptic curves and its application to watermarking. In: Fifth conference on algebraic geometry, number theory, coding theory and cryptography, University of Tokyo (2003)

Query by Humming System Through Multiscale Music Entropy

Trisiladevi C. Nagavi and Nagappa U. Bhajantri

Abstract Query by humming (QBH) is one of the most active areas of research under music information retrieval (MIR) domain. QBH employs meticulous approaches for matching hummed query to music excerpts existing within the music database. This paper proposes QBH system based on the estimation of multiscale music entropy (MME). The proposed technique exploits the statistical reliability through the MME for music signals approximation. Further, the Kd tree is employed for indexing MME feature vectors of music database leading to reduced search space and retrieval time. Later, MME feature vectors are extracted from humming query for recognition and retrieval of the corresponding song from music database. The experimental results demonstrate that the proposed MME and Kd tree-based QBH system provides higher discrimination capability than the existing contemporary techniques.

Keywords Entropy · Kd tree · Multiscale music entropy (MME) · QBH

1 Introduction

The contemporary personalized ways of music processing techniques are changing the nature of music dissemination allowing ardent music lovers to get access to large music collections widespread over genres and artists. Thus, various researchers have contributed in distinct ways for the growth of music

T.C. Nagavi (✉)
Deptartment of Computer Science and Engineering,
Sri Jayachamarajendra College of Engineering, Mysore, India
e-mail: tnagavi@yahoo.com

N.U. Bhajantri
Department of Computer Science and Engineering,
Government Engineering College, Chamarajanagar, India
e-mail: bhajan3nu@gmail.com

© Springer India 2015
L.C. Jain et al. (eds.), *Intelligent Computing, Communication and Devices*,
Advances in Intelligent Systems and Computing 309,
DOI 10.1007/978-81-322-2009-1_17

dissemination leading to the invention of MIR systems. The MIR research is exploring new technique referred as QBH system for music database search using humming a short music excerpt.

So far, most of the QBH research efforts [1–7] have focused on building systems based on dynamic programming [7], hierarchical filtering [1], iterative deepening melody alignment [2, 6], time series alignment [3], dynamic time warping (DTW) [4], wavelet transforms, and envelope transforms [5] approaches. Although these techniques exhibit good experimental performance, there is a need of consistency estimation in understanding the nature of a music signal. Also to conquer the inadequacy of using big knowledge base, music indexing is being used.

The growing scientific and commercial interest in audio and music processing is seen recently in the works of various MIR applications. They are based on pitch [8, 9], rhythm [10], and note [5] estimation. In this paper, we are proposing new music melody estimation approach based on multiscale entropy (ME). Also we present, Kd tree-based music melody database indexing structure for efficient and reduced space searching.

This paper is organized as follows. In Sect. 2, comprehension of related contributions is presented. The brief view of the proposed QBH system is described in Sect. 3. The entropy estimation procedure, Kd tree-based indexing, query processing, and searching are presented in Sects. 4, 5, and 6, respectively. The results and discussions are portrayed in Sect. 7. Finally, we draw the conclusions in Sect. 8.

2 Comprehension of Related Contributions

The music signals often exhibit complex structure of regularity. Through the regularity estimation, it is possible to perceive the characteristics of the music signals. In the past, several applications are developed based on regularity estimation through entropy such as heart rate analysis [11], music beat tracking [12], traffic flow analysis [13], human gait dynamics [14], and measuring predictability in networks [15].

The effectiveness of these applications motivates us to exploit the potential of regularity estimation techniques for developing QBH systems. Several methods [12, 16–19] on entropy estimation have been developed to detect specific pattern in music signals. The task of beat tracking is popularly used mechanism in music information retrieval (MIR). Davies and Plumbley [12] have explored a new method for the evaluation of beat tracking systems.

Another work [16] focuses on music retrieval based on emotion recognition. The system enhances maximum entropy with Gaussian and Laplace priors to model features for music emotion recognition. Authors were successful in getting better results for maximum entropy with priors. An interdisciplinary research [17] was proposed to explore the concept of entropy in jazz improvisation. In order to measure the improvisation, they adapted standard melodic, harmonic, and rhythmic entropy analysis techniques.

Subsequently, the research work [18] addresses matching of music performances of the same piece of music, making it suitable for monitoring applications. The entropy-based audio finger printing was employed to represent music in the form of symbolic strings. The system was able to correctly identify Pop music with DTW, edit distance (ED), and longest common subsequence (LCS) distance measures in less than a second per comparison.

An information-based approach [19] was proposed for meaningful musical events analysis and recognition. The entropy derived from a predictive model of music corresponds to important feature of music. Also, neural network model was employed to estimate instantaneous entropy for music. Even the relationship between listener's subjective sense of tension and entropy is considered in the work.

The majority of the music processing systems have focused on developing holistic single sample entropy-based techniques for capturing regularity in the music signals. Nevertheless, they cannot capture the long-term structures and large amount of information present in the music signals. In regard to this disadvantage, we are proposing MME technique, which uses single sample entropies of a music signal at multiple scales to uncover several types of structure.

3 Proposed Query by Humming System

In order to illustrate the proposed system of QBH, let us consider distinguished training and testing phases. Each of these phases perform different operations on the input signal such as preprocessing, framing, entropy estimation, Kd tree-based indexing, and searching. Through the simplified block diagram of a general system as shown in Fig. 1, the proposed QBH system's steps are discussed.

3.1 Preprocessing

Preprocessing is applied on the MP3 songs database to remove nonessential information, for down sampling and converting to mono channel [20].

3.2 Framing

In framing, the music signal is split into several frames, such that each frame is analyzed in short time instead of analyzing the entire signal at once [20]. In our experiments, the frame length is set to 25 with 10 ms overlap between two adjacent frames.

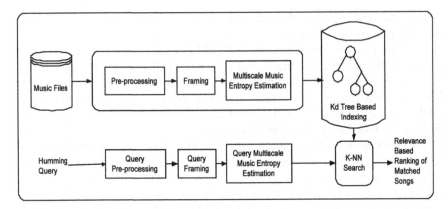

Fig. 1 Block diagram of the proposed QBH system

4 Entropy Estimation

The music signal data points often exhibit complex fluctuations containing information about the underlying regularity or consistency. Hence an important problem in music processing is determining whether the music signal has a regularity component. In this section, we discuss the entropy estimation procedure which can be quite effectively used to quantify the amount of regularity.

4.1 Shannon Entropy

Shannon's information theory defined a quantitative measure of information associated with an event A, which is a set of outcomes of some experiment. If $P(A)$ is the probability of an event A, then information associated with A is given by

$$i(A) = \log_b \frac{1}{P(A)} = -\log_b P(A) \tag{1}$$

So, the amount of information associated with low probability event is more and vice versa. If we have a set of independent events A_i, which are sets of outcomes of experiment S, then the average information associated with the experiment is given by

$$H(S) = \sum_{i=1}^{N} P(A_i) i(A_i) = -\sum_{i=1}^{N} P(A_i) \log_b P(A_i) \tag{2}$$

Here, the quantity $H(S)$ is an information entropy or Shannon entropy of S. So entropy is defined as measure of mean information content or information consistency or information regularity [21]. The choice of the logarithm base in Eq. (2) determines the unit of information. In this work, we have used the bits units obtained by the use of logarithm base 2.

4.2 Multiscale Entropy

Conventional single scale entropy measures, purely quantify the regularity of data samples. Significant structural richness or complexity which assimilates correlations over multiple spatiotemporal scales can be derived through multiscale analysis [22]. So, ME is adopted as a new method for measuring the complexity of data samples.

The Fig. 2a describes the pictorial representation of coarse graining baseline procedure for scale 1, 2, and 3. For a given data set, $X = \{x_1, x_2, x_3, \ldots, x_N\}$, multiple coarse grained sequences are constructed by averaging the data points within nonoverlapping windows of increasing length. Each element of the coarse grained sequences, y_j^δ, is generated according to the equation:

$$y_j^\delta = \frac{1}{\delta} \sum_{i=(j-1)\delta+1}^{j\delta} x_i \tag{3}$$

where δ represents scale level $1-l$ and $1 <= j < = N/\delta$. The length of the coarse grained sequence is $M = N/\delta$. For scale level $\delta = 1$, the coarse grained sequence is same as the original sequence x_i where $i = 1-N$.

We then render the entropy measurement e_δ of each coarse grained sequence y_j^δ of length M for scale level $\delta = 1-l$ as defined in the following equation:

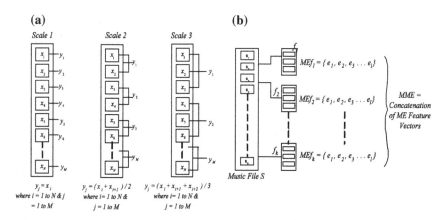

Fig. 2 **a** Schematic illustration of the coarse graining procedure for *scale 1, 2,* and *3.* **b** MME estimation procedure

$$e_\delta = -\sum_{j=1}^{M} P\left(y_j^\delta\right) \log_b P\left(y_j^\delta\right) \tag{4}$$

where $P\left(y_j^\delta\right)$ indicates probability of data point y_j at scale level δ and b is logarithm base. In this work, we have used the bits units obtained by the use of logarithm base 10. Now, ME is defined as the feature vector of entropies e_δ for scale levels $\delta = 1$–l as shown in Eq. (5):

$$\mathrm{ME}(X, \delta) = \overset{t}{\underset{\delta=1}{II}} e_\delta \tag{5}$$

where X is an original data set.

4.3 Multiscale Music Entropy

In this section, we reproduce the analysis of ME with reference to music signals. Given a music signal database

$$D = \{S_1, S_2, \ldots S_R\} \tag{6}$$

and a query Q, the aim is to find all the music signals in D that contain the specified query Q as well as feature vector similar to that of Q. One dimensional music signal S is defined as a sequence of values:

$$S = [s_1, s_2, \ldots s_N] \tag{7}$$

where N is the number of samples in S. Primary objective is to represent music melody in the form of ME.

Each music signal S is split into several frames, as defined in Eq. (8), so that it is possible to analyze each frame in short time.

$$S = [f_1, f_2, \ldots f_k] \tag{8}$$

where $f_1 = [s_1, s_2, \ldots s_{250}], f_2 = [s_{251}, s_{252}, \ldots s_{500}], \ldots f_k$. So, for every frame f, $\mathrm{MEf}_{i, \text{ feature}}$ vector of entropies e_δ for scale levels $\delta = 1$–l are constructed by following the procedure discussed in Sect. 4.2. Same is defined in Eq. (9):

$$\mathrm{MEf}_i = \overset{t}{\underset{\delta=1}{II}} e_\delta \tag{9}$$

where $i = 1$–k.

Finally feature vectors, MEf, of frames, $f_1 - f_k$, are concatenated to form the feature vector of one music signal referred as MME, expressed through equation as:

$$\text{MME} = \overset{t}{\underset{i=1}{II}} \text{MEf}_i \tag{10}$$

So, this process is repeated for all music signals and knowledge base of MME is built during training phase. MME construction system for database D is depicted in Fig. 2b and steps are shown in Algorithm 1.

Algorithm 1 : Procedure to construct Multiscale Music Entropy (MME) for music database

Input: a music database D, number of music signals R, number of entropy scale levels $\delta = 1$ to l and number of samples in each music signal N

Output: Multiscale Music Entropy MME_D data set

/* Framing */
for $i = 1 : R$ do
split music signal S_i of database D into several frames
$S_i = [f_1, f_2, ... f_k]$ /* k= number of frames */
end

/*Generating coarse grained sequence & estimating entropy for each sequence*/
for $z = 1 : R$ do
for $t = 1 : k$ do
for $\delta = 1 : l$ do
for $j = 1 : N/\delta$ do
multiple coarse grained sequences are generated for frame, f_t, of each music signal
$y_j^\delta = \frac{1}{\delta} \sum_{i=(j-1)\delta+1}^{j\delta} x_i$
/* x= original sequence in frame, f_t */
end
then render the entropy measurement of each coarse grained sequence
$e_\delta = - \sum_{j=1}^{M} P(y_j^\delta) \log_b P(y_j^\delta)$
where $P(y_j^\delta)$ indicates probability of data point y_j at scale level δ , $M = \frac{N}{\delta}$ and b is logarithm base
end

/*Concatenation of all entropies of coarse grained sequence to form MEf for every frame*/
$MEf_t = \overset{l}{\underset{\delta=1}{\|}} e_\delta$
end

/* Concatenation of all MEf of every frame to form MME for every music signal*/
$MME = \overset{k}{\underset{i=1}{\|}} MEf_i$
insert generated MME to the result data set MME_D
end
return the result data set MME_D

---

5 Kd Tree-Based Indexing

The Kd tree is a binary tree, which iteratively splits the entire input space into partitions. Each internal node represents splitting condition and leaves of this node denote subsets of the partition. Thus, the root node represents the entire input space, while the leaf nodes explicitly record the data points of smallest partitions [23]. The tree construction process becomes accustomed to the local characteristics of input space and the sizes of partitions at the same phase are not necessarily equal.

Our implementation employs the multiscale music entropy-balanced (MMEB) Kd tree for music database indexing. The tree indexing utilizes the MME statistical properties of music data for efficient and accurate searches. The MMEB Kd tree is an extension of statistical Kd tree which optimizes the multidimensional decision tree based on balancing split decisions. This balancing policy is adopted during tree building process so that the average and worst case search depth is reduced.

To construct a tree from a MME knowledge base, top down recursive procedure is employed. The common variation of splitting a subset of partition is the median point μ of the dimension d. This method of splitting guarantees a balanced tree and leads to reduced search depth. The cost of building a tree from N data points is O (NlogN). The steps of building MMEB Kd tree are shown in Algorithm 2.

Algorithm 2 : Procedure to construct Multiscale Music Entropy Balanced (MMEB) Kd tree for indexing music database

Input:number of music signals R in music database and Multiscale Music Entropy feature vectors $\{ MME_i \} \in Knowledge\ Base,\ where\ i\ =\ 1\ to\ R$

Output: A MMEB Kd Tree $\{K\}$. Every internal node has a split condition, $\{d, \mu\}$
  where d is the dimension and μ is the median to split on.
  Here all feature vectors,$\{ MME_i \} \in Knowledge\ Base,\ where\ i\ =\ 1\ to\ R,$
  belong to the left or right child based on split condition, $\{ d,\ \mu \}$.
  The leaf nodes have a list of indices to the features that ended up in that node

/* Build Function */
1. assign all the feature vectors ,$\{ MME_i \} \in Knowledge\ Base,\ where\ i\ =\ 1\ to\ R$ to the root node of K
2. every node in the tree is visited using *Breadth First Order*
3.*for* for all the feature vectors ,$\{ MME_i \} \in Knowledge\ Base,\ where\ i\ =$ $1\ to\ R$,dimension $d = 1, 2, 3, ..Z\ do$
compute its *median* μ from the points in that node
choose a *dimension d* at random
choose the split condition$\{ d,\ \mu \}$
for for all points that belong to this node: *do*
if $\{ MME_i[d] \} \leq median\ \mu$
  assign $\{ MME_i[d] \}$ to *left* [node]
  else
  assign $\{ MME_i[d] \}$ to*right* [node]
end if
end
end
4. step 3 is performed recursively until the subdivision depth exceeds threshold

---

6 Query Processing and Searching

Query processing basically involves query preprocessing, framing, and entropy estimation as discussed in Sects. 3 and 4. The melody database is a set of MME feature vectors indexed by MMEB Kd tree. The user hums different proportions of the query melody which is segmented into frames of 25 ms. Subsequently, each frame is resolved into query multiscale music entropy (QMME) feature vector.

Query vector QMME is searched in MMEB Kd tree using K-Nearest Neighbor (K-NN) algorithm to find those music signals whose MME feature vectors are close to QMME. Tree is recursively searched from the root to leaves. In concurrence with neighborhood range, search is carried out to locate those branches, which are close to the branches where the query resides. The cost of searching MMEB Kd tree with N data points is O (logN). The steps of K-NN searching are shown in Algorithm 3.

7 Results and Discussions

In order to evaluate the proposed method, several experiments are conducted with datasets and interesting trends are observed in the performance. The performance evaluation is done through standard yardsticks such as mean reciprocal rank (MRR), mean of average (MoA), and Top X Hit Rate.

The training database chosen for the experiments consists of 1,000 Indian devotional monophonic songs. These are further segmented to 1,350 fragments. Then a subset of 100, 200, 500, 800, 1,000, and 1,350 fragments are employed during training phase. The query database consists of total 200 hums from ten participants. The humming query is searched in the music database and QBH system's performance analysis is evaluated with different metrics such as MRR, MoA, and Top X Hit Rate [6, 7, 24].

Algorithm 3 : Procedure to search Multiscale Music Entropy Balanced (MMEB) Kd tree using K-NN

Input: MMEB Kd tree $\{K\}$, number of music signals R in music database, Multiscale Music Entropy feature vectors $\{ MME_i \} \in Knowledge Base$, where $i = 1 to R$ used to build tree, a query feature vector, $QMME$ and k to decide number of nearest neighbors

Output: A set of relevant music signals $\{A_i\}$ where $i = 1 to k$ which are nearest neighbors with their distances $\{B_i\}$ where $i = 1 to k$ to the query feature vector, $QMME$

/* K-NN Search Function */
1. given a humming query Q, extract its $QMME$ features
2. search the Kd Tree K to find the nearest neighbors $\{A_i\}$ where $i = 1 to k$ with their distances $\{B_i\}$ where $i = 1 to k$
3. initialize a priority queue, PQ, with the root nodes of the K tree by appending
   $branch = (node, val) with val = 0$
4. the priority queue PQ is indexed by $val [branch]$ and it returns the branch with smallest val
5. initialize $counter = 0, vector = [ ]$
6. for $counter <= max no of backtracking steps do$
   retrieve the top branch from PQ
   descend the tree defined by branch till leaf by adding unexplored branches on the way to PQ
$for i = 1 to length (QMME feature vector) do$
   if $QMME(i) == vector(i)$ then /*Matching condition*/
      $result \leftarrow result \cup QMME(i)$
   end if
   if $vector(i) == *$ then /*Partial query matching */
      perform match with $K.left, vector and result$
      perform match with $K.right, vector and result$
   else
      if $vector(i) < QMME(i)$ then
      perform match with $K.left, vector and result$
      else
      perform match with $K.right, vector and result$
      end if
   end if
end
end
7. locate the k nearest neighbors to PQ in vector and return the sorted vector
   $\{A_i\}$ where $i = 1 to k$ and their distances $\{B_i\}$ where $i = 1 to k$

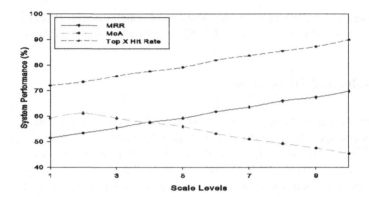

Fig. 3 QBH system's performance analysis with *MRR, MoA,* and *Top X Hit Rate*

7.1 Performance Analysis

MRR metric estimates any system which generates list of possible responses to a query. The probability of the target song coming in first position is estimated through MRR [20]. In the proposed strategy, MRR is obtained in the range 51.50–69.72 % for different scale levels. Also the performance of the system in terms of MRR goes on increasing with an increase in the scale level as depicted in Fig. 3. This is an indication that system's potential ability of discriminating songs increases with higher scale levels of MME estimation.

Similarly mean of accuracy (MoA) metric estimates the mean or average rank at which desired song is found for every query. The proposed strategy yields MoA in the range 59.23–45.32 % with scale levels one to ten. Experimental results demonstrate that MoA decreases with higher scale levels of MME estimation as depicted in Fig. 3. This is an indication that mean or average rank of the retrieved song decreases with increase in scale levels of MME estimation.

The third performance measure the Top X Hit Rate is employed to calculate approximately the percentage of successful queries. The different scale level of MME estimation procedure has significant effect on Top X Hit Rate and same is portrayed in Fig. 3. The Top X Hit Rate varies between 71.91 to 89.90 % for *X* value 10 and 1–10 scale levels of MME estimation. More humming queries return the desired song in the top 10 list when the scale level of MME estimation is increased from 1 to 10. The proposed QBH system based on MME exhibits significant performance in terms of accuracy, MRR and MoA metrics.

8 Conclusions

This paper is an attempt to exploit the advantages of Multiscale Entropy Analysis technique for estimating regularity and complex structure of music signals. Here, MME feature vectors are employed to represent music melody for music search through QBH system. Also, music melody is indexed through MMEB Kd tree to support fast and accurate search in the database to produce effective music retrievals. Proposed approach yields sensible performance by returning the desired song within the top 10 hits 89.90 % of the time and as the top hit 69.72 % of the time on a database with 1,000 songs and 1,350 fragments. Exhaustive exploration of the possibility of combining Multiscale Entropy Analysis techniques is to be considered in future work.

References

1. Addis, A., Armano, G., Vargiu, E.: Using the progressive filtering approach to deal with input imbalance in large-scale taxonomies. In: Proceedings of LSHC Workshop of ECIR (2010)
2. Jang, J.S.R., Lee, H.R.: Hierarchical filtering method for content based music retrieval via acoustic input. In: Proceedings of the 9th ACM Multimedia Conference, Canada, pp. 401–410 (2001)
3. Adams, N.H., Bartsch, M.A., Shifrin, J.B., Wakefileld, G.H.: Time series alignment for music information retrieval. In: Proceedings of 5th ISMIR, pp. 303–311 (2004)
4. Selina, C., Eamonn, K., David, H., Michael, P.: Iterative deepening dynamic time warping for time series. In: Proceedings of 2nd SIAM International Conference on Data Mining (2002)
5. Zhu, Y., Shasha, D.: Warping indexes with envelope transforms for QBH. In: Proceedings of the ACMSIGMOD International Conference on Management of Data, California, pp. 181–192 (2003)
6. Adams, N., Marquez, D., Wakefileld, G.: Iterative deepening for melody alignment and retrieval. In: Proceedings of ISMIR, pp. 199–206 (2005)
7. Jang, J.S.R., Lee, H.R.: An initial study on progressive filtering based on DP for QBSH. In: Proceedings of 7th IEEE Pacific-Rim Conference on Adv. in MIP, China, pp. 971–978 (2006)
8. Raju, M.A., Sundaram, B., Preeti Rao:. Tansen: a query-by-humming based music retrieval system. In: Proceedings of the National Conference on Communications (NCC) (2003)
9. Shifrin, J., Pardo, B., Meek, C., Birmingham, W.: HMM based musical query retrieval. In: Proceedings of 2nd ACM/IEEE-CS Joint Conference on DL, Oregon, USA, pp. 295–300 (2002)
10. Jeon, W., Ma, C.: Efficient search of music pitch contours using wavelet transforms and segmented DTW. In: Proceedings of IEEE Internationl Conference on ICASSP, Prague, pp. 2304–2307 (2011)
11. Thuraisingham, R.A., Gottwald, G.A.: On multiscale entropy analysis for physiological data. Technical report (2006)
12. Davies, M.E.P., Plumbley, M.D.: On the use of entropy for beat tracking evaluation. In: Proceedings of IEEE International Conference on ASSP, Honolulu, HI, pp. 1305–1308 (2007)
13. Yan, R.Y., Zheng, Q.H.: Multi-scale entropy based traffic analysis and anomaly detection. In: Proceedings of 8th International Conference on ISDA, Kaohsiung, Taiwan, pp. 151–157 (2008)
14. Costa, M., Peng, C.K., Goldberger, A.L., Hausdorff, J.M.: Multiscale entropy analysis of human gait dynamics. Phys. A Stat. Mech. Appl. Technical report (2003)
15. Riihijarvi, J., Wellens, M., Mahonen, P.: Measuring complexity and predictability in networks with MEA. In: Proceedings of IEEE INFOCOM, Rio de Janeiro, pp. 1107–1115 (2009)
16. He, H., Chen, B., Guo, J.: Emotion recognition of pop music based on maximum entropy with priors. In: Proceedings of 13th Pacific-Asia Conference, PAKDD, Thailand, pp. 788–795 (2009)

17. Simon, S.J.: Measuring Information in Jazz Improvisation. Technical report School of Library and Information Science, University of South Florida, South Florida (2007)
18. Ibarrola, A.C., Chavez, E.: On musical performances identification, entropy and string matching. In: Proceedings of MICAI, Springer, Advances in AI LNCS, pp. 952–962 (2006)
19. Cox, G.: On the relationship between entropy and meaning in music: an exploration with recurrent neural networks. In: Proceedings of the Annual Meeting of the Cognitive Science Society (2010)
20. Trisiladevi, C.N., Nagappa, U.B.: Perceptive analysis of QBS system through query excerption. In: Proceedings of the 2nd International Conference on CCSEIT, ACM, India, pp. 580–586 (2012)
21. Sayood, K.: Introduction to Data Compression, 3rd edn. Elsevier (2006)
22. Costa, M., Goldberger, A.L., Peng, C.K.: Multiscale entropy analysis of biological signals. Technical report (2005)
23. Aly, M., Munich, M., Perona, P.: Distributed Kd-trees for retrieval from very large image collections. In: Proceedings of BMVC, Dundee, UK (2011)
24. Jang, J.S.R., Lee, H.R.: A general framework of progressive filtering and its application to query by singing/humming 16:350–358 (2008)

An Empirical Analysis of Training Algorithms of Neural Networks: A Case Study of EEG Signal Classification Using Java Framework

**Sandeep Kumar Satapathy, Alok Kumar Jagadev
and Satchidananda Dehuri**

Abstract With the pace of modern lifestyle, about 40–50 million people in the world suffer from epilepsy—a disease with neurological disorder. Electroencephalography (EEG) is the process of recording brain signals that generate due to a small amount of electric discharge in brain. This may occur due to the information flow among several neurons. Therefore, in every minute, analysis of EEG signal can solve much neurological disorders like epilepsy. In this paper, a systematic procedure for analysis and classification of EEG signal is discussed for identification of epilepsy in a human brain. The analysis of EEG signal is made through a series of steps from feature extraction to classification. Feature extraction from EEG signal is done through discrete wavelet transform (DWT), and the classification task is carried out by MLPNN based on supervised training algorithms such as backpropagation, resilient propagation (RPROP), and Manhattan update rule. Experimental study in a Java platform confirms that RPROP trained MLPNN to classify EEG signal is promising as compared to back-propagation or Manhattan update rule trained MLPNN.

1 Introduction

Analysis and measurement of EEG signal are used for examination of human brain state and function [2, 4]. Like positron emission tomography (PET) and magnetic resonance imaging (MRI), EEG method has still important place due to its

S.K. Satapathy (✉) · A.K. Jagadev
Department of Computer Science and Engineering,
Institute of Technical Education and Research, SOA University,
Bhubaneswar, India
e-mail: sandeepkumar04@gmail.com

S. Dehuri
Department of Systems Engineering, Ajou University,
San 5, Woncheon-Dong, Suwon, South Korea
e-mail: satchi.lapa@gmail.com

© Springer India 2015
L.C. Jain et al. (eds.), *Intelligent Computing, Communication and Devices*,
Advances in Intelligent Systems and Computing 309,
DOI 10.1007/978-81-322-2009-1_18

excellent temporal resolution, low price, and availability. Human brain consists of several neurons [3]. When there will be any flow of information into brain, these neurons hit each other. During this procedure, electricity is generated, which is very small in amount and transient in nature. This is also called as a non-stationary signal, because the frequency of this signal is not fixed for a certain amount of time. Biologically, it is believed that during the activation of brain cells, the synaptic currents are produced within the dendrites. As a result of that, these currents generate a magnetic field measureable by a machine and a secondary electrical field over the scalp measurable by EEG systems [1].

The structure of this paper is organized as follows: The first part deals with the introduction to the EEG systems, followed by various EEG recording methodologies; the second part shows some of the previous findings and related work in this domain. The third part of the layout introduces the discrete wavelet transform (DWT) technique for EEG signal decomposition, followed by the fourth segment that introduces various propagation training algorithms. Lastly, the experimental study and result analysis show the comparison of various training algorithms used in MLPNN.

2 Related Work

There are many methods proposed for epileptic seizure detection, and some of them are discussed below. Guler et al. [5] have found an accuracy of 93.63 % by applying the MLPNN along with feature extraction techniques such as DWT and Lyapunov exponents. Naderi et al. [6] have proposed a very efficient classifier model using recurrent neural network based on Bartlett's idea of splitting the data into segments and finding the average of their periodogram. Guler et al. [7] have used DWT and Lyapunov exponents as feature extraction techniques, and they have used RNN to design classifier model for which they got an accuracy of 96.79 %. Subasi et al. [8] have also proposed an efficient classification technique for epileptic seizure detection using SVM.

3 Discrete Wavelet Transform for EEG Signal Decomposition

EEG signal is non-stationary signal, and it is not easy to get transient and distinct features from the signal without any proper methodology applied. The wavelet transform decomposes the signal into different scales with different levels of resolution by dilating the mother wavelet. It is defined as shown in Eq. (1):

$$\varphi_{a,b}(t) = \frac{1}{\sqrt{|a|}} \varphi\left(\frac{t-b}{a}\right) \tag{1}$$

where $a, b \in R, a \neq 0$ measures the degree of compression and the time location of the wavelet, respectively.

DWT signal is very much similar to sub-band coding and pyramidal coding or multiresolution analysis that was proposed by Crochiere et al. [9]. The DWT uses multiresolution filter banks and special wavelet filters for the analysis and reconstruction of signals. The resolution of the signal (that is a measure of the amount of detail information in the signal) is determined by the filtering, and the scale is determined by upsampling and downsampling [10, 11]. It is also called as the Mallat algorithm or Mallat tree decomposition [11]. This algorithm includes decomposition of a signal into the approximations and details.

4 Training Algorithms for MLPNN

In this paper, we have discussed three different training algorithms used in multilayer perceptron neural network. They are backpropagation, resilient propagation (RPROP), and Manhattan update rule.

4.1 Backpropagation

It is the most widely used propagation training algorithm for feed-forward neural networks [12]. This algorithm is different from other algorithms in terms of the weight-updating strategies. In backpropagation, generally weight is updated by the following Eq. (2).

$$wij(k+1) = wij(k) + \Delta wij(k), \tag{2}$$

where in regular gradient decent,

$$\Delta wij(k) = -\eta \frac{\partial E}{\partial wij}(k), \tag{3}$$

with a momentum term

$$\Delta wij(k) = -\eta \frac{\partial E}{\partial wij}(k) + \mu \Delta wij(k-1). \tag{4}$$

4.2 Manhattan Update Rule

The basic problem with backpropagation training algorithm is to determine the degree to which weights are changed. Manhattan update rule only uses the sign of the gradient, and magnitude is discarded. If the magnitude is zero, then no change is made to the weight or threshold value. If the sign is positive, then the weight or threshold value is increased by a specific amount defined by a constant. If the sign is negative, then the weight or threshold value is decreased by a specific amount defined by a constant. This constant must be provided to training algorithm as a parameter.

4.3 Resilient Propagation

RPROP, short form for resilient propagation [13], is a supervised training algorithm for feed-forward neural network. Instead of magnitude, it takes into account only the sign of the partial derivative or gradient decent and acts independently on each weight. The advantage of RPROP algorithm is that it requires no setting of parameters before using it. The weight updating is performed according to the following equations. Equation 1 from above is same for the RPROP for weight update.

$$
\Delta wij(k) = \begin{cases} +\Delta ij(k), \text{if } \frac{\partial E}{\partial wij}(k) > 0 \\ -\Delta ij(k), \text{if } \frac{\partial E}{\partial wij}(k) < 0 \\ \quad 0, \text{Otherwise} \end{cases} \tag{5}
$$

$$
\Delta ij(k) = \begin{cases} \eta + *\Delta ij(k-1), Sij > 0 \\ \eta - *\Delta ij(k-1), Sij < 0 \\ \quad \Delta ij(k-1), \text{Otherwise} \end{cases} \tag{6}
$$

where $Sij = \frac{\partial E}{\partial wij}(k-1) * \frac{\partial E}{\partial wij}(k)$, $\eta_+ = 1.2$ and $\eta_- = 0.5$.

5 Experimental Study, Results, and Analysis

Classification of EEG [14, 15] brain signal is done for the identification of different epileptic seizures using MLPNN with three training algorithms. The design of MLPNN model consists of two phases. First is training phase, and second is testing phase. For our work, the training algorithms such as backpropagation, RPROP, and Manhattan update rule are used and their performances are compared.

A detailed step-by-step procedure for EEG signal analysis and classification is mentioned below:

Phase I. *EEG signal decomposition*
This is the signal analysis phase where the raw EEG signal is collected from publicly available database. This dataset generally consists of five sets (A, B, C, D, and E). Each set is having 100 single channels of EEG recording. In our work, we have considered two sets A (normal patient) and E (epileptic patient). These signals are decomposed using DWT [16, 17] with Daubechies method of order 2 and up to level 4 (Fig. 1).

Phase II. *Feature Extraction*
After signal decomposition, a set of statistical features are extracted such as min, max, mean, and standard deviation. Altogether a complete dataset of dimension 200×20 is constructed with 200 samples and 20 features.

Phase III. *Java Platform and Parameter Setup*
A complete Java framework has been designed to carry out all the experimental analysis. It is created using basic core Java packages such as util, io, awt, and swing. Specific package has been created where each package contains methods that take required arguments for execution. Backpropagation training takes two parameters to execute such

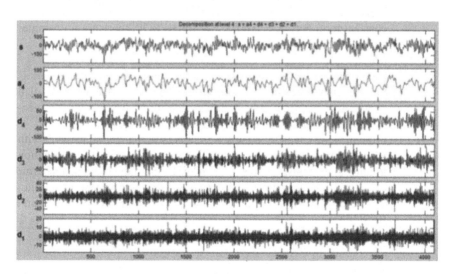

Fig. 1 Single-channel EEG from normal person with four detailed and one approximation coefficient after decomposition using DWT

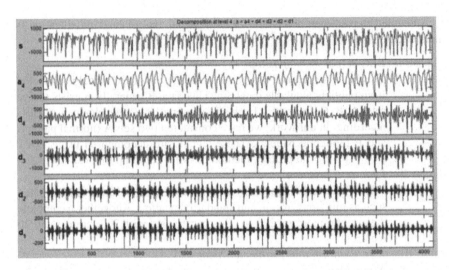

Fig. 2 Single-channel EEG from epilepsy-affected person with four detailed and one approximation coefficient after decomposition using DWT

as learning rate ($\eta = 0.7$) and momentum coefficient ($\mu = 0.8$). Similarly, Manhattan update rule takes only one parameter, which is the constant term (say $\alpha = 0.001$). But for RPROP, there are no parameters required (Fig. 2).

Phase IV. *Classification*

A classifier model is designed using a MLPNN [18, 19]. The network is trained using three different propagation training algorithms. A detailed comparison between three training algorithms using different measurement parameters is shown in Table 1 given below.

From the above table of comparison, it is clear that neural network with RPROP provides the best result as compared to other propagation training algorithms. Figures 3, 4, and 5 illustrate the mean square error (MSE) during training of the MLPNN using three different training algorithms such as backpropagation, Manhattan update rule, and RPROP.

Table 1 Comparison of three different training algorithms used in MLPNN for EEG signal classification study

Classifier type	Signal analysis tool	Total accuracy	Specificity	Sensitivity	PPV	NPV	F1 score	FDR	FPR
MLPNN with backpropagation	DWT with DB2, level 4	94.5	94.1	95	94	95	94.5	6	5.9
MLPNN with Manhattan update rule	DWT with DB2, level 4	96	96.9	95	97	95	96.04	3	3.06
MLPNN with RPROP	DWT with DB2, level 4	100	100	100	100	100	100	0	0

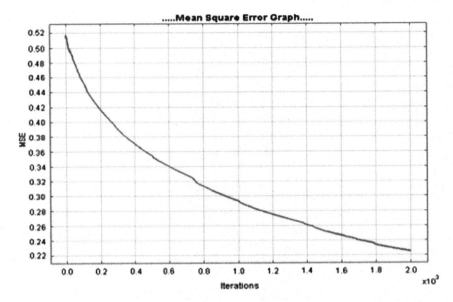

Fig. 3 Mean square error graph for MLPNN with backpropagation using maximum number of epochs 2000

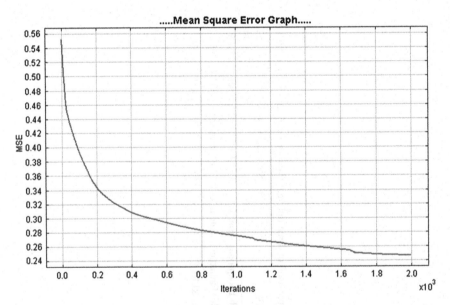

Fig. 4 Mean square error graph for MLPNN with Manhattan update rule with maximum number of epochs 2000

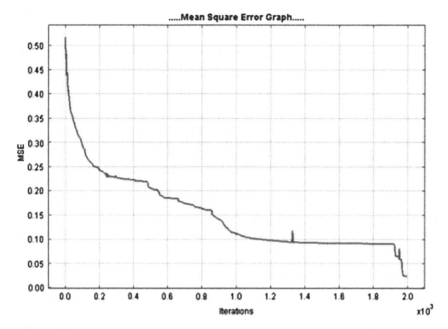

Fig. 5 Mean square error graph for MLPNN with RPROP with maximum number of epochs 2000

6 Summary and Conclusion

We have clearly presented a comparison of three different training algorithms used for designing artificial neural network model to analyze the EEG signal effectively [20]. A Java-based framework has been designed to create a neural network model trained with three training algorithms. It is proved that for this specific domain, neural network with RPROP is most efficient as compared to other training algorithms.

References

1. Sanei, S., Chambers, J.A.: EEG Signal Processing. Wiley Publications, New York (2008)
2. Niedermeyer, E., Lopes da Silva, F.: Electroencephalography: Basic Principles, Clinical Applications, and Related Fields. Lippincot Williams & Wilkins, 2004
3. Towle, V.L., Bolaños, J., Suarez, D., Tan, K., Grzeszczuk, R., Levin, D.N., Cakmur, R., Frank, S.A., Spire, J.P.: The spatial location of EEG electrodes: locating the best fitting sphere relative to cortical anatomy. Electroencephalogr. Clin. Neuro. Physiol. **86**(1), 1–6 (2003)

4. Aurlien, H., Gjerde, I.O., Aarseth, J.H., Karlsen, B., Skeidsvoll, H., Gilhus, N.E.: EEG background activity described by a large computerized database. Clin. Neurophysiol. **115**(3), 665–673 (2004)
5. Guler, I., Beyli, E.D.U.: Multi-class support vector machines for EEG-signals classification. IEEE Trans. Inf Technol. Biomed. **11**(2), 117–126 (2007)
6. Naderi, M.A., Homayoun, M. N., Analysis and classification of EEG signals using spectral analysis and recurrent neural networks. In: International Conference on Biomedical Engineering (ICBME), pp. 1–4 (2010)
7. Guler, N.F., Ubeyli, E.D., Gule, I.: Recurrent neural network employing Lyapunov exponents for EEG signal classification. Expert Syst. Appl. **29**, 506–514 (2005)
8. Subasi, A., Ismail Gursoy, M.: EEG signal classification using PCA, ICA, LDA and support vector machines. Expert Syst. Appl. **37**, 8659–8666 (2010)
9. Durand, S., Froment, J.: Artifacts Free signal denoising with wavelets. In: IEEE, published in the 2001 International Conference on Acoustics, Speech and Signal Processing, vol. 6, pp. 3685–3688 (2001)
10. Ocak, H.: Optimal classification of epileptic seizures in EEG using wavelet analysis and genetic algorithm. Sig. Process. **88**, 1858–1867 (2008)
11. Sifuzzaman1, M., Islam1, M.R., Ali, M.Z.: Application of wavelet transform and its advantages compared to fourier transform. J. Phys. Sci. **13**, 121–134 (2009)
12. Subasi, A., Ercelebi, E.: Classification of EEG signals using neural network and logistic regression. Comput. Methods Programs Biomed. **78**, 87–99 (2005)
13. Riedmiller, M., Braun, H.: A direct adaptive method for faster backpropagation learning: the RPROP algorithm. In: IEEE Internal Conference on Neural Networks, vol 1, pp. 586–591 (1993)
14. Arab, M.R., Suratgar, A.A., Martínez Hernández, V.M., Ashtiani, A.R.: Electroencephalogram signals processing for the diagnosis of petit mal and grand mal epilepsies using an artificial neural network. J. Appl. Res. Technol. **8**(1), 120–129 (2010)
15. Hagan, M.T., Menhaj, M.B.: Training feedforward networks with the Marquardt algorithm. IEEE Trans. Neural Netw. **5**(6), 989–993 (1994)
16. Adeli, H., Zhou, Z., Dadmehr, N.: Analysis of EEG records in an epileptic patient using wavelet transform. J. Neurosci. Methods **123**, 69–87 (2003)
17. Acir, N., Oztura, I., Kuntalp, M., Baklan, B., Guzelis, C.: Automatic detection of epileptiform events in EEG by a three-stage procedure based on artificial neural networks. IEEE Trans. Biomed. Eng. **52**(1), 30–40 (2005)
18. D'Alessandro, M., Esteller, R., Vachtsevanos, G., Hinson, A., Echauz, A., Litt, B.: Epileptic seizure prediction using hybrid feature selection over multiple intracranial EEG electrode contacts: a report of four patients. IEEE Trans. Biomed. Eng. **50**(5), 603–615 (2003)
19. Nigam, V.P., Graupe, D.: A neural-network-based detection of epilepsy. Neurol. Res. **26**(1), 55–60 (2004)
20. Subasi, A.: Automatic detection of epileptic seizure using dynamic fuzzy neural networks. Expert Syst. Appl. **31**, 320–328 (2006)

Relating Perceptual Feature Space and Context Drift Information in Query by Humming System

Trisiladevi C. Nagavi and Nagappa U. Bhajantri

Abstract The advancement in the field of music signal processing insists on effective music information retrieval (MIR) techniques. Query by humming (QBH) system is one of the active research areas under MIR. In this paper, we propose a QBH system for automatically retrieving the desired song based on humming query and human perceptual features. In the proposed system, five perceptual features corresponding to four perceptual properties are extracted. Further, the temporal relationship of these features is estimated through the transfer entropy (TE). The trajectory of TE of the target music database is analyzed to find the match for humming query. Series of experiments are conducted to evaluate the effectiveness of the system with 1,200 songs target database and 200 humming queries. The results show that the proposed method is robust in finding desired song automatically with hummed query as input.

Keywords Context drift · Music information retrieval (MIR) · Perceptual features · Query by humming (QBH) · Transfer entropy (TE)

1 Introduction

The music information retrieval (MIR) systems deal with automatic music information processing and retrieval for applications like radio or disco jockey playlist generation, online music access, frame-up of personal and public music

T.C. Nagavi (✉)
Department of Computer Science and Engineering, S.J. College of Engineering, Mysore, India
e-mail: tnagavi@yahoo.com

N.U. Bhajantri
Department of Computer Science and Engineering, Government Engineering College, Chamarajanagar, India
e-mail: bhajan3nu@gmail.com

© Springer India 2015
L.C. Jain et al. (eds.), *Intelligent Computing, Communication and Devices,*
Advances in Intelligent Systems and Computing 309,
DOI 10.1007/978-81-322-2009-1_19

collections. The query by humming (QBH) is a music retrieval technique based on hum of a small piece of music under MIR systems. The characteristic features of music melody are categorized into the following types: statistical, acoustical, and perceptual. Since the music data contains lot of perceivable information, we feel perceptual features are the best to discriminate among different music samples.

A great deal of research [9–13] is carried out in the field of QBH, but there is no significant amount of literature [1–4] toward capturing the perceptual aspects. Most of the researches based on perceptual features [1–3] extract pitch, loudness, and intensity features. Recently, spectral and harmonicity [4] features have become popular because of the rich information content and discrimination capacity. Therefore, in this paper, we are proposing QBH system based on enhanced perceptual features such as spectrum centroid (SC), spectrum spread (SS), pitch class profile (PCP), and harmonicity. Then, in order to capture the dynamics and context switching of a music signal, transfer entropy (TE) estimation is carried out.

The rest of the paper is organized as follows. Existing work on content-based music retrieval using perceptual features is elaborated in Sect. 2. The overview of the proposed framework is described in Sect. 3. Perceptual features used for representing music melody are discussed in detail in Sect. 4. The TE estimation based on perceptual features is presented in Sect. 5. Experimental results are discussed in Sect. 6. Finally, Sect. 7 concludes the paper along with the future work.

2 Comprehension of Related Contributions

Few studies [1–4] have explored the techniques of music retrieval from perceptual features. These systems in general presented useful strategies to extract and build music content analysis systems and also reviewed different perceptual feature-based methods for the music retrieval.

A Web-based QBH system was built using pitch and rhythmic information using dynamic programming classifier. The system evaluation yielded 43.44 and 75.63 % accuracy for exact and top 10 matches, respectively [1]. In another work [2], music segmentation framework based on perceptual features cutting points detection is presented. Variety of combinations of feature extraction and machine learning algorithms are employed for classifying music into perceptual categories by the authors [3]. Results were over the baseline for different combination of feature set and learning algorithms. Authors feel that combining issues like cultural, usage pattern, and listening habits with audio features may yield enhanced accuracy. Based on SC, spectral flux, spectral roll-off, and zero crossing rate multiple feature vectors content-based audio retrieval system is proposed in another work [4] and obtained 60–70 % accuracy.

Literature review based on music content representation and retrieval shows lack of perception-based models. After the review of the preexisting work in this domain, it is evident that there are some isolated attempts in exploiting perceptual features for QBH applications. Here, we are presenting a unique approach for QBH system based on perceptual features and TE to represent different aspects of music melody.

3 Envisioned Query by Humming System

There exists substantial research [5–7] to provide evidence that the majority of the people are more acutely sensitive to the perceptual features in preference to structures of the music. Researchers [5–7] enumerated perceptual features in psychoacoustics, particularly pitch, loudness, timbre, and beat. The prominence between many of these characteristics as well as individuals perception of hearing is also mentioned. Due to this fact, we envisioned a music retrieval system based on humming and perceptual features. Figure 1 details the overview of the proposed framework.

The preprocessing exercised on music database removes nonessential information such as noise, down-samples music signals to 8 kHz, and converts each music signal to monochannel [9]. Then, framing splits music signal into several frames, such that each frame is analyzed in short time instead of analyzing the entire signal at once [9].

The envisioned QBH system performs music knowledge base generation followed by music retrieval and analysis. Each one of these stages of development accomplishes specific operations on the input signal which include preprocessing,

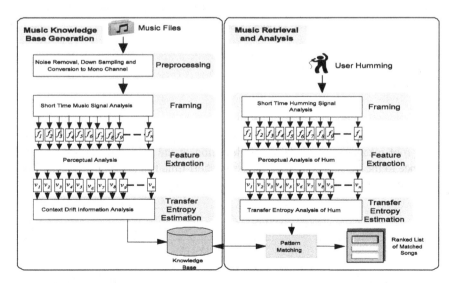

Fig. 1 Envisioned query by humming (QBH) system

framing, perceptual analysis, and context drift information analysis. Eventually, pattern matching is carried out to fetch the preferred song in response to hummed query.

4 Perceptual Analysis

This section emphasizes the set of prophesy features that manifest the music perception appropriately either in frequency or time domain. They are SC, SS, PCP, and harmonicity.

4.1 Spectral Centroid

The SC is referred as the center of gravity of the magnitude spectrum of the short-time Fourier transform (STFT) and computed for every short-time frame of music signal. It is a common approximation of brightness and a point where most of the energy is concentrated which is correlated with dominant frequency of the signal. The following equation is used for estimating the SC:

$$SC_i = \frac{\sum_{k=1}^{N/2} f(k)|X_i(k)|}{\sum_{k=1}^{N/2} |X_i(k)|} \tag{1}$$

where N is the total number of the fast Fourier transform (FFT) points, $X_i(k)$ is the power of the kth FFT point in the ith frame, and $f(k)$ is the corresponding frequency of the FFT point. Each frame produces singular SC value.

4.2 Spectral Spread

The SS describes the shape property of the spectrum which is obtained through STFT. This metric helps to distinguish pure tones from noise. Further, it is defined as the root mean square value of the deviation of the magnitude spectrum with reference to the SC and computed for every short-time frame of music signal. The following equation precisely represents SS:

$$SS_i = \sqrt{\frac{\sum_{k=1}^{N/2} \left((f(k) - SC_i)^2 |X_i(k)| \right)}{\sum_{k=1}^{N/2} |X_i(k)|}} \tag{2}$$

Like SC, SS also is a singular value generated from each frame.

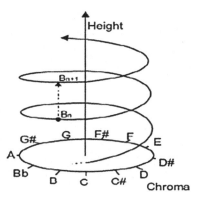

Fig. 2 Pitch class profile (PCP) or chroma

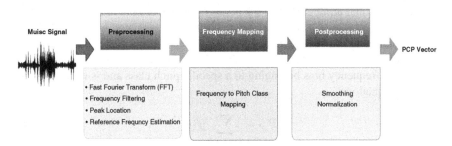

Fig. 3 Block diagram of computing pitch class profile (PCP) or chroma vector

4.3 Pitch Class Profile or Chroma

The pitch-related music descriptors which subsume tonal information for music perception are useful for high semantic description. Perception of pitch has 2 aspects: One is height and another is chroma. Since PCP is one of the best multi-pitch estimators leading to the tonal description, we are using PCP in this system. Other 2 estimators are constant Q profile and harmonic PCP. PCP also referred as chroma represents the energy distribution of the music signal across a predefined 12 pitch classes. As depicted in Fig. 2, each pitch class represents 1 note of the 12 tone equal temperament. Our system extracts PCP from every frame for music structure analysis. The general block diagram for extracting PCP is depicted in Fig. 3.

4.3.1 Preprocessing

PCP uses the FFT to convert music signal into frequency domain. Then, frequency filtering is applied so that we use desired frequency band between 100 and 5,000 Hz. Peak location of local maxima of the spectrum is performed. Then, estimation of reference tuning frequency is performed. The 440 Hz is considered as the standard reference frequency for frequency deviation analysis.

4.3.2 Frequency Mapping

Frequency to pitch class mapping divides the spectrum into regions. Further, the amplitudes of every region are added and divided by the number of bins inside the region, which produces a histogram. Then, the histogram is folded, by bringing tones of similar class to the same chroma bin, producing PCP vector of size 12; every index signifies the intensity of the one note. The PCP is expressed precisely as follows:

$$p(k) = [12 \times \log_2((k/N) \times (f_{sr}/f_{ref}))] \mathrm{mod} 12 \tag{3}$$

where N is the total number of the FFT points or bins, k is a bin in the FFT with the range $0 \leq k \leq N - 1$, f_{ref} is the reference frequency associated with PCP[0], and f_{sr} is the sampling rate. Here, each FFT bin is mapped to its adjacent note. Calculation of the values of the PCP elements is performed by summing the magnitude of all frequency bins belonging to a specific pitch class and is formulated in the below equation:

$$PCP_{[p]} = \sum_{k:p(k)=p} |f(k)|^2 \tag{4}$$

where k is a bin in the FFT with $0 \leq k \leq N - 1$, $f(k)$ is the frequency bin belonging to specific pitch class, and $p = 0$–11. Each frame generates a vector of 12 PCP values.

4.3.3 Postprocessing

After estimating the PCP vectors, smoothing is applied for introducing the robustness against fast fluctuations and variations. Finally, the estimated features are normalized framewise by dividing through the highest value to remove dependency on global loudness. Then, sequences of PCP values are obtained.

4.4 Harmonicity

Harmonicity property establishes distinction among periodic also referred as harmonic and non-periodic also referred as inharmonic noise like signals. Frequencies at integer multiples of the fundamental frequency are referred as harmonics. Harmonic structure of a music signal can be described through harmonic spectral centroid (HSC) and harmonic spectral spread (HSS). They are based on fundamental frequency estimation and recognition of harmonic peaks in the spectrum. These descriptors indicate statistical properties of the harmonic frequencies as well as their amplitudes.

4.4.1 Harmonic Spectral Centroid

The HSC is estimated as the amplitude-weighted average of the harmonic frequencies within a frame, expressed through equation as follows:

$$\text{HSC}_i = \frac{\sum_{h=1}^{M} f_i(h) A_i(h)}{\sum_{h=1}^{M} A_i(h)} \tag{5}$$

where M is the total number of harmonic frequencies, $A_i(h)$ is the amplitude of the hth harmonic frequency in the ith frame, and $f_i(h)$ is the corresponding harmonic frequency. Each frame produces singular HSC value, which represents the harmonic peak describing the brightness and sharpness of the spectrum.

4.4.2 Harmonic Spectral Spread

The HSS is the power-weighted root mean square deviation of the harmonic frequencies from the HSC, more precisely defined through equation as follows:

$$\text{HSS}_i = \frac{1}{\text{HSC}_i} \sqrt{\frac{\sum_{h=1}^{M} P_i^2(h)[f_i(h) - \text{HSC}_i]^2}{\sum_{h=1}^{M} P_i^2(h)}} \tag{6}$$

where M is the total number of harmonic frequencies, P_i^2 is the power of the hth harmonic frequency in the ith frame, and $f_i(h)$ is the corresponding harmonic frequency. Every frame generates single HSS value signifying the bandwidth of the harmonic frequencies also referred as the variation of harmonic frequencies.

5 Context Drift Information Analysis

The music signal often exhibits complex fluctuations containing information about the underlying consistency and dynamism. Hence, an important problem in music processing is detecting and locating these complex fluctuations in music signal. The context drifting is a process of context switching from one situation to another. Drifts may appear instantaneously or gradually [8]. Music signal also experiences context drifting while it moves from one position to the other.

In this section, we discuss the TE estimation procedure which can be quite effectively used to quantify the amount of music signal complex fluctuations and context drifting. To begin with, the perceptual features are computed from music signal as discussed in Sect. 4. Then, complex fluctuations and context drift information from perceptual space is estimated using TE measure, because it is reliable, efficient, as well as a better predictor.

5.1 Perceptual Feature Space

The TE estimation in the context of perceptual feature space is applied on frame of feature instances. We define frame of perceptual feature instances as a sequence comprising sequentially arranged values. Now, X is defined as the perceptual feature vector for frames $i = 1$ to R as shown in equation below:

$$X_{i=1}^{R} = \left\|_{j=1}^{t} s_j \right. \tag{7}$$

where $s_j \in \{(SC_i) \| (SS_i) \| (Sone_i) \| (PCP_i) \| (HSC_i) \| (HSS_i) \| (HSD_i) \| (HSV_i)\}$ and t is the total number of features per frame. So, this process is iterated for all music signals and knowledge base of X is developed.

5.2 Transfer Entropy Estimation

In this subsection, we employee TE estimation approach that automatically fine-tune to complex fluctuations and context drifts based on Shannon's entropy [8]. The TE captures dynamics and temporal relationship of a music signal and is based on rates of entropy change. It is used as symmetric measure of temporal relationship between two frames of perceptual features. The TE is mathematically represented as

$$TE = \sum_{j=1}^{t} P(s_{j+1},\ s_j,\ r_j) \log_b \frac{P(s_{j+1} \mid s_j, r_j)}{P(s_{j+1} \mid s_j)} \tag{8}$$

where s_j is jth feature in current frame, r_j is jth feature in the next frame, P denotes the probability of having the status in the following parentheses, and vertical bar in the parentheses indicates the conditional probability. We have used logarithms with base 2, and our units are in bits.

6 Results and Discussions

In order to match the humming query with the music database, we have employed a similarity function which independently takes into account signal frequency in addition to positional information [11]. Given a song S of music database D and humming query Q, feature vector TE_s extracted from song is matched with query TE_Q by means of the similarity measure:

$$\text{Similarity}(TE_s, TE_Q) = \sum_{i=1}^{R} \min(TE_{s_i}, TE_{Q_i}) \times \frac{\left(\sqrt{2} - d(TE_{s_i}, TE_{Q_i})\right)}{\sqrt{2}} \quad (9)$$

where $d(TE_{s_i}, TE_{Q_i}) = \sqrt{\sum_{i=0}^{R} (TE_{s_i}, TE_{Q_i})^2}$ is a Euclidean distance function and R is number of frames.

So as to evaluate the proposed system, various experiments are accomplished with datasets and significant trends are noticed in the performance. The training database chosen for the investigations encompasses 1,200 Indian devotional monophonic MP3 songs. However, these are further segmented to 1,495 fragments. Then, a subset of 100, 200, 500, 800, 1,000, 1,300, and 1,495 fragments are used during training phase. The detail specifications as well as figures used for establishing training database are shown in Table 1.

The humming query database comprises of total 200 hums from ten individuals. They were furnished with the lyrics of the fragments and advised to hum various proportions effectively. The performance evaluation is carried out with the aid of conventional yardsticks such as mean reciprocal rank (MRR), mean of average (MoA), and Top X hit rate, and results are tabulated in Table 2.

In the proposed strategy, the potential for the desired song appearing in top one position is predicted through the MRR. The MRR values were obtained in the range 52.12–70.80 % for various number of song fragments. The MRR goes on increasing with decreasing number of song fragments as tabulated in Table 2. This indicates that the system's ability of song recognition increases with decreasing number of song fragments.

Similarly, MoA metric provides the mean or average rank at which target is located for all queries. The proposed system facilitates MoA in the range 60.21–46.23 % for number of song fragments 1,495–100. To be precise, MoA decreases with increasing number of song fragments as shown in Table 2. That is, the mean or average rank of the retrieved song decreases with decreasing size of the database. Moreover, the likelihood of the target song arriving in initial five rankings tends to increase.

The third performance measure the Top X hit rate is used to indicate the percentage of fruitful queries. The number of song fragments employed has considerable effect on Top X hit rate as portrayed in Table 2. The Top X hit rate is in the range 72.23–90.13 % for number of song fragments 1,495–100. The

Table 1 Summary of training database

Category	Number of songs	Number of fragments	Age	Singers	M	F
Database I	1,200	1,495	20–50 years	39 professionals	22	17

Table 2 Performance analysis using number of song fragments

Performance	Number of song fragments						
	1,495	1,300	1,000	800	500	200	100
MRR	52.12	54.20	57.19	60.64	63.24	67.88	70.80
MoA	60.21	58.14	56.98	54.34	51.84	48.22	46.23
Top X hit rate	72.23	76.20	78.23	81.29	85.11	87.10	90.13

Table 3 Percentage of songs precisely retrieved using ranks and number of song fragments

Rank	Number of song fragments						
	1,495	1,300	1,000	800	500	200	100
1–2	58.90	59.20	61.99	64.53	65.94	67.81	70.03
3–4	56.10	57.14	61.98	62.34	64.39	65.48	68.22
5–6	53.77	55.12	58.55	59.11	61.44	62.12	65.32
7–8	51.11	53.00	55.11	57.35	59.91	60.21	62.14
9–10	49.87	51.14	52.99	54.77	56.14	58.40	59.33

observation is discriminating tendency deteriorates slowly when number of fragments are increased from 100 to 1,495.

Substantially more result analysis is carried out to learn about the likelihood of the desired song occurring in top 1–10 ranks. The rankings are tabulated in Table 3. Significantly better rankings are noted for top 1–2 ranks and number of song fragments 100. The proposed system is proficient at matching the query precisely when song fragments are less because of the availability of more discriminating information. Besides that, our results ensure that the desired song appears with in top 5 ranks normally.

7 Conclusions

Conventionally used music content recognition features do not rely upon human perception model. In this work, we propose to exploit the advantages of enhanced perceptual features and TE for music search through QBH system. Proposed perceptual features provide increased music retrieval reliability in comparison with conventionally used features, because these are derived based on a mathematical model of the human ear. Further, TE estimation is performed with reference to perceptual feature space to capture dynamics of the music retrieval system. Our experiments yield optimized performance by returning the target song within top ranks 72.23–90.13 % of the time and as the top rank 52.12–70.80 % of the time with various number of song fragments. In future, the potential of perceptual analysis technique is to be investigated with expanded melody database.

References

1. Cao, L., Hao, P., Zhou, C.: Music radar: a web-based QBH system. Technical Article
2. Jian, M.H., Lin, C.H., Chen, A.L.P.: Perceptual analysis for music segmentation. In: Proceedings of the SPIE, Storage and Ret. Methods and Application for Multimedia, vol. 5307, pp. 223–234 (2004)
3. Pohle, T., Pampalk, E., Widmer, G.: Evaluation of frequently used audio features for classification of music into perception categories. In: Proceedings of the 4th International WS on CBMI (2005)
4. Nandedkar, V.: Audio retrieval using multiple feature Vec. In: IJEEE, vol. 1(1) (2011)
5. Zhang, X., Ras, Z.W.: Analysis of sound features for music timbre recognition. In: Proceedings of the IEEE CS International Conference on MUE, Seoul, Korea, pp. 3–8, (2007)
6. Mitrovic, D., Zeppelzauer, M., Breiteneder, C.: Features for content based audio retrieval. Thesis appeared in advances in computers, vol. 78, pp. 71–150 (2010)
7. Sezgin, C., Gunsel, B., Kurt, G.: A New Perceptual Feature Set for Audio Emotion Recognition. Technical Report, Department of Electronic and Communication Engineering, Istanbul Technical University, Turkey (2010)
8. Vorburger, P., Bernstein, A.: Entropy-based concept shift detection. In: Proceedings of the ICDM, IEEE Computer Society, pp. 1113–1118 (2006)
9. Trisiladevi, C.N., Nagappa, U.B.: Perceptive analysis of QBS system through query excerption. In Proceedings of the 2nd International Conference on CCSEIT, ACM, India, pp. 580–586 (2012)
10. Guo, Z., Wang, Q., Liu, G., Guo, J.: A QBH system based on locality sensitive hashing indexes. Elsevier J. Sig. Process. 93(8), 2229–2243 (2013)
11. Trisiladevi, C.N., Nagappa, U.B.: Progressive filtering using multiresolution histograms for QBH system. In: Proceedings of the 1st ICMCCA, Springer LNEE, India, vol. 213, pp. 253–265 (2013)
12. Jang, J.S.R., Lee, H.R.: A general framework of progressive filtering and its application to QBSH. IEEE Trans. Audio Speech Lang Process 16(2), 350–358 (2008)
13. Qin, J., Lin, H., Liu, X.: Query by humming systems using melody matching model based on the Genetic Algorithm. J. S/W 6(12), 2416–2420 (2011)

First-Fit Semi-partitioned Scheduling Based on Rate Monotonic Algorithm

Saeed Senobary and Mahmoud Naghibzadeh

Abstract Semi-partitioned scheduling is a new approach for allocating real-time tasks to processors such that utilization is enhanced. Each semi-partitioned approach has two phases, partitioning and scheduling. In partitioning phase, tasks are assigned to the processors. In this phase, some tasks are probably split into several subtasks and each assigned to a different processor. The second phase is the policy to determine how to schedule tasks on each processor. The main challenge of semi-partitioned scheduling algorithms is how to partition and split tasks by which they are safely scheduled under the identified scheduling policy. This paper proposes a new semi-partitioned scheduling algorithm called SRM-FF for real-time periodic tasks over multiprocessor platforms. The scheduling policy used within each processor is based on rate monotonic algorithm. The partitioning phase of our proposed approach includes two sub-phases. Task splitting is done only in the second sub-phase. In the first sub-phase, processors are selected by a first-fit method. The use of first-fit method makes SRM-FF create lower number of subtasks in comparison to previous work hence the number of context switches of subtasks and overhead due to task splitting are reduced. The feasibility of tasks and subtasks which are partitioned by SRM-FF is formally proved.

Keywords Embedded systems · Real time scheduling · Rate monotonic · Semi-partitioned technique

S. Senobary (✉)
Imam Reza International University, Mashhad, Iran
e-mail: s.senobary@imamreza.ac.ir

M. Naghibzadeh
Department of Computer Engineering, Ferdowsi University of Mashhad, Mashhad, Iran
e-mail: naghibzadeh@um.ac.ir

© Springer India 2015
L.C. Jain et al. (eds.), *Intelligent Computing, Communication and Devices*,
Advances in Intelligent Systems and Computing 309,
DOI 10.1007/978-81-322-2009-1_20

1 Introduction

Multiprocessor real-time scheduling theory has been studied since the late 1960s and early 1970s [1]. These scheduling algorithms are classified as global scheduling and partition scheduling. In global scheduling, there is only one queue for the entire system and tasks can run on different processors. This class of scheduling algorithms have high overhead. On the other hand, in partition scheduling, each processor has a separate queue and each task may only run on one processor. The problem of partition scheduling is very similar to bin-packing which is known to be NP-Hard [2]. In recent years, some papers introduce a new class called semi-partitioned scheduling by which the processor utilization is enhanced [3–9]. A semi-partitioned scheduling algorithm has two phases, partitioning and scheduling. In partitioning phase tasks are assigned to the processors. In this phase, some tasks are probably split into several subtasks and each assigned to a different processor. The second phase is the policy to determine how to schedule tasks on each processor. The main challenge of semi-partitioned scheduling algorithms is how to partition and split tasks by which they are safely scheduled under the identified scheduling policy.

In this paper, we present a new semi-partitioned scheduling algorithm called SRM-FF for periodic tasks on multiprocessor platforms. The scheduling policy which is used in SRM-FF is based on rate monotonic algorithm. This algorithm is known to be an optimal algorithm for real-time systems [1]. Since 1973, some papers have tried to improve the performance of rate monotonic algorithm in some special cases. Such as the yielding-first rate monotonic scheduling approach which raises the processor utilization up to 100 % in some special cases [10].

The partitioning phase of our proposed approach includes two sub-phases. In the first sub-phase, most tasks are entirely assigned to the processors in which these processors are safely scheduled under rate monotonic algorithm. In the second sub-phase of partitioning phase of SRM-FF, the remaining tasks are assigned to the remaining processors. If a task cannot be entirely assigned to a processor, it is split into two subtasks. The first subtask is assigned to the processor and the second subtask is put back in front of the queue. Such tasks are called split-task while other tasks which are entirely assigned to a processor are called non-split task.

In the first sub-phase of partitioning phase of SRM-FF, processors are chosen by a first-fit method. This heuristic method has the best result among all allocation methods [11]. First-fit method makes SRM-FF to create lower number of subtasks in comparison to previous work and hence the number of context switches and the overhead due to task splitting are reduced. The feasibility of tasks which are partitioned by SRM-FF is formally proved.

The rest of this paper is organized as follows. In Sect. 2, basic concepts are introduced. Sections 3 and 4 relate to our proposed approach and evaluations. Finally, conclusion is discussed in Sect. 5.

2 Basic Concepts

Tasks in our proposed system are periodic and their deadline parameters (i.e. relative deadline) are assumed to be equal to their periods. A request of task τ_i $i = \{1, ..., n\}$, is called a job. Every job of task τ_i is modeled by two parameters C_i and T_i in which C_i is the worst case execution time and T_i is its period. Utilization of task τ_i is defined by $U_i = \frac{C_i}{T_i}$. Every job of task τ_i must be completed before the next job of the same task arrives. Response time of a job is the time span from job arriving up to its execution completion. Liu and Layland [1] proved that the worst response time of task τ_i occurs when it simultaneously requests with all the higher priority tasks. Task τ_i is feasible if its worst response time, R_i, is lower than or equal to its period ($R_i \leq T_i$). In this system, we want to assign task set $\Gamma = \{(C_1, T_1), (C_2, T_2) ... (C_n, T_n)\}$ to m processors, $\{M_1, ..., M_m\}$. The total utilization of task set Γ is equal to: $U(\Gamma) = \sum_{i=1}^{n} \frac{C_i}{T_i}$.

The average processor utilization of task set Γ with m processors is: $U_m(\Gamma) = \frac{U(\Gamma)}{m}$. The processor utilization of M_a, $a = \{1, ..., m\}$, is the total utilization of tasks set Γ' which is assigned to M_a. The worst response time of task τ_i under rate monotonic can be calculated by Formula 1 [12]. Suppose tasks are sorted in ascending order of their periods. Now, the worst response time of task τ_i is:

$$R_i^o = C_i$$
$$R_i^{k+1} = C_i + \sum_{j<i} \left\lceil \frac{R_i^k}{T_j} \right\rceil * C_j \tag{1}$$

This iteration is terminated when $R_i^{k+1} = R_i^k$. If $R_i^k \leq T_i$, then, task $\tau_i(C_i, T_i)$ is feasible under rate monotonic algorithm. SRM-FF uses response time analysis to check feasibility of tasks on a specific processor. The remaining processor utilization of a processor is another concept which is defined here. Suppose task set $\Gamma' = \{(C_1, T_1), (C_2, T_2) ... (C_n, T_{n'})\}$, is assigned to processor M_a, the remaining processor utilization of M_a is defined by:

$$\eta(M_a) = 1 - U(\Gamma') \tag{2}$$

3 SRM-FF Algorithm

In this section, we introduce our proposed approach which is called SRM-FF. In SRM-FF tasks are sorted in ascending order of their priority, i.e. $i < j$ τ_j has higher priority over τ_i. Therefore, current task which is in front of the queue is the lowest priority task. The partitioning phase of SRM-FF is demonstrated in Algorithm 1. As mentioned, this phase includes two separate sub-phases. In the first sub-phase, processors are chosen by first-fit method. In the first-fit method, it is tried to accommodate tasks into the first processor as much as possible. In the other words, If the worst response time of all tasks in the first processor (including the current task) is lower than or equal to their periods, then, the current task which is in front of the queue is added to task set of this processor, Line 7 from Algorithm 1.

If an infeasible task exist, then the current task is assigned to the next processor. Based on the first-fit method, this processor is empty. This process repeats for other processors as well. It will be stopped until no empty processor exist in the queue of processors, or no other task exist in the queue of tasks. Therefore, at the end of the first sub-phase of partitioning phase of SRM-FF, some tasks are entirely assigned to some processors. As mentioned, such tasks are called non-split tasks. It is easy to derive following property from the first sub-phase of partitioning phase of SRM-FF.

Lemma 1 *At least one task is entirely assigned to each processor.*

Proof Proof follows the assigning approach which is used by SRM-FF.   □

Considering Lemma 1, the lowest priority task on each processor is a non-split task. In the second sub-phase of partitioning phase of SRM-FF, the remaining tasks are assigned to the processors which are not filled yet. In this sub-phase, the processor which has the highest remaining processor utilization is chosen for assigning the current task. In this work, processors are filled parallel.

If the current task, i.e. τ_i, cannot entirely be assigned to a processor, it is split into two subtasks, τ_{i1} (C_{i1}, T_i) and τ_{i2} $(C_i - C_{i1}, T_i)$. The worst case execution time of the first subtask is obtained by which all tasks in the host processor, including the first subtask, will be safely scheduled under rate monotonic algorithm. Binary search between $[0, C_i]$ can be used for finding the suitable value for the worst case execution time of subtask τ_{i1}. The second subtask, τ_{i2}, is put back in front of the queue.

Algorithm 1. Partitioning phase of SRM-FF

Input: task set $\Gamma=\{\tau_1 ...,\tau_n\}$ and list of processors $\{M_1, ..., M_m\}$.	15. *Pick a processor, i.e. M_a, which* $\eta(M_a)$ *is maximum*
// Sort all tasks in ascending order of their priority.	16. *If all tasks in M_a including τ_i are feasible **then***
// Sub-phase 1	17. *Assign τ_i to processor M_a.*
1. $M_a = 1$	18. *If τ_i is subtask **then***
2. **For each** *task τ_i in Γ **do***	19. *Remove M_a from the queue of processors*
3. **If** *M_a is equal to m **then***	20. **End if**
4. **Break for.**	21. **Else**
5. **End if**	22. *Split τ_i into two subtasks τ_{i1} and τ_{i2}.*
6. **If** *all tasks in M_a including τ_i are feasible **then***	23. *Find the best value for C_{i1}.*
7. *Assign τ_i to processor M_a.*	24. *Assign τ_{i1} to processor M_a.*
8. **Else**	25. *Put back τ_{i2} into front of the queue.*
9. $M_a = M_a + 1$.	26. *Remove M_a from queue of processors*
10. *Assign τ_i to processor M_a.*	27. **End if**
11. **End if**	28. **Else** *// not exist a processor*
12. **End for**	29. *Return "cannot assign task set Γ to m processors".*
	30. **End if**
// Sub-phase 2	31. **End for**
	32. **Return** *"task set Γ is successfully assigned to m processors"*
13. **For each** *task τ_i in remaining tasks **do***	
14. **If exist** *a processor in the queue of processors **then***	

After adding the first subtask to a processor it is removed from queue of processors, Line 26 from Algorithm 1. Based on the sorting method we have the following lemma.

Lemma 2 *The highest priority task on each processor is split.*

Proof Proof follows the sorting and splitting methods of SRM-FF.   □

Considering the task splitting method of SRM-FF, a task is probably split into several subtasks. According Lemma 2, each subtask, i.e. $\tau_{ij} j = \{1, ..., q - 1\}$ τ_{iq} is the last subtask of task τ_i, is the highest priority task on its own processor and starts its execution as soon as it requests a job.

The overhead of task splitting is measured in some previous works [5, 7]. It is necessary to synchronize subtasks of a split-task to rely on their measurement. Therefore, we define the concept of release time for each subtask of a split-task.

Release time of subtask $\tau_{ij} j = \{1, ..., q\}$ is defined by: $\mu(\tau_{ij}) = \sum_{\ell=1}^{j-1} R_{i\ell}$.

Where $R_{i\ell}$ determines the worst response time of subtask $\tau_{i\ell}$ in its own processor. It should be mentioned that, the release time of non-split tasks is assumed to be zero. Considering Lemma 2, the worst response time of each subtask, except the last one, is equal to its worst case execution time. The release time of last subtask of a split-task, i.e. τ_{iq}, is equal to $\sum_{\ell=1}^{q} C_{i\ell}$ which is larger than its worst case execution time, because

$$T_i \geq C_i$$
$$C_i = C_{i1} + C_{i2} + \ldots + C_{iq}$$
$$T_i - (C_{i1} + C_{i2} + \ldots + C_{iq-1}) \geq C_{iq}$$

Therefore, the last subtask of a split-task has enough time to complete its execution if starts its execution as soon as its release time is over. To guaranty this situation, after adding the last subtask of a split-task to a processor, this processor is removed from the queue of processors hence no other task or subtask is assigned to the processor, Line 19 from Algorithm 1. Therefore, the following lemmas are stated to guaranty feasibility of subtasks and non-split tasks.

Lemma 3 *If a task, i.e. τ_i, by partitioning phase of SRM-FF is split into several subtasks, then, all subtasks of split-task τ_i can meet their deadline parameters.*

Proof Considering the second sub-phase of partitioning phase of SRM-FF, the highest priority task on each processor is split. After that, these processors are removed from the queue of processors hence no other task or subtask is assigned to them. Therefore, based on Lemma 2 each subtask starts its execution as soon as its release time is over. Thus, all subtasks of a split-task can safely be scheduled under rate monotonic algorithm.   □

Lemma 4 *Non-split tasks which are partitioned using SRM-FF, can meet their deadline parameters.*

Proof Lemma 4 can be proved by noting that whenever a task is assigned to a processor, the feasibility of the task set on that processor is assured based on the admission control which is used by SRM-FF. Lines 6 and 16 from Algorithm 1.□

Thus, it is formally proved that any task which are successfully partitioned by SRM-FF and scheduled under rate monotonic policy can meet its deadline parameters.

4 Evaluations

In this section, the performance of our proposed approach in comparison to one previous work is investigated. RM-TS is another approach which uses rate monotonic policy to propose a semi-partitioned scheduling algorithm for multi-processor platforms [4]. Worst-fit method is used in partitioning phase of RM-TS for allocating tasks to the processors. This approach splits the highest priority task on each processor. Tasks are divided into two groups, light and heavy. First, heavy tasks which satisfy a specific condition are pre-assigned. Then, the remaining tasks are assigned to the processors. Response time analysis is used as admission control of RM-TS. Therefore, the average processor utilization of SRM-FF is so close to average processor utilization of RM-TS. But, the number of subtasks which are created by SRM-FF is lower than that of RM-TS.

Same as other researches, in RM-TS, a subtask can start its execution, if the prior subtasks are finished. Therefore, this process imposes some overhead to the system [5, 7]. Considering paper [7] the overall overhead due to task splitting is almost equal to 65 μs.

Therefore, as the number of subtasks which are created by an approach becomes low, the overhead due to task splitting can be reduced. To evaluate this issue, we generate 1,000 task sets while the total utilization of each task set is a random value between [0.7 × α, α]. α determines the maximum total utilization of each task set and its value will be one of the values of the following set.

$$\alpha \in \{4, 8, 32, 64\} \tag{3}$$

Period of each task is a random value between [5, 1,000] and utilization of each task is also randomly generated between [0.01, 1]. Distribution of our random function is uniform. The number of processors are assumed to be variable so that each approach can successfully partition such task sets. The average processor utilization of each approach is demonstrated in Table 1. Considering Table 1, the average processor utilization of SRM-FF is so close to that of RM-TS. The total number of created subtasks is demonstrated in Fig. 1a. As shown in Fig. 1a, the total number of subtasks which are created by SRM-FF is lower than that of RM-TS.

Table 1 Average processor utilization of each approach

Maximum utilization of each task set	4	8	16	32	64
SRM-FF	0.81	0.83	0.84	0.841	0.84
RM-TS	0.81	0.83	0.84	0.847	0.846

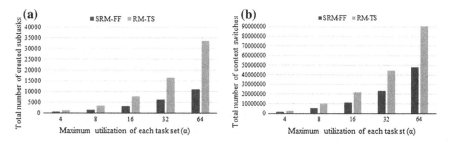

Fig. 1 Experimental result. **a** Total number of created subtasks. **b** Total number of context switches

Table 2 Difference of two approaches in term of number of context switches

Maximum utilization of each task set	4	8	16	32	64
Difference of two approaches as percentage	30.7	45.4	49	47.4	47

The main motivation which makes SRM-FF has this result is the usage of first-fit method. After partitioning phase, the execution of each task set is simulated under rate monotonic policy. The total number of context switches are calculated during each simulation. These context switches occur when a job of a subtask completes its execution and the job of next subtask of the same task starts its execution on another processor. Results of this test are illustrated in Fig. 1b. Difference of number of context switches in SRM-FF compared with RM-TS is shown in Table 2. Considering Table 2 and Fig. 1b, the total number of context switches which are created by SRM-FF is lower than that of RM-TS. For instance, the total number of context switches in SRM-FF is 49 % lower than that of RM-TS, when α is equal to 16.

5 Conclusion

In this paper, a new semi-partitioned scheduling algorithm called SRM-FF is proposed by which the number of created subtasks is reduced in comparison to previous work. This superiority occurs because SRM-FF uses first-fit method to allocate tasks over processors. The simulation results demonstrate that the number of context switches and the overhead due to task splitting can be reduced by declining the number of created subtasks.

In our future works, to improve the processor utilization, some manners will be presented to change the partitioning phase of SRM-FF. Improving our evaluations like measurement the real overhead due to task splitting in a practical system is one of our other purposes for future works well.

References

1. Liu, C.L., Layland, J.W.: Scheduling algorithms for multiprogramming in a hard-real-time environment. J. ACM, 43–73 (1973)
2. Garey, M.R., Johnson, D.S. (eds.): Computers and Intractability: A Guide to the Theory of NP-Completeness (1979)
3. George, L., Courbin, P., Sorel, Y.: Job vs. portioned partitioning for the earliest deadline first semi-partitioned scheduling. J. Syst. Architect. **57**(5), 518–535 (2011)
4. Guan, N., et al.: Parametric utilization bounds for fixed-priority multiprocessor scheduling. In: Parallel and Distributed Processing Symposium (IPDPS). IEEE, Shanghai, pp. 261–272 (2012)

5. Kandhalu, A., et al.: pCOMPATS: period-compatible task allocation and splitting on multi-core processors. In: 18th Real Time and Embedded Technology and Applications Symposium (RTAS). IEEE, Beijing, pp. 307–316 (2012)
6. Kato, S., Yamasaki, N.: Semi-partitioned fixed-priority scheduling on multiprocessors. In: 15th Real-Time and Embedded Technology and Applications Symposium (RTAS), IEEE, San Francisco, CA, pp. 23–32 (2009)
7. Lakshmanan, K., Rajkumar, R.R., Lehoczky, J.P.: Partitioned fixed-priority preemptive scheduling for multi-core processors. In: Euromicro conference on real-time systems (ECRTS), Dublin, pp 239–248 (2009)
8. Naghibzadeh, M., et al.: Efficient semi-partitioning and rate-monotonic scheduling hard real-time tasks on multi-core systems. In: 8th IEEE International Symposium on Industrial Embedded Systems (SIES), Porto, pp. 85–88 (2013)
9. Guan, N., Stigge, M., Yi, W., Yu, G.: Fixed-priority multiprocessor scheduling with Liu and Layland's utilization bound. In: Real-Time and Embedded Technology and Applications Symposium (RTAS), 2010 16th IEEE, pp. 165–174 (2013)
10. Naghibzadeh, M., Kim, K.H.K.: The yielding-first rate-monotonic scheduling approach and its efficiency assessment. Comput. Syst. Sci. Eng. **18**, 173–180 (2003)
11. Lauzac, S., Melhem, R., Mosse, D.: An improved rate-monotonic admission control and its applications. IEEE Trans. Comput. **52**(3), 337–350 (2003)
12. Lehoczky, J., Sha, L., Ding, Y.: The rate monotonic scheduling algorithm: exact characterization and average case behavior. In: Real Time Systems Symposium, Santa Monica (1989)

A Novel Approach for Data Hiding in Color Images Using LSB Based Steganography with Improved Capacity and Resistance to Statistical Attacks

Savita Badhan and Mamta Juneja

Abstract In modern communication, security, privacy, and integrity of the secret data being transmitted is the most important issue and steganographic techniques help in accomplishing this task. In this paper, we have proposed a data hiding scheme using steganography and cryptography in spatial domain. RGB 24 bit bitmap image carries the secret DES encrypted message that is embedded by utilizing the different pixel intensities of the cover RGB image. At receiver side, message is received and message is extracted by applying the reverse of embedding process. The proposed system results in high embedding capacity and also successfully resists statistical attacks like RS analysis, histogram analysis, and chi-square analysis.

Keywords Steganography · Cryptography · Variable least significant bit embedding · Steganalysis

1 Introduction

Steganography [1] helps in achieving secret data communication between two entities. It is a word with Greek origin meaning "secret communication." For many years, it has captured the attraction of researchers. Steganography may be defined as the technique of hiding confidential data in such a way that no one apart from the authorized parties is aware of the transmission taking place. Various

S. Badhan (✉) · M. Juneja
Department of Computer Science and Engineering,
University Institute of Engineering and Technology,
Punjab University, Chandigarh, India
e-mail: savitabadhan23@gmail.com

M. Juneja
e-mail: er_mamta@yahoo.com

© Springer India 2015
L.C. Jain et al. (eds.), *Intelligent Computing, Communication and Devices*,
Advances in Intelligent Systems and Computing 309,
DOI 10.1007/978-81-322-2009-1_21

steganography techniques [2, 3] have been developed in both spatial domain and transform domain. In modern digital world of fast and high-speed communication, there is high requirement of safe and secure communication that is achieved with both cryptography [4] and steganography. Now, cryptography alone is not sufficient as a number of techniques have been developed that find the loopholes in the cryptography system, thereby cracking them. Both techniques have different security abilities. The main goal of steganography is keeping the message's existence hidden. In contrast, the goal of cryptography is scrambling the message in an open environment in such a way that makes it meaningless and unreadable until and unless the decryption key is available. So Steganography along with cryptography is the best solution for information hiding. In steganography, the security is very much influenced by the choice the cover image used for hiding the secret message. Large images are the most desirable for steganography because they have more space to hide the data. In this paper, stress is on image steganography. Digital color images are typically referred to as 24-bit RGB images that may be in any format maybe bmp, jpeg, png, etc. Generally, main requirements of any image steganographic system are high capacity, high imperceptibility, and robustness and temper resistance. High capacity means the ability to hide large amount of information into the system. Robustness is that the system must remain intact even if modifications like cropping, compression, filtering, noise addition, etc., are applied on it. High imperceptibility is maintaining the high quality of the stego-image. Temper resistance means that it should be difficult to modify and detect the message with various attacks like statistical and visual attacks once it has been embedded. In spatial domain, most common methods used to hide information within image include [2, 3] least significant bit (LSB) insertion, multiple base notational system (MBNS), and masking and filtering steganography. As the technology is advancing every day, so is the advancement in the techniques that aim to break the stego-systems by using various algorithms that are particularly designed for this purpose. So a lot of research has been made to aid in discovering the steganalysis techniques for the perception of hidden data. Steganalysis is a relatively new technique that evolved in late 1990s. It is the art of discovering the hidden files in the carriers of any stego-system. It aims to find the existence of the message, content, and length of the message and even more details like detecting the message used for embedding, encryption algorithm used, etc. Various steganalysis techniques are also available to detect steganography like histogram analysis [5], chi-square analysis [5], and RS analysis [6].

2 Proposed Work

In this section, we describe our proposed method, which is then compared with the method proposed by Hussain et al. [7]. In this paper, the work proposed by Hussain has been further modified for increasing capacity security and temper resistance. In proposed work, the carrier cover image is a 24-bit RGB bitmap color image. And the secret data used is the text message. Also a matrix is maintained for those pixels that utilize 8, 5, 3, and 2 bits to store data. Before embedding the secret message file into the cover image, the process of encryption has been used to encrypt the secret data using data encryption standards (DES) [8] algorithm for increasing the security of the system rather than just applying the 8-bit XOR operation on the secret data, thereby making it less vulnerable to attacks. After this, the encrypted message is embedded into the cover image using the process embedding algorithm. The embedding algorithm divides the pixel intensities of the color image into different pixel intensity ranges ranging from high-intensity pixel values to low-intensity pixel values. Other modification proposed in the work is that in the entire image wherever the value of any two red, green, or blue values is zero, the remaining one's all the pixel value bits, i.e., 8 bits, are utilized for embedding secret data. After embedding, the stego-image so formed is sent to the receiver. The original message at receiver side is extracted from the stego-image using the decryption process. In addition to being more secure, the proposed work also successfully resists to the statistical attacks like RS analysis, chi-square analysis, and histogram analysis. Proposed work architecture and the embedding algorithm are described in Sects. 2.1 and 2.2, respectively.

2.1 Proposed Work Architecture

The proposed work architecture depicts the methodology that is followed in designing the proposed steganographic system. It gives general idea of the work methodology and the steps involved (Fig. 1).

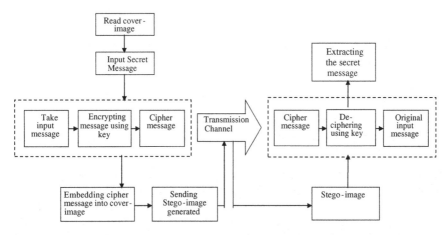

Fig. 1 Proposed work architecture

2.2 Proposed Algorithm

A. Embedding module:

Step 1:   Read cover image for embedding secret text message.
Step 2:   Input encrypted secret message.
Step 3:   Embedding secret message inside cover image using proposed embedding scheme.
Step 4:   Sending generated stego-image to the receiving party.

B. Extraction module:

Step 1:   Receiving stego-image.
Step 2:   Entering the stego-key for deciphering the stego-image.
Step 3:   Applying reverse embedding procedure for extraction of original text message.

2.3 Proposed Embedding Scheme

The proposed embedding scheme is described in Table 1 in which different pixel intensity ranges have been utilized for hiding data accordingly. In Table 1, pixel intensity represents the pixel value, data bit to embed represents the number of message bits to be embedded into the cover image, matrix entry maintains a matrix which denotes the 5 LSB are embedded and utilize bits represents the total number of bits embedded into a pixel, and x represents don't care bits (either 0 or 1).

3 Experimental Results and Comparison Analysis

The experimental results have been evaluated based on two criteria: first on the basis of imperceptibility and payload capacity and second on the basis of resistance to stastical attacks.

3.1 Evaluation Based on Stego-Image Quality and Payload Capacity

Imperceptibility is the factor that is used to evaluate the stego-image quality. It results in high value when the difference between the generated stego-image and chosen cover image is less. For evaluating stego-images based on this criteria, peak signal-to-noise ratio (PSNR) and mean square error (MSE) values are calculated. Value for PSNR should be high, and for MSE, it should be low. Second

Table 1 Our proposed work

S.no. bits	Pixel intensity	Data bit to embed	Matrix entry	Utilized
1	240–255	0	–	4
2	240–255	1	1	5
3	231–239	0	1	3
4	231–239	1	–	4
5	224–230	x	1	5
6	192–223	0	1	3
7	192–223	1	1	2
8	192–198	x	1	5
9	51–191	0	1	2
10	51–191	1	–	1
11	32–50	x	–	1
12	16–31	0	1	3
13	16–31	1	1	2
14	0–15	x	–	4
15	$R = 255, G = 255, B = 255$	x	1	8
16	$R = 255, G = 255, B = 255$	x	1	8
17	$R = 255, G = 255, B = 255$	x	1	8

Table 2 Value of MSE, PSNR, percentage of pixels, and changed bytes percentage

Cover image	Embedded data bytes	Percentage of used pixel in image	Percentage of changed bytes	PSNR	MSE
Our method using Abraham Lincoln's letter					
Lena	1785	30.0464	58.3755	0.0038	43.13
Baboon	1785	47.9922	54.6375	0.0041	44.47
Pepper	1785	17.7162	53.1245	0.0093	45.05
Mypic	1785	31.5437	55.4325	0.0155	48.23
Hussain's method using Abraham Lincoln's letter					
Lena	1785	0.9666	55.5906	65.7180	0.0174
Baboon	1785	1.1237	51.3975	69.6609	0.0070
Pepper	1785	0.8867	59.9455	59.5402	0.0723

thing that is considered is the payload capacity which is defined as the capacity of the image to hide details within it without any distortion to the original image. Value for payload should be as high as possible. Four-colored bitmap cover images lena, baboon, pepper, and mypic are used, each of size 512 × 512 in our experiment. Twenty-four-bit RGB color image is used as cover image. Text message is Abraham Lincoln's letter to his son's teacher that is to be hidden into the cover image. They were compared to work done by Hussain, and the results obtained are shown in Table 2.

(a)   (b)

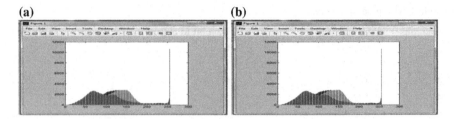

Fig. 2 Histogram analysis results for mypic.bmp. **a** Cover image histogram. **b** Stego-image histogram

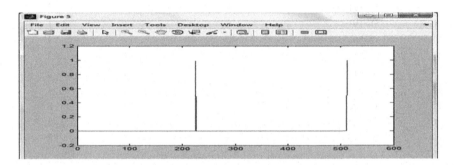

Fig. 3 Chi-square analysis for mypic.bmp

3.2 Evaluation Based on Resistance to Stastical Attacks

Results on "Mypic" for histogram analysis, chi-square analysis, and Rs analysis have been shown in Figs. 2, 3 and 4, respectively.

A. Histogram Analysis

The results of histogram analysis technique [5] are shown in Fig. 2. As there are no differences found in histograms of both the original and stego images so the proposed system could not be attacked.

B. Chi-Square Analysis

The results of chi-square analysis technique proposed in [5] are shown in Fig. 3. And, the proposed system successfully sustains this attack as the graph obtained fulfills the required range.

C. RS Analysis

The results of RS analysis technique proposed in [6] are shown in Fig. 4a, b and Table 5. And, the proposed system successfully sustains this attack. Table 5 shows the values of RS analysis results for before embedding and after embedding. Figure 5a shows the graph obtained according to obtained values, and Fig. 5b

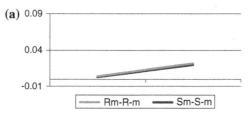

(a)

(b)

```
File   Edit   Format   View   Help
[RS Analysis](RBG-disjoint groups):
Occupation[Red Channel]:24.21036%=0.24210bpp(bitsperpixel)
   [Red] Approximate size:6159.11499bytes.
Occupation[Green Channel]:23.45963%=0.23460bpp(bitsperpixel)
   [Green] Approximate size:5968.12866bytes.
Occupation[Blue Channel]:22.23282%=0.22233bpp(bitsperpixel)
   [Blue] Approximate size:5656.02905bytes.

After Embedding:
Occupation[Red Channel]:24.21036%=0.24210bpp(bitsperpixel)
   [Red] Approximate size:7933.25000bytes.
Occupation[Green Channel]:23.45963%=0.23460bpp(bitsperpixel)
   [Green] Approximate size:7687.25000bytes.
Occupation[Blue Channel]:22.23282%=0.22233bpp(bitsperpixel)
   [Blue] Approximate size:7285.25000bytes.
```

Fig. 4 a RS analysis plot for mypic.bmp. **b** RS analysis results for mypic.bmp

shows the results for occupation of each channel for disjoint groups and after embedding.

The values obtained for Rm–R–m and Sm–S–m show that the difference between them is less than 10 % that indicates the temper resistance of the proposed system. So these attacks are successfully resisted.

4 Conclusions

The main advantage of our proposed work is that it better utilizes the pixel intensity ranges by utilizing lower ranges in and also those intensity area pixels of the colored component where pixel intensity of any of the color component red, green, or blue is zero. It has resulted in good statistical result for PSNR, MSE, and for the percentage of pixels utilized. Also it is less vulnerable to statistical attacks like histogram analysis, chi-square analysis, and RS analysis.

References

1. Provos, N., et al.: Hide and seek: an introduction to steganography. IEEE Secur. Priv. Mag. **1**, 32–44 (2003)
2. Hamid, N., et al.: Image steganography techniques: an overview. Int. J. Comput. Sci. Secur. **6**(3), 168–187 (2012)

3. Kruus, P., et al.: A survey of steganography techniques for image files. Adv. Secur. Res. J. 5(1), 41–52 (2003)
4. Manferdeli, J.L., Wagner, D.A.: Cryptography and Cryptanalysis. Springer, Heidelberg (2013)
5. Westfeld, A., Pfitzmann, A.: Attacks on steganographic systems. In: Lecture Notes in Computer Science. Springer, Berlin, pp. 61–75 (2000)
6. Fridrich, J., Goljan, M: Practical steganalysis of digital images—state of the art. In: Proceedings of SPIE, Security and Watermarking of Multimedia Contents IV, E.J. Delp III and P.W. Wong, pp. 1–13, 2002
7. Hussain, M., et al.: Pixel intensity based high capacity data embedding method. In: IEEE (2010)
8. DES: Wikipedia. http://en.wikipidea.org/wiki/DES

A Comparative Study of Correlation Based Stereo Matching Algorithms: Illumination and Exposure

N.S. Bindu and H.S. Sheshadri

Abstract Stereo matching is one of the most active research areas in computer vision for an accurate estimation of disparity. Many algorithms for computing stereo algorithm have been proposed, but there has been very little work on experimentally evaluating algorithm performance, especially using real-time scenario. Many researchers have been undergone past from many decades to find an accurate disparity, but still it is not an easy task to choose an appropriate algorithm for the required real-time application. To overcome from this problem, we proposed an experimental comparison of several different cross-correlation-based stereo algorithms and also introduce an objective that evaluates a set of six known correlation-based stereo algorithms. An evaluation of correlation-based stereo matching algorithm results will be very useful for selecting the appropriate stereo algorithms for a given application. Here, we make use of two stereo pairs: Aloe and Cloth from Middlebury stereo datasets. This work mainly focuses on the evaluation of robustness to change in illumination and exposure.

Keywords Disparity estimation · Stereo matching · Similarity measure · Illumination · Exposure

N.S. Bindu (✉)
Department of Electronics and Communication Engineering,
Vidyavardhaka College of Engineering, Mysore, India
e-mail: bindu.ns5@gmail.com

H.S. Sheshadri
Department of Electronics and Communication Engineering,
PES College of Engineering, Mandya, India
e-mail: hssheshadri@gmail.com

© Springer India 2015
L.C. Jain et al. (eds.), *Intelligent Computing, Communication and Devices*,
Advances in Intelligent Systems and Computing 309,
DOI 10.1007/978-81-322-2009-1_22

1 Introduction

During the past few years, the estimation of disparity field of an image sequence has been playing an increasingly important rule in a large number of applications of a computer vision area. Some examples of such applications can be found in video coding, 3D reconstruction from stereo image pairs, object recognition, and motion estimation. Stereo matching has been one of the most active research topics in computer vision [1]. Stereo matching algorithms aim at extracting 3D structure of a particular scene by finding the correct correspondence between the images from a different viewpoints. Finding an accurate correspondence in any of the stereo algorithms is not an easy and simple task. During the correspondence matching, there occur a number of difficulties, and some of them are occluded regions, object boundaries, and textureless regions. A matching cost should be computed at each and every single pixel for all the disparities under consideration. Approaches to all the stereo correspondence problems or stereo matching algorithms can be classified into two categories: area based and feature based. In area-based method, the disparity estimation at a given pixel is based on similarity measurement performed in a finite window, whereas in feature-based methods, global cost functions will be defined to solve an optimization problem. As far as the real-time applications are concerned, area-based stereo matching is the most powerful tool compared to feature-based stereo matching algorithm. Some of the feature-based stereo matching algorithms are sum of absolute difference (SAD), sum of squared difference (SSD), normalized cross-correlation (NCC), ZSAD, ZSSD, ZNCC, etc. In this work, we are particularly evaluating a set of cross-correlation-based stereo matching algorithms suitable for real-time application, which generally have low computational complexity and less storage requirement. This paper is organized as follows. Section 2 presents the related work.

2 Related Work

As correspondence algorithm is a major aspect in detection and tracking, research has been done from many decades. All the correspondence algorithms belong to two main categories as mentioned in Sect. 1. In this work, we are mainly concentrating on work done on the correlation-based methods. In the recent years, area-based approaches have been received a very good attention mainly because of its attribute-like flexibility and also the computational cost as it is a major role in stereo correspondence. Scharstein et al. [1] done a detailed study on various matching algorithms that have been presented. Many different correlation-based similarity measures and comparisons have been done in [2], but some parameters that are essential for the efficiency are missing from the list. Hseu et al. [3] has presented other similarity measures and comparative analysis is given. In [4, 5], the author has analyzed some of area-based algorithms among various algorithms and also the feature-based algorithms. Here, the author has considered some parameters which are going to affect the performance. The contribution to this work will make us help in choosing the more efficient algorithm by considering different aspects of parameters. In [6], the author has proposed a method that estimates the computation time of a correlation-based algorithm. In [6], the author has not considered any other parameters. This survey comes to a conclusion by saying that only a very few and less attempts are made for comparison and also characterizing the various aspects of correlation-based matching algorithms and its application. In [7, 8], the author has evaluated the execution time and evaluation of various correlation-based stereo algorithms.

3 Area-Based Stereo Matching

As we mentioned above, area-based similarity measure is the most powerful, and also, it is more efficient methods of a correlation-based algorithms. This work mainly contributes and focuses on cross-correlation-based stereo algorithms. Various cross-correlation-based stereo algorithms have been proposed, and among them, we will discuss only a few powerful stereo algorithms as shown in Eqs. (1–6). We consider two dissimilarity measure functions called SSD and SAD. Also, we consider the extended versions of SAD and SSD in zero mean and also the least square. NCC is one of the most commonly used and standard window-based matching techniques. It matches pixel of interest of two windows. The normalization compensates differences in both gain and bias within the window, and also, it is the optimal method which compensates Gaussian noise. As NCC is a standard window-based matching technique, we include this similarity measure function. In the equation below, $I_1(i,j)$ represents the coordinates of the reference image and $I_2(i,j)$ represents the coordinates of the search image. A proper correlation window is selected.

$$\text{SAD} = \sum_{(i,j)\in W} |I_1(i,j) - I_2(x+i, y+i)| \tag{1}$$

$$\text{SSD} = \sum_{(i,j)\in W} (I_1(i,j) - I_2(x+i, y+i))^2 \tag{2}$$

$$\text{NCC} = \frac{\sum_{(i,j\in W)} (I_1(i,j)I_2(x+i, y+j))}{\sqrt{[2]\sum_{(i,j)\in W} I_1^2(i,j) \cdot \sum_{(i,j)\in W} I_2^2(x+i, y+j)}} \tag{3}$$

$$\text{ZNCC} = \frac{\sum_{(i,j)\in W} (I_1(i,j) - \bar{I}_1(i,j)) \cdot I_2(x+i, y+j) - \bar{I}_2(x+i, y+j)}{\sqrt{[2]\sum_{(i,j)\in W} (I_1(i,j) - \bar{I}_1(i,j))^2 \cdot \sum_{(i,j)\in W} (I_2(x+i, y+j) - \bar{I}_2(x+i, y+j))^2}}$$

$$\tag{4}$$

$$\text{ZSAD} = \sum_{(i,j) \in W} |I_1(i,j) - \bar{I}_1(i,j) - I_2(x+i, y+j) + \bar{I}_2(x+i, y+j)| \qquad (5)$$

$$\text{ZSSD} = \sum_{(i,j) \in W} (I_1(i,j) - \bar{I}_1(i,j) - I_2(x+i, y+j) + \bar{I}_2(x+i, y+j))^2 \qquad (6)$$

4 Experiment Results

4.1 Method of Evaluation

Here we consider a pair of left and right of Aloe stereo image, Cloth stereo image and also its ground truth disparity which is taken from one of most often used datasets called Middlebury stereo datasets. Here, the disparity range is considered to be from 0 to 70 pixels. The main contribution of this work is to evaluate the above-said algorithms by considering various parameters; one includes disparity, robustness to change in illumination, and also robustness to change in exposure. Figures 1 and 2 show the estimated disparities of various correlation-based algorithms for an Aloe and Cloth stereo image pair.

4.2 Robustness to Change in Illumination

In almost all the paper, most of the authors make an assumption saying that the corresponding pixels will have similar color values. But the above-said statement does not hold for the case of stereo input image that has different corresponding color values. In this work, we consider a few most popular algorithms to check the robustness to change in illumination as well as robustness to change in exposure. As mentioned above, we have already used Aloe and Cloth [9] stereo input images. In this work, we will set the index value of exposure to be as 1, and also, we will vary the index value of illumination to be from 1 to 3. Here, we also consider a fixed window size of 9. Figures 3 and 4 show the output disparity. From the output disparity, we can clearly observe that a small change in light source seems to be a very realistic and very challenging task in stereo matching algorithms. As it is mentioned in literature review, compared to NCC and ZNCC the SAD, SSD are very sensitive to change in light sources. But NCC and ZNCC are very insensitive to the changes in light sources. One of the major drawbacks of NCC and ZNCC is that they create a fattening effect on the object boundaries.

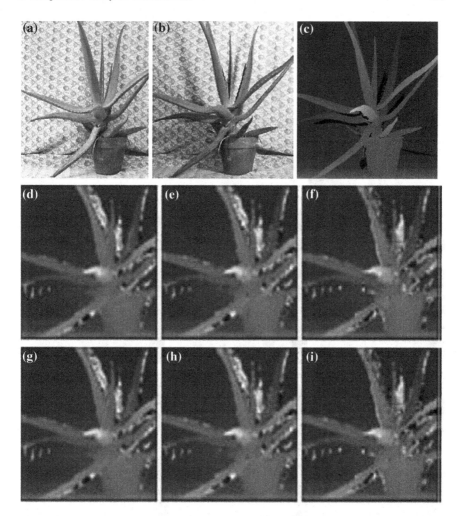

Fig. 1 Estimated disparities of various correlation algorithms for an Aloe stereo image pair.
a Left. **b** Right. **c** Ground. **d** SAD. **e** SSD. **f** NCC. **g** ZSAD. **h** ZSSD. **i** ZNCC

4.3 Robustness to Change in Exposure

In this work, we consider a few most popular algorithms to check the robustness to change in illumination as well as robustness to change in exposure. As mentioned above, we have already used Aloe and Cloth [9] stereo input images. Here also, we consider a stereo image pair which has a fixed illumination index value to be 1, and we slowly vary the exposure index value from 0 to 2. The results of these above-said constraints are shown in Figs. 5 and 6 in terms of output disparity. The changes made in exposure will seriously affect SAD and SSD stereo matching

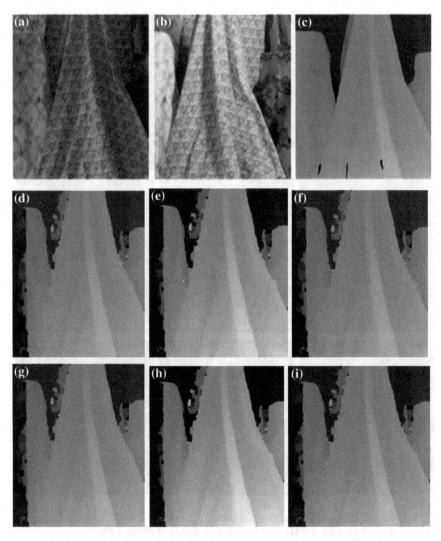

Fig. 2 Estimated disparities of various correlation algorithms for an Cloth stereo image pair.
a Left. **b** Right. **c** Ground. **d** SAD. **e** SSD. **f** NCC. **g** ZSAD. **h** ZSSD. **i** ZNCC

algorithm, and as a result, this produces either very dark or light outputs in the
output disparity. By this result, it is slightly very difficult to identify the objects'
edge features in the output disparity. NCC and ZNCC are still having a robustness
to above-said changes, i.e., change in light sources.

Fig. 3 Results of Aloe stereo algorithms with varying illumination. **a** Left. **b** Right. **c** Ground. **d** SAD. **e** SSD. **f** NCC. **g** ZSAD. **h** ZSSD. **i** ZNCC

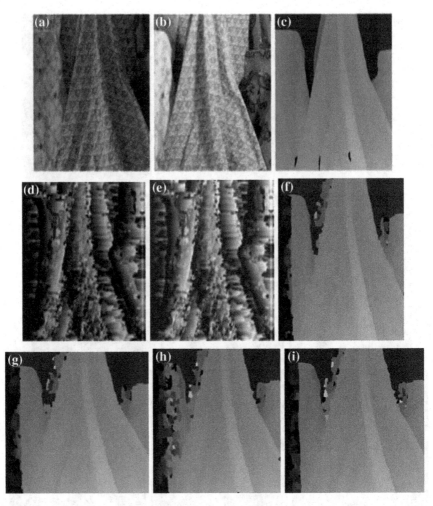

Fig. 4 Results of Cloth stereo algorithms with varying illumination. **a** Left. **b** Right. **c** Ground. **d** SAD. **e** SSD. **f** NCC. **g** ZSAD. **h** ZSSD. **i** ZNCC

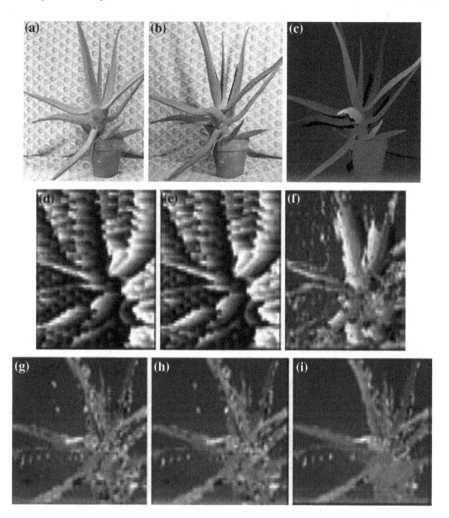

Fig. 5 Results of stereo algorithms with varying exposure. **a** Left. **b** Right. **c** Ground. **d** SAD.
e SSD. **f** NCC. **g** ZSAD. **h** ZSSD. **i** ZNCC

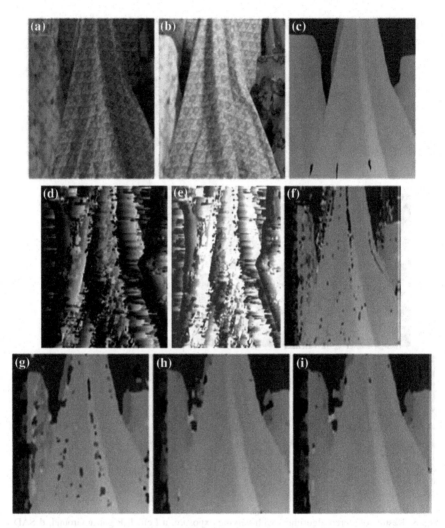

Fig. 6 Results of stereo algorithms with varying exposure. **a** Left. **b** Right. **c** Ground. **d** SAD.
e SSD. **f** NCC. **g** ZSAD. **h** ZSSD. **i** ZNCC

5 Conclusion

This paper presents a comparative analysis of several area-based similarity measure functions. As far as real-time applications are considered, the correlation-
based stereo algorithms play an important role as it is one of the most popular and
very efficient and effective algorithms. In this work, we have considered 6 correlation-based stereo algorithms, namely SAD, SSD, NCC, ZSAD, ZSSD, and
ZNCC, and also, analysis has been conducted considering illumination and
exposure. The use of suitable window size of 9 has created a very interesting fact

in finding a robustness to change in illumination and exposure in our algorithm. We can conclude that selection of suitable window size plays an important role in deciding the final disparity. SAD-based stereo algorithms are comparably less expensive compared to all other existing methods. It is observed that NCC with large window size is most efficient and it can be used in the real-time applications such as object detection and object tracking systems.

Acknowledgement The authors would like to thank the anonymous reviewers for their constructive comments. Also, I would like to thank my guide Dr. H.S. Sheshadri and my friend U. Ragavendra for their support. This research was supported in part by PESCE, Mandya, India.

References

1. Scharstein, D., Szeliski, R., Zabih, R.: A taxonomy and evaluation of dense two-frame stereo correspondence algorithms. Int. J. Comput. Vis. **47**, 7–42 (2002)
2. Aschwanden, P., Guggenbuhl, W.: Experimental results from a comparative study on correlation type registration algorithms. In: Forstner, W., Ruweidel, S. (eds.) Robust Computer Vision, pp. 268–282. Wichmann, Karlsruhe (1992)
3. Hseu, H., Bhalerao, A., Wilson, R.: Image matching based on the co-occurrence matrix. Technical report, University of Warwick, Coventry, UK
4. Faugeras, O., et al.: Qualitative and quantitative comparison of some area and feature based stereo algorithms. In: Fostner, W., Ruweides, S. (eds.) Robust Computer Vision, pp. 1–26. Wichmann, Karlsruhe (1992)
5. Arsenio, A., Marques, J.S.: Performance analysis and characterization of matching algorithms. In: Proceedings of the 5th International Symposium on Intelligent Robotic Systems, Stockholm, Sweden (1997)
6. Sun, C.: Multi-resolution rectangular sub regioning stereo matching using fast correlation and dynamic programming techniques. CMIS report no. 98/246 (1998)
7. Bindu, N.S., Sheshadri, H.S.: An evaluation of correlation based stereo matching algorithms by considering various parameters. In: Proceedings of International Conference on Recent Trends in Signal Processing, Image Processing and VLSI, Bangalore (2014)
8. Bindu, N.S., Sheshadri, H.S.: A comparative study and evaluation of stereo matching costs for radiometric differences. In: Proceedings of International Conference on Recent Trends in Signal Processing, Image Processing and VLSI, Bangalore (2014)
9. Middlebury Stereo Vision.: http://www.vision.middlebury.edu/stereo/

A Hybrid Motion Based-Object Tracking Algorithm for VTS

Prasenjit Das and Nirmalya Kar

Abstract This paper presents a hybrid motion path estimation-based object tracking algorithm for virtual touch screen (VTS) system using a projector, a camera, and LEDs as cursor indicator. The position of hand in current frame is estimated based on its position in previous frames. At the estimated hand position, a square region of interest (ROI) is considered to search for exact location of hand indicator. We also devised a modified full search (FS) algorithm to apply on entire frame. Efficiency of the algorithm is evaluated based on hand indicator detection rate with respect to frame rate and distance between the camera and projection surface. Performance comparison with other algorithms using FS on entire frame or fixed size ROI shows that implementation of the proposed algorithm significantly improves real-time performance by increasing accuracy and reducing computational time.

Keywords Direction vector · Full frame search · Motion estimation · Object tracking · Region of interest · Virtual touch screen

1 Introduction

Object tracking in live video streams has many applications such as surveillance cameras, vehicle navigation, perceptual user interface, and augmented reality [1]. The idea of virtual touch screen (VTS) is first introduced by Tosas based on the contour tracking framework proposed by Blake and Isard [2]. Several vision-based

P. Das (✉) · N. Kar
Department of Electronics and Communication Engineering, National Institute of
Technology, Agartala, India
e-mail: pj.cstech@gmail.com

N. Kar
e-mail: nirmalya@nita.ac.in

© Springer India 2015
L.C. Jain et al. (eds.), *Intelligent Computing, Communication and Devices*,
Advances in Intelligent Systems and Computing 309,
DOI 10.1007/978-81-322-2009-1_23

interactive systems have been proposed in [3, 4]. To detect and track a user's hand, a shape and hand-color-based model-based tracking algorithm was discussed in [5]. Some researchers [6] use mechanical devices that directly measure motions and spatial positions of a hand. Some of the hand tracking algorithms use artic-ulated hand model. A good overview of vision-based full DOF hand motion estimation algorithm is provided in [7]. In Sminchisescu and Triggs [8] proposed an articulated human motion estimation technique with covariance scaled sam-pling. Some algorithms use multiple cues to reduce ambiguity or relieve the related ill-posed problem [9, 10]. Zhao and Dai proposed a hand tracking approach in [11], which can auto-initialize a hand tracker-based monocular vision, and recover the pose of an articulated hand model. We developed a motion path estimation-based tracking algorithm for VTS to achieve maximum real-time response.

2 The VTS and Its Challenges

A VTS has three components: (1) a camera and (2) a projector connected to a computing device. Using signals generated in the computing device, the projector creates a graphical user interface on any ordinary surface which can be manipu-lated by a set of LEDs attached to a user's fingers.

The biggest challenge for achieving real-time performance is dependency on efficient processing of most of the frames captured by the camera, which in turn depends on 3 independent factors—distance between the camera and projection surface (d), frame rate (f_{rate}), and hand movement speed (s).

3 Motion-Based Object Tracking Scheme

The fingertip LED detection process has two parts—(1) LED detection and (2) tracking. The system analyzes each frame and tries to detect a pixel blob with size greater than a predefined threshold size (T_{Blob}) and each pixel in this blob has luminance value greater than T_{Final} [12]. In the subsequent frames, LED blob is searched in a reduced estimated ROI. The entire process is shown in the form of a flowchart in Fig. 1.

The forementioned tracking algorithm uses following sub-algorithms.

Full Frame Search (FFS). FFS is applied on the entire frame if the LED location is not known beforehand. The traditional full search (FS) applies row-wise scan from top to bottom. But instead, we apply a spiral scan method which scans the boundary pixels first and continues toward center of the frame until an illu-minant blob is found or all the pixels are scanned (whichever first).

Region of Interest (ROI) Estimation. The algorithm steps are as follows.

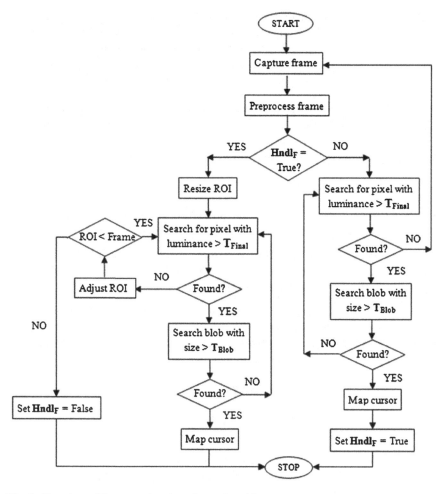

Fig. 1 Flowchart of finger motion detection and tracking

- Let $p_c(x_c, y_c)$ and $p_{c-1}(x_{c-1}, y_{c-1})$ be the location of LED in current frame and previous frame. We calculate the average speed by the following.

$$a = \left\lceil \frac{1}{p} \sum_{k=0}^{p-1} d_k \right\rceil \tag{1}$$

where,

$$d_k = \sqrt{\left(x_{c-k} - x_{c-(k+1)}\right)^2 + \left(y_{c-k} - y_{c-(k+1)}\right)^2} \tag{2}$$

- We consider an error margin of 20 pixels. So $(a + 20)$ is half of ROI side.

- The motion path estimation-based search algorithm requires the ROI to be divided into 3×3 block matrix. So we add a rounding value (r) to a so that, $(2 \times (a + 20) + r) \bmod 3 = 0$. So the final ROI side length is,

$$l = 2 \times (a + 20) + r, \quad \text{where } r = \{0, 1, 2\} \tag{3}$$

- The calculated ROI is centered at the last known location of the cursor.
- If the LED is not detected in the initial ROI, then its side length is incremented by $2 \times (l/3)$ pixels (see Fig. 2b).
- The ROI can increase up to two times, if it does not cross frame boundary i.e.,

$$\text{width}_{\text{ROI}} < \text{width}_{\text{frame}} \quad \text{and} \quad \text{height}_{\text{ROI}} < \text{height}_{\text{frame}}$$

- When LED is found, we calculate its distance (d') from its location in previous fame. We store this distance for average speed calculation in next frame.

Motion Path Estimation. This algorithm makes search inside ROI less exhaustive yet more accurate by predicting the blocks with highest probability of having the LED. The steps are explained below.

- We take two locations $p_1(x_1, y_1)$ and $p_2(x_2, y_2)$ of LED from two consecutive frames and calculate the angle (Θ_1) between them.
- Similarly, we calculate Θ_2 between $p_2(x_2, y_2)$ and $p_3(x_3, y_3)$ in next frame.
- Now we calculate the deviation of angle and note the sign.

$$\text{dev\_ang}_1 = |\Theta_1 - \Theta_2| \tag{4}$$

- We calculate the angle deviation for p frames and calculate their average.

$$\text{Avg}_{\text{ang}} = \frac{1}{p - 2} \sum_{k=1}^{p-2} \text{dev\_ang}_k \tag{5}$$

- This average angle is added to the angle (Θ_c) between the locations of LED in current frame (center of ROI) and previous frame denoted by $p_c(x_c, y_c)$ and $p_{c-1}(x_{c-1}, y_{c-1})$, respectively.

$$\text{Ang}_F = \Theta_c + \text{Avg}_{\text{ang}} \times \text{sgn}, \quad \text{for sgn} = \{-1, 1\} \tag{6}$$

- We get a direction vector (dv) at point pc which makes Ang_F angle with positive horizontal vector (x axis) as shown in Fig. 3a.
- We apply a FS on the central block and the blocks pointed by the direction vector and within $\pm 45°$ range. The $\pm 45°$ range is the error margin. The same pattern is applied on the incremented ROI (see Fig. 3b).

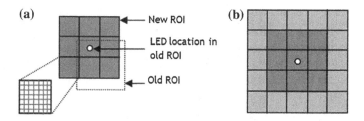

Fig. 2 ROI estimation. **a** Estimated new ROI. **b** Incremented ROI

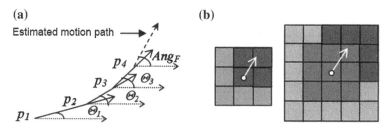

Fig. 3 Motion path estimation. **a** *dv* estimation. **b** Primary search region in initial ROI and incremented ROI

4 Simulation Results

We applied the proposed motion detection algorithm on the existing VTS system. The simulation results are compared with two more algorithms—one using search in entire frame (i.e., ROI = frame size) and the other uses dynamically calculated fixed size ROI. Figure 4 shows motion detection rate by all the three. Fixed ROI-based algorithm has higher, but less stable detection rate than FS algorithm. The proposed algorithm has highest yet stable detection rate.

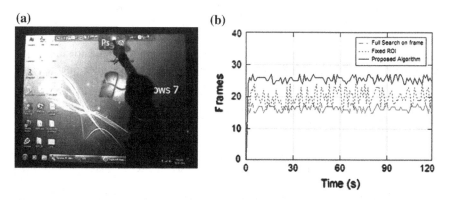

Fig. 4 Comparison of motion detection rate

Table 1 Accuracy comparison based on frame rate

captured frames	Full search on frame		Fixed size ROI		Proposed algorithm	
	Detected frames	Accuracy (%)	Detected frames	Accuracy (%)	Detected frames	Accuracy (%)
259	145	55.98	176	67.95	211	81.47
272	151	55.51	181	66.54	214	78.68
248	138	55.64	179	72.18	190	76.61
277	136	49.1	193	69.68	245	88.44
285	143	50.18	184	64.56	236	82.69

Table 2 Accuracy comparison based on distance taken on first 300 frames

d	Full search on frame		Fixed size ROI		Proposed algorithm	
	Detected frames	Accuracy (%)	Detected frames	Accuracy (%)	Detected frames	Accuracy (%)
2 m	159	53	187	62.3	223	74.3
4 m	174	58	172	57.3	234	78
7 m	175	58.3	203	67.67	209	69.6

Table 1 gives a comparative accuracy analysis from five tests each of 10 s duration. The proposed algorithm has the highest detection rate with 81.58 % on average.

Next we compare the accuracy by varying the distance (d) between the camera and the projection surface. Results in Table 2 show that, with an average 73.97 % accuracy, the proposed algorithm outperforms the other algorithms.

5 Conclusion and Future Work

In this paper, we have presented a motion-based object tracking algorithm for virtual touch screen system (VTSS). This tracking algorithm is specifically optimized for all the HCI applications using LEDs or any bright color blob as hand indicator. Although this algorithm is efficient for tracking the indicator for cursor movement, but it is not optimized for detecting gestures using other fingers. This will be the focus of our future work.

References

1. Yilmaz, A., Javed, O., Shah, M.: Object tracking: a survey. ACM Comput. Surv. (CSUR) **38**(4), Article 13 (2006)
2. Isard, M., Blake, A.: Condensation—conditional density propagation for visual tracking. Int. J. Comput. Vis. **28**(1), 5–28 (1998)

3. Kölsch, M., Turk, M., Höllerer, T.: Vision-based interfaces for mobility. In: First Annual International Conference on Mobile and Ubiquitous Systems: Networking and Services (MobiQuitous'04), pp. 86–94 (2004)

4. Wilson, A.D.: Play anywhere: a compact interactive tabletop projection-vision system. In: Proceedings of the 18th Annual ACM Symposium on User Interface Software and Technology, Seattle, WA, USA, pp. 83–92 (2005)

5. Kurata, T., Okuma, T., Kourogi, M., Sakaue, K.: The hand mouse: GMM hand-color classification and mean shift tracking, In: Proceedings of 2nd International Workshop on RATFG-RTS2001, pp. 119–124 (2001)

6. Fels, S.S., Hinton, G.E.: Glove-talk: a neural network interface between a data-glove and a speech synthesizer. IEEE Trans. Neur. Netw. 4, 2–8 (1993)

7. Erol, A., Bebis, G., Nicolescu, M., Boyle, R.D., Twombly, X.: A review on vision based full DOF hand motion estimation. In: Proceedings of Computer Vision and Pattern Recognition (CVPR'05), San Diego, CA, USA, pp. 75–75 (2005)

8. Sminchisescu, C., Triggs, B.: Estimating articulated human motion with covariance scaled sampling. Int. J. Robot. Res. 22(6), 371–393 (2003)

9. Lu, S., Metaxas, D.N., Samaras, D., Oliensis, J.: Using multiple cues for hand tracking and model refinement. In: Proceedings of the 2003 IEEE Computer Society Conference on Computer Vision and Pattern Recognition (CVPR'v03), pp. 443–450 (2003)

10. Liang, W., Yun-de, J., Tang-li, L., Lei, H., Xin-xiao, W.: Hand motion tracking using simulated annealing method in a discrete space. J. Beijing Inst. Technol. 16(1), 61–66 (2007)

11. Yong-jia, Z., Shu-ling, D.: A robust and fast monocular-vision-based hand tracking method for virtual touch screen. In: Proceedings of 2nd International Congress on Image and Signal Processing (CISP'09), pp. 1–5 (2009)

12. Das, S., Rudrapal, D., Jamatia, A., Kumari, L.: Preprocessing and screen-cursor mapping for a virtual touch screen on a projected area. Int. J. Comput. Sci. Eng. (IJCSE) 3, 2420–2429 (2011)

Side Lobe Level and Null Control Design of Linear Antenna Array Using Taguchi's Method

Rakesh Kumar, Saumendra Ku. Mohanty and B.B. Mangaraj

Abstract This paper presents a global electromagnetic optimization technique using Taguchi's method (TM) for synthesis of linear antenna array (LAA). TM was developed on the basis of orthogonal array (OA) concept. In this paper, Taguchi method yields better lower side lobe levels (SLL) and desired null control pattern which is achieved with less number of iterations compared to traditional optimization techniques such as particle swarm optimization (PSO) and Cuckoo search (CS).

Keywords LAA · OA · Array factor · Excitation magnitude · SLL · Null control

1 Introduction

Linear antenna array (LAA) [1] optimizations have received high attention in the field of electromagnetic for many applications. Two of the examples are minimization of side lobe levels (SLLs) [2] and null placement control [3]. They are usually accomplished by Schelkunoff and Chebyshev methods [4–7]. In this paper, we have used a fast optimization technique, Taguchi's method (TM) [8–11], to find the optimized excitation magnitudes of 20 elements symmetrical LAA. As particle swarm optimization (PSO) [4, 6] depends on number of particles in the search space, which forms that much number of sets of optimizing parameters, so iterations required to converge the fitness function and number of times fitness

R. Kumar (✉) · S.Ku. Mohanty
School of Electronics, ITER, SOA University, Bhubaneswar, Odisha, India
e-mail: rakeshiniter@gmail.com

S.Ku. Mohanty
e-mail: saumendramohanty@soauniversity.ac.in

B.B. Mangaraj
Dept. of ETC, VSSUT, Burla, Odisha, India
e-mail: bbmangaraj@yahoo.co.in

© Springer India 2015
L.C. Jain et al. (eds.), *Intelligent Computing, Communication and Devices*,
Advances in Intelligent Systems and Computing 309,
DOI 10.1007/978-81-322-2009-1_24

function evaluated is much more than TM. Again, as Cuckoo search (CS) [12] depends on number of nests and discovery rate of alien eggs, the number of iterations required is less than PSO but still much more than TM. So by using orthogonal array (OA) [9], Taguchi method reduces the number of iterations and converges to the optimal solution quickly.

2 Taguchi's Optimization Method

To initialize the problem, TM depends on number of input parameter. According to our problem, by relating to Fig. 1, ten excitation magnitudes should be optimized. So OA should have ten columns to present those ten parameters or elements. Each parameter has three levels such as 1, 2, and 3 to define the nonlinear effect. By OA approach, only 27 numbers of experiments are required [10]. To represent OA, a simple notation is used as OA (N, k, s, t) where N represents number of experiments or number of rows, k represents number of parameter to be optimized, s represents number of levels which are selected from set $S = \{1, 2, 3\}$, and t represents strength or tuple $(0 \le t \le k)$.

The array factor in azimuth plane is [6]

$$AF(\theta) = 2 \sum_{n=1}^{10} a(n)\cos[\beta d(n) \cos(\theta)] \tag{1}$$

where β is the wave number, θ is the azimuth angle, $a(n)$ is the excitation magnitude of element, and $d(n)$ is the position of elements. Here, we have assumed uniform phase excitation, i.e., $\varphi_n = 0$.

Our optimization aim is to minimize the fitness value which is a function of SLL and controlling of nulls by providing non-uniform excitation magnitude to individual element of LAA, which is done by varying $a(n)$. So fitness function is given by [6],

$$\text{fitness} = \sum_i \frac{1}{\Delta\theta_i} \int_{\theta_{li}}^{\theta_{ui}} |AF(\theta)|^2 d\theta + \sum_k |AF(\theta_k)|^2 \tag{2}$$

where θ_{ui} and θ_{li} represent the spatial regions in which SLL minimizes, θ_k is the direction of nulls, and $\Delta\theta_i = \theta_{ui} - \theta_{li}$. First term of the above equation solves for minimization of SLL, and second term solves for controlling of nulls.

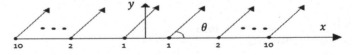

Fig. 1 Geometry of 20 elements uniformly spaced symmetrical LAA

2.1 Designing of Input Parameter Using OA

After assigning the numerical values to the three levels (1, 2, and 3 of OA) for each input parameter, experiments are conducted. Here, the excitation magnitude will be optimized in the range of [0, 1]. For the first iteration, value of level 2 represented as notation $a(n)_1^2$, assumed as the center of computational range [10].

$$\text{i.e. } a(n)_1^2 = \frac{\min + \max}{2} = \frac{0 + 1}{2} = 0.5 \tag{3}$$

where max and min represent upper and lower bound of optimization range, respectively, and $a(n)$ represents nth input excitation magnitude. Value of level 1 represented as notation $a(n)_1^1 = a(n)_1^2 - LD_1$. Value of level 3 represented as notation $a(n)_1^3 = a(n)_1^2 + LD_1$. The level difference is calculated as

$$LD_1 = \frac{\max - \min}{\text{number of levels} + 1} = \frac{1 - 0}{3 + 1} = \frac{1}{4} = 0.25 \tag{4}$$

2.2 Building of Response Table

Each experiment or row will provide 10 non-uniform level values which will act as excitation magnitude to calculate fitness values, and from these values, we can calculate signal to noise ratio (S/N) by the formula [10, 11]

$$\eta = -20 \log (\text{fitness}) dB \tag{5}$$

Fitness and corresponding S/N ratio are used to build a response table by averaging the (S/N) ratio of the same level m for each parameter n and level m by formula [7]

$$\tilde{\eta}(m, n) = \frac{S}{N} \sum_{i, \text{OA}(i,n)=m} (\eta_i) \tag{6}$$

2.3 Optimal Level Values Identification and Confirmation Experiment

Each column of response table [10] provides the largest (S/N) ratio which indicates the corresponding optimal level, and its value is called as optimal value. Hence, for ten parameters, we will have ten optimal levels and their corresponding values. By combining these optimal level values, a confirmation test is conducted to produce a fitness value which is acted as the fitness value of the current iteration.

2.4 Reducing the Optimization Range

If the termination criteria are failed according to current iteration, the process is repeated for next iteration by using following equation:

$$a(n)_{i+1}^2 = a(n)_i^{opt}$$ (7)

The optimal level values of current iteration regarded as central values (i.e. level 2) for the next iteration. For next iteration, $LD_{i+1} = RR \times LD_i$. Where the value of RR is set between 0.5 to 1 and high value of RR provides slower convergence. Here we have considered the RR value as 0.9.

2.5 Checking of Level Values and Termination Criteria

If the level values are higher from maximum range, then it will be set to maximum and if the level values are lower to the minimum range, then it will be set to minimum.

By increasing number of iteration, the LD of each element is decreased. So fitness value of current and next iteration is close to each other due to the closeness between the level values. Below equation is used to terminate the optimization procedure.

$$\frac{LD_i}{LD_1} < \text{converged value}$$ (8)

The converged value can be between 0.001 and 0.01. If optimization goal is achieved or Eq. (8) is satisfied, the optimization process will terminate.

3 Experimental Results

MATLAB Simulation Parameters:

Number of Elements: 20 (non-uniformly excited amplitude), Inter elemental spa ing: 0.5 λ, Frequency: 300 MHz, Azimuth Angle: [0°, 180°]. After the termination of optimization process, we found ten optimized excitation magnitudes such as [0.9968 0.9523 0.7795 0.6469 0.8221 0.7836 1.000 0.7844 0.5450 0.2828]. The first SLL of conventional array pattern has been lowered from −13.26 to −22.1 dB of about 8.84 dB, and nulls are achieved around −42 dB at the desired directions. We have split the MATLAB figure into two distinct figures column-wise (radiation pattern and fitness curve) for better analysis (Fig. 2). The simulation results of sidelobe suppression and desired null control using Taguchi's method compared with conventional Dolph-Tschebyscheff array is presented in Fig. 2.

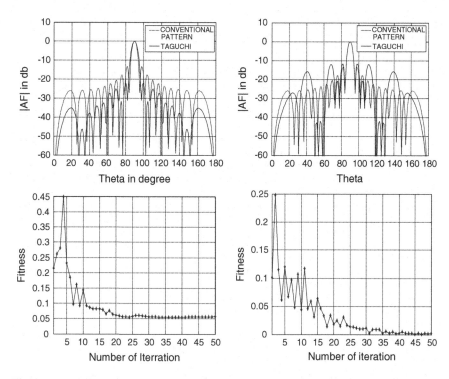

Fig. 2 The regions of suppressed SLL are [1°, 84°] and [96°, 180°] and desired nulls at [50°, 60°] and at [120°, 130°] of 20 elements symmetrical LAA with their fitness convergence curve column-wise represented

4 Conclusion

The TM is applied in this paper to optimize the excitation magnitudes for suppressing the SLL and placement of nulls in the desired direction with dramatically reduced number of iterations nearly 36 iterations for SLL suppression and 46 iterations for null control pattern. TM is compared with the conventional LAA pattern, where elements are excited with uniform amplitude. Easy implementation and quick convergence to desired goal are the important characteristics of TM. Compared to PSO, CS, and other evolutionary algorithms, Taguchi method is easier to implement and requires minimum iterations to achieve the desired goal. So TM will most likely be an increasingly attractive option, in the field of antennas and electromagnetics.

References

1. Wolfgang, H.K.: Basic array theory. Proc. IEEE **80**, 127–139 (1992)
2. Balanis, C.A.: Antenna theory: analysis and design, 3rd edn. Wiley, Singapore (2005)
3. Jazi, S.A.: A new formulation for the design of Chebyshev arrays. IEEE Trans. Antennas Propagat. **42**, 439–443 (1994)
4. Boeringer, D.W., Werner, D.H.: Particle swarm optimization versus genetic algorithms for phased array synthesis. IEEE Trans. Antennas Propag. **52**, 771–779 (2004)
5. Yan, K.K., Lu, Y.: Sidelobe reduction in array-pattern synthesis using genetic algorithm. IEEE Trans. Antennas Propag. **45**, 1117–1122 (1997)
6. Khodier, M.M., Christodoulou, C.G.: Linear array geometry synthesis with minimum side lobe level and null controlling using particle swarm optimization. IEEE Trans. Antennas Propag. **53**, 2674–2679 (2005)
7. Lee, B.G.: Smart antennas: linear array synthesis including mutual coupling effect. The University of Queensland, School of Information Technology and Electrical Engineering, Bachelor of Engineering Thesis, pp. 27–30 (2001)
8. Taguchi, G., Chowdhury, S., Wu, Y.: Taguchi's quality engineering handbook. Wiley, New York (2005)
9. Hedayat, A.S., Sloane, N.J.A., Stufken, J.: Orthogonal arrays: theory and applications. Springer, New York (1999)
10. Weng, W.C., Yang, F., Elsherbeni, A.Z.: Linear antenna array synthesis using Taguchi's method: a novel optimization technique in electromagnetic. IEEE Trans. Antennas Propag. **55**, 68–73 (2007)
11. Chou, T.Y.: Applications of the Taguchi method for optimized package design. In: Proceedings of IEEE 5th Topical Meeting Electrical Performance of Electronic Packaging, Napa, CA, pp. 14–17 (1996)
12. Stevanetic, B.B., Kolundzija, B.M.: Comparison of differential evolution and cuckoo optimization for antenna array problems. In: IEEE International Symposium Antennas Propagation Society (APSURSI), Chicago, pp. 1–2 (2012)

A Novel Approach for Jamming and Tracking of Mobile Objects

Aradhana Misra, Nirmal Kumar Rout and Kaustav Dash

Abstract An adaptive algorithm called the least mean square (LMS) is employed to study jamming and tracking of mobile objects. The concepts of antenna array, adaptive antenna array and smart antenna have been employed in the proposed method. The focus is on the adaptive jamming technique. The LMS adaptive algorithm is used to reject a particular direction signal. The proposed approach is also used to track the moving object by projecting the main beam towards the target, then estimate its next position and track the object for a known value of Azimuthal and elevation angular position. While estimating the new position of the target by comparing the previous and present value of echo signal, main signal get interference from noise signal. Here, the main objective is to make an algorithm that can track the target, estimate its new position and increase signal to noise ratio of the system. Detailed simulation has been carried out to validate the proposed method using the MATLAB.

Keywords Jamming · Tracking · LMS · Adaptive algorithm · Adaptive filters

1 Introduction

In recent years, a growing field of research in adaptive signal processing has resulted in a variety of adaptive automation whose characteristics in few ways resemble certain characteristics of living being and biological adaptive process. In

A. Misra (✉) · N.K. Rout
School of Electronics Engineering, KIIT University, Bhubaneswar, India
e-mail: aradhana.misra6@gmail.com

N.K. Rout
e-mail: routnirmal@rediffmail.com

K. Dash
GITAM, Bhubaneswar, India
e-mail: dash.kaustav@gmail.com

© Springer India 2015
L.C. Jain et al. (eds.), *Intelligent Computing, Communication and Devices*,
Advances in Intelligent Systems and Computing 309,
DOI 10.1007/978-81-322-2009-1_25

217

our proposed work, we have considered adaptive algorithm (least mean square, *LMS*) to jam signals coming from particular direction and train the system so that the main lobe is in the direction of desired signal. We also use the LMS concept to track a moving object by steering the main lobe in its direction. We generate the radiation pattern for the same and plot the learning/error curve. We carry out the simulations to find that for fast convergence with large number of iteration, LMS is a suitable algorithm [1].

2 Smart Antenna

We have considered the concept of smart antennas in our analysis of the system. The basic principle of smart antennas is illustrated in Fig. 1. They are most often realized with either switched beam or fully adaptive array antennas. An array of antenna consists of two or more antennas spatially arranged and electrically interconnected to produce a directional radiation pattern. A smart antenna system consists of an antenna array, an associated RF hardware and a computer controller that changes the array pattern in response to the radio frequency environment, in order to improve the performance of a communication or radar system [2, 3].

3 Algorithm Applied in the System

Least mean square algorithm is a very popular adaptive signal-processing algorithm. The LMS is popular mainly because of its simplicity and easy computation. It is used in variety of practical application [4]. It provides stable and robust performance against different signal, and it has MSE behaviour.

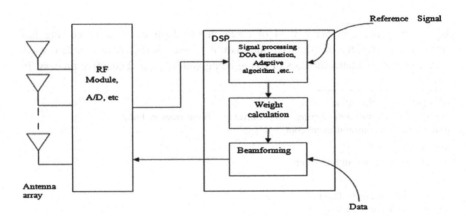

Fig. 1 Block diagram of smart antenna

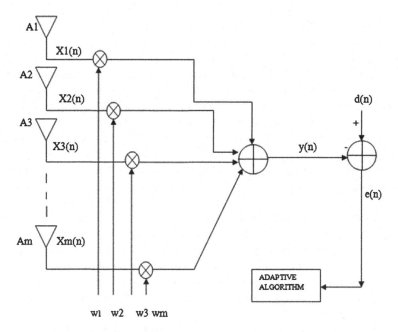

Fig. 2 Block diagram of adaptive filter

The LMS is the most used algorithm in adaptive filtering. It is a gradient descent algorithm; it adjusts the adaptive filter taps modifying them by an amount proportional to the instantaneous estimate of the gradient of the error surface [5]. The block diagram used in our adaptive filter is shown in Fig. 2.

In case of general, the LMS algorithm and the weight vector coefficient for the finite impulse response (FIR) filter are updated as in following equation:

$$w(n + 1) = w(n) + \mu e(n)y(n) \tag{1}$$

where $w(n)$ is the weight vector coefficient, which can be expressed as following equation:

$$w(n) = [w0(nw1(n)...wm(n)] \tag{2}$$

where the filter length is $(m + 1)$. μ is known as step factor (which is convergence parameter of the LMS algorithm). $e(n)$ is the output error, which is represented as follows:

$$e(n) = d(n) - y(n) \tag{3}$$

$y(n)$ being the filter output and $d(n)$ is the reference signal [1, 6].

4 Simulation Results

The adaptive filter with MATLAB is simulated, and the results prove its performance is better than the use of a fixed filter designed by conventional methods.

Figure 3 shows that it is possible to block signals coming from directions of about 10, 80° and get the desired signal coming from 50°. For jamming, four signals with direction of arrival 50, 80, 120, and 10° angle have been considered. Number of antenna array elements is taken as 9 with frequency of operation of 0.5 MHz. Total iteration for LMS adaptive algorithm is 1,000. Then, steering vector matrix is generated and also, the signals s1, s2, s3, and s4 are generated (considering value as random integral). To get the complex envelope of the above signals, their PSK modulation is taken, and then, complex Gaussian noise is generated and added at the time of reception. X is the input vector which is equal to the received signal S convolved with steering matrix A along with the additive noise; $X = A*S$ + noise. Initial weight vector is taken to be zero. It is found out that the main lobe is formed at 50° for s1 and null are at undesired position (80, 120, 10). So it is concluded that using adaptive LMS algorithm, a constructive interference occurs to generate the desired signal and destructive interference occurs at undesired directions.

Figure 4 shows the tracking of the moving object. Number of antenna array elements is taken to be 9 with frequency of operation of 0.5 MHz. Number of iterations taken is 10,000. The step factor (μ) is considered to be 0.00000001. The LMS filter coefficients are initialized to zero. Then, steering vector matrix is generated and also, the signals s1, s2, s3, and s4 are generated (considering value as random integral). To get the complex envelope of the above signals, their PSK modulation is taken, and then, complex Gaussian noise is generated and added at the time of reception. X is the input vector which is found out. Filter coefficients are found out by using the LMS algorithm.

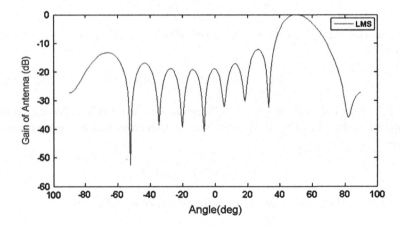

Fig. 3 Simulation result for jamming (main lobe occurs in the direction of 50°)

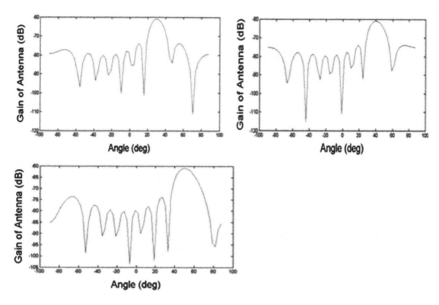

Fig. 4 Simulation result for tracking (main lobe occurs in the direction of about 30, 40, and 50°)

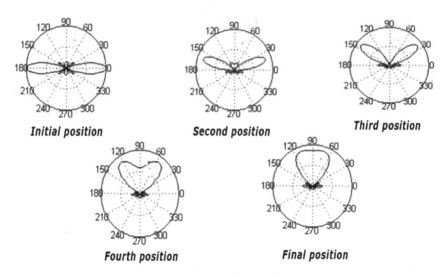

Fig. 5 Radiation patterns while tracking the moving object

Figure 5 shows the different radiation patterns. It gives an idea how the main beam follows the moving object. The initial position of the main lobe is 0°. Then, it traverses in the direction of the moving object. Finally, it is at 60° position.

Fig. 6 Mean square error
(MSE) verses iterations curve

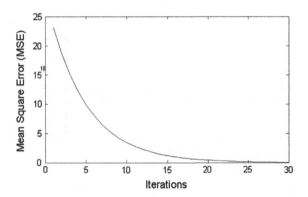

Figure 6 shows how the system is adapting itself and learns by training itself. We also find that there is a gradual decrease in the mean square error, and it is converging in nature. The number of iterations taken is 30.

5 Conclusion

This paper has investigated adaptive LMS algorithm and has successfully implemented it in jamming and tracking. It is found that adaptive array technique used here is better than the mechanical steering of antenna. In case of jamming, the null is steered towards the undesired signal. In order to track a moving object, the main lobe is steered in the direction of the target and its information is collected on timely basis. The approach taken to track a moving object was to project our main beam towards the target, then estimate its future position and track the object.

Considering the objective of studying the feasibility of implementing adaptive arrays in MATLAB, this effort met with success. MATLAB provides the necessary tools and functions to implement most adaptive arrays. The amount of effort needed to perform the simulations is greatly reduced during this study.

References

1. Widrow, B., McCool, J.M., Larimore, M., Johnson, C.R.: Stationary and nonstationary learning characteristics of the LMS adaptive filter. Proc. IEEE **64**, 1151–1162 (1976)
2. Godara, L.C.:Smart Antenna. CRC Press, Boca Raton (2004)
3. Balanis, C.A.: Antenna Theory, vol. 2. Wiley, New York (1997)
4. Kaur, M.K.K., Mittal, R.: Improvement in capacity and signal strength using LMS algorithms. Int. J. Comput. Appl. **1**(5), 103–107 (2007)
5. Haykin, S.: Adaptive Filter Theory. Prentice-Hall, Englewood Cliffs (1991)
6. Raymond, K.E., Johnston, H.: A variable step size LMS algorithm. IEEE Trans. Signal Process. **40**(7), 1633–1642 (1992)

Performance Analysis of Low-Complexity Multiuser STBC MC-CDMA System

Sadananda Behera and Sarat Kumar Patra

Abstract In this paper, a low complexity and efficient multiuser space time block code multicarrier code division multiple access (STBC MC-CDMA) for downlink wireless communication system is proposed. STBC MC-CDMA provides diversity gain to improve transmission efficiency of mobile wireless systems where both STBC encoder and decoder are in time domain thus reducing the complexity at the receiver side. Proposed STBC MC-CDMA scheme achieves a diversity order of 2 without channel state information (CSI) at the transmitter under flat fading conditions without bandwidth expansion. In this paper, the STBC MC-CDMA is compared with the STBC OFDM scheme under Rayleigh fading channel and AWGN channel using zero forcing (ZF) linear detection scheme and as anticipated the proposed STBC MC-CDMA outperforms STBC OFDM. Simulation results verify this.

Keywords STBC · MC-CDMA · CSI · Linear detection schemes · Rayleigh fading channel

1 Introduction

Multicarrier code division multiple access (MC-CDMA) also known as OFDM-CDMA is a promising technology for 4G wireless communication systems [1, 2]. MC-CDMA spreads the data in frequency direction, thus increasing the frequency diversity gain [1]. Since the symbol rate is lower than the original rate, the effect of

S. Behera (✉) · S.K. Patra
Department of Electronics and Communication Engineering,
National Institute of Technology Rourkela, Rourkela 769008, Odisha, India
e-mail: sadanandabehera07@gmail.com

S.K. Patra
e-mail: skpatra@nitrkl.ac.in

© Springer India 2015
L.C. Jain et al. (eds.), *Intelligent Computing, Communication and Devices*,
Advances in Intelligent Systems and Computing 309,
DOI 10.1007/978-81-322-2009-1_26

223

ISI is eliminated [3], thus making the complex equalizer design much easier. So MC-CDMA provides a flexible system design, since the spreading factor (SF) is not necessarily to be taken equal to the number of subcarriers [1, 2].

Space–time block code (STBC) is a special form of MIMO and originally employed for 2 transmit and 1 receive antenna by Alamouti [4]. Since STBCs are originally proposed only for flat fading channels, application of this scheme to frequency selective channel is challenging. Hence, integration with multicarrier techniques such as OFDM and MC-CDMA is essential which convert frequency selective channel to several flat fading channels eliminating ISI. In Ref. [5], STBC OFDM is explained in which both STBC encoders and decoders are in time domain. In this proposed STBC MC-CDMA, the equalization is carried out in time domain. Time domain equalization is preferred over frequency domain equalization because it is simple to implement and less complex.

This paper is organized as follows. Following this section, mathematical model of STBC MC-CDMA is explained. The communication system model is discussed in Sect. 3. Section 4 contains the simulation results, and Sect. 5 presents the concluding remarks.

Notation: $(\cdot)^T, (\cdot)^*, (\cdot)^H$ represent transpose, complex conjugate, and Hermitian operation, respectively.

2 Mathematical Model

In this section, a brief description about the STBC encoding scheme for two transmitting antennas and one receive antenna is given.

2.1 STBC Encoding Scheme

In the first time instant, antenna 1 transmits $X1$ and antenna 2 transmits X_2 simultaneously, while in second time instant, antenna 1 transmits $-X_2^*$ and antenna 2 transmits X_1^* [4]. The coding rate of this STBC is one since in two symbol periods only two signals are transmitted. Since the STBC codes are orthogonal, the symbols from different antennas can be distinguished at the receiver by simple zero forcing (ZF) equalization [1]. At the receiver, we can mathematically model the received signals as follows:

Received signal at 1st time slot:

$$y(1) = [h_1 \quad h_2]\begin{bmatrix} X_1 \\ X_2 \end{bmatrix} + n(1) \tag{1}$$

Received signal at 2nd time slot:

$$y(2) = [h_1 \quad h_2] \begin{bmatrix} -X_2^* \\ X_1^* \end{bmatrix} + n(2) \tag{2}$$

where h_1 and h_2 are flat fading channel coefficients and $n(1)$ and $n(2)$ are noise components which is generally additive white Gaussian noise (AWGN). In this paper, we have assumed that the channel is assumed to be constant over the two time slots.

3 Communication System Model

The baseband processing at the transmitter is presented at Fig. 1. The information bits of different users are digitally modulated and then spread using orthogonal Walsh–Hadamard (W–H) codes. All the spread signals are summed up to produce the CDMA signal and then passed through STBC encoder. Each of these STBC encoded outputs is passed through individual OFDM modulators. After IFFT operation, the cyclic prefixes (CP) are added and then transmitted through channel.

3.1 STBC Decoding Scheme

At the receiver, first the CP is removed and then the channel equalization is carried out in time domain as shown in Fig. 2. In the conventional STBC MC-CDMA [6, 7], equalization is implemented in frequency domain which increases the

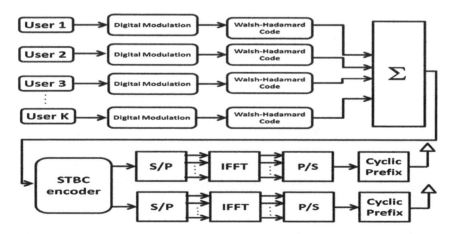

Fig. 1 Transmitter block diagram of STBC MC-CDMA system

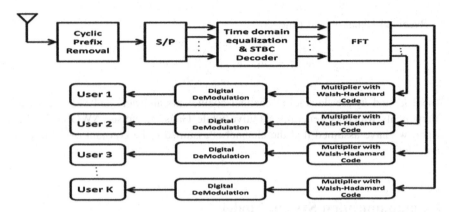

Fig. 2 Receiver block diagram of STBC MC-CDMA system

complexity by processing the channel coefficients which are already present in time domain into frequency domain. For detection of signal, simple ZF equalizer can be used. Now, the received signal vectors from (1) and (2) as

$$\begin{bmatrix} y(1) \\ y(2) \end{bmatrix} = \begin{bmatrix} h_1 & h_2 \\ -h_2^* & -h_1^* \end{bmatrix} \begin{bmatrix} X_1 \\ X_2 \end{bmatrix} + \begin{bmatrix} n(1) \\ n(2)^* \end{bmatrix} \tag{3}$$

For detection of signals X_1 and X_2, zero forcing receiver is employed as under

Let, $H = \begin{bmatrix} h_1 & h_2 \\ h_2^* & -h_1^* \end{bmatrix}$ and $\bar{Y} = \begin{bmatrix} y(1) \\ y(2)^* \end{bmatrix}$, then we can define $W = (H^H H)^{-1} H^H$

To estimate the transmitted symbols, we can employ

$$\tilde{X} = \begin{bmatrix} X_1 \\ X_2 \end{bmatrix} = W\bar{Y} = (H^H H)^{-1} H^H \begin{bmatrix} y(1) \\ y(2)^* \end{bmatrix} \tag{4}$$

After the channel effects are nullified according to above equations, the data are sent to the FFT block and the signals are despread by using the same W–H code assigned to the specified user followed by the detection of the signals to retrieve all the user information bits.

4 Results and Discussion

Figures 3 and 4 present that the STBC MC-CDMA performs better than STBC OFDM under AWGN and Rayleigh channel. For BER of 10^{-3}, STBC MC-CDMA achieves a SNR gain of around 6 dB in AWGN and Rayleigh fading channel. In Fig. 5, it is observed that as the number of users increase, the BER performance degrades.

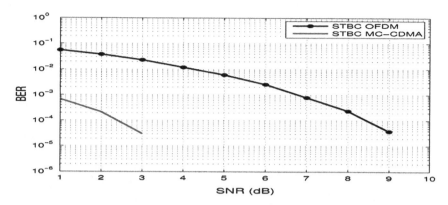

Fig. 3 BER performance of STBC OFDM and STBC MC-CDMA under AWGN channel for single user

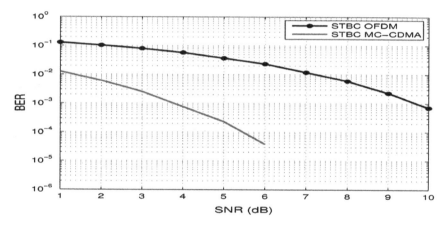

Fig. 4 BER performance of STBC OFDM and STBC MC-CDMA under Rayleigh fading channel for single user

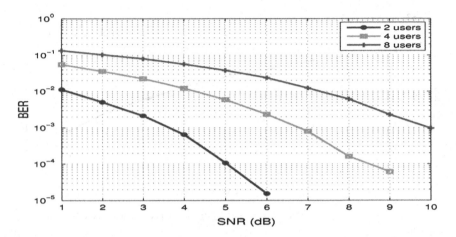

Fig. 5 BER performance for 2, 4, and 8 users for SF = 4 of STBC MC-CDMA under Rayleigh fading channel

5 Conclusions

In this paper, a low-complexity multiuser STBC MC-CDMA scheme is proposed. The proposed scheme is compared with STBC OFDM. Also the simulation results for different users for STBC MC-CDMA are presented. Multiuser BER performances are also given which ensures that BER performance degrades as the number of users increase. It is observed that STBC MC-CDMA performs better than STBC OFDM under AWGN and Rayleigh fading channel.

References

1. Fazel, K., Kaiser, S.: Multi-carrier and spread spectrum systems: From OFDM and MC-CDMA to LTE and WiMAX. Wiley (2008)
2. Hara, S., Prasad, R.: Overview of multicarrier CDMA. IEEE Commun. Mag. **35**(12), 126–133 (1997)
3. Yang, L.-L.: Multicarrier communications. Wiley (2009)
4. Alamouti, S.: A simple transmit diversity technique for wireless communications. IEEE J. Sel. Areas Commun. **16**(8), 1451–1458 (1998)
5. Hassib, M.D., Singh, M., Ismail, M., Nordin, R.: Efficient and low complexity STBC-OFDM scheme over fading channel. In: 18th Asia-Pacific Conference on Communications (APCC), 2012. IEEE, pp. 402–406 (2012)
6. Chen, J.-D., Ueng, F.-B., Chang, J.-C., Su, H.: Performance analysis of OFDM-CDMA receivers in multipath fading channels. IEEE Trans. Veh. Technol. **58**(9), 4805–4818 (2009)
7. D'Orazio, L., Sacchi, C., Fedrizzi, R., De Natale, F.G.: An adaptive minimum-BER approach for multi-user detection in STBC-MIMO MC-CDMA systems. In: Global Telecommunications Conference, 2007. GLOBECOM'07. IEEE, pp. 3427–3431 (2007)

On the Design an Enhanced Bandwidth of Elliptical Shape CPW-Fed Fractal Monopole Antenna for UWB Application

Satyabrata Maiti, Naikatmana Pani, Thiyam Romila Devi and Amlan Datta

Abstract This paper presents a design of compact elliptical-shaped CPW-fed planer UWB fractal antenna. A novel planer UWB antenna using a fifth iteration elliptical fractal shape is presented in this paper. The frequency characteristics of antenna consist of UWB properties in the range 2.0–16 GHz that corresponds to the impedance bandwidth of 140 %. The antenna has nearly good omnidirectional radiation pattern and peak gain of 4.9 dBi. The group delay profile of the proposed antenna lies within 1 ns. The areas of applications are medical imaging, wireless communication, and vehicular radar.

Keywords Fractal geometry · Ultrawide band · Coplanar waveguide · Impedance bandwidth

1 Introduction

In February 2002, the frequency band between 3.1 and 10.6 GHz was assigned as the ultrawide band (UWB) usable frequency by the Federal Communication Commission (FCC) [1], USA, since it provides high data rate at a very high speed [2, 3]. This has increased the demand for smaller size antenna having broadband features. One of the many technological challenges of an ultra wide band system lies in the high level of integration that UWB products require at low cost and low

S. Maiti (✉) · N. Pani · T.R. Devi · A. Datta
School of Electronics Engineering, KIIT University, Bhubaneswar, India
e-mail: Satyabratamaiti12@gmail.com

N. Pani
e-mail: naikatman@gmail.com

T.R. Devi
e-mail: romilath@gmail.com

A. Datta
e-mail: amlandatta01@gmail.com

© Springer India 2015
L.C. Jain et al. (eds.), *Intelligent Computing, Communication and Devices*,
Advances in Intelligent Systems and Computing 309,
DOI 10.1007/978-81-322-2009-1_27

power consumption. However, antenna design is a challenging task in UWB systems due to a 140 % impedance bandwidth. Various designs of UWB antenna have been accounted for, where subwavelength structures as SRR and electromagnetic bandgap structures are used to create notch bands. The printed monopole antennas have been developed in current years catering ultrawide band range [4–6]. Various matching techniques are reported to increase the bandwidth and thereby depletion of size. Optimization of feed gap, beveling of ground planes [7], feed gap etc. are used to increase the bandwidth and hence to obtain UWB [8, 9]. Currently, the self-recursive nature of the fractal geometries has been utilized to design electrically smaller ultrawide band antennas. A novel approach to obtain a multiband miniaturized antenna was to include fractal geometry.

2 Antenna Design and Parametric Study

The geometry of the proposed antenna structure is designed on a substrate of $\varepsilon_r = 4.3$, thickness 1.53 mm with a dimension of 45×48 mm^2 ($W_{sub} \times L_{sub}$), and loss tangent waveguide coplanar wave guide feed. The impendence bandwidth of the designed antenna covers the range from 2.0 GHz to 16 GHz unlike a simple elliptical Co planar waveguide feed monopole of same size whose operating bandwidth ranges from 3.1 GHz to 14 GHz.

The initial height d_g of the two ground planes is taken to be 14.5 mm Fig. 1a and then beveled to the height as depicted in Fig. 1b. At first, the planar antenna was designed such that it covers the entire UWB range. The proposed antenna structure is shown in Fig. 2. The effect of various parameters of the feed gap and the radiating patch gap between ground planes is studied. It can be that there is a shift to lower frequency, for S_{11} dB better than 10 dB, as iteration increases.

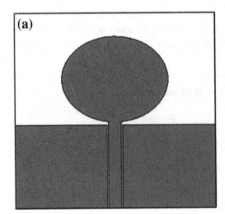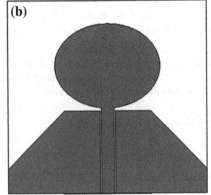

Fig. 1 a Initial elliptical monopole. **b** Construction of beveled ground

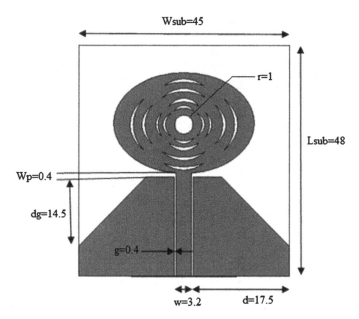

Fig. 2 Proposed elliptical fractal antenna

Introduction of the fractal shape enhances the effective electrical path of the surface current which in turn increases the effective impedance bandwidth [10–12]. This fine-tunes the desired impedance bandwidth frequency range of UWB antenna.

The gap between the patch and the ground plane, w_p, and the gap between the feed line and the ground plane, g, are the two most important parameters which determine the UWB characteristics of the antenna as shown in Fig. 3a, b. By varying these two parameters, the antenna is made to cover the entire UWB range from 3.1 to 10.6 GHz. It is observed that the bandwidth of the antenna increases as the gap g decreases. So the optimized value of the gap, g, is fixed at 0.5 mm. Then, elliptical patch contains fifth iterative structure. In the first iteration, a horizontal ellipse having its major axis = 15 mm and minor axis = 10.5 mm is intersected with the vertical ellipse having same dimension. Then, horizontal ellipse and vertical ellipse having major axis = 9.5 mm and minor axis = 8 mm, respectively, are subtracted from it. The same process is repeated for each iteration using the scaling down of 1.0 on both axes.

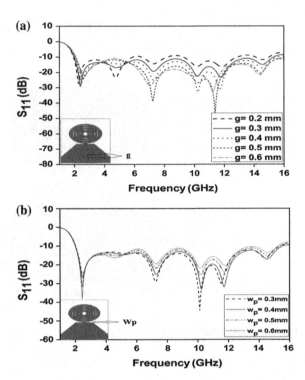

Fig. 3 **a** Simulated result of proposed antenna gap variation between feed and ground and **b** simulated result of proposed antenna gap variation between patch and ground

3 Results and Discussion

3.1 Return Loss

The proposed antenna is evaluated by the finite integration method by using time domain solver of CST microwave studio. The designed antenna has a compact size of 45×48 mm^2. The simulated characteristic of the designed antenna is shown in Fig. 4, and it is noticeable that the impedance bandwidth ranges from 2 to 16 GHz.

3.2 Current Distribution

The current distribution at four frequencies, 3.0, 5.5, 7.5, and 10 GHz, is shown in Fig. 5. Antenna behaves as a radiating slot which can be formed between the ground plane and radiating patch. The current distribution at 5.5 GHz is shown in Fig. 5 that shows that it results in standing wave due to the concentration of current near the slot.

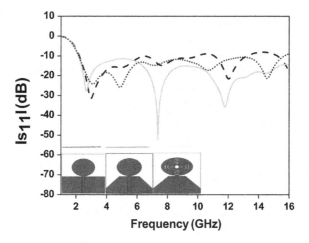

Fig. 4 Comparison of s_{11} with and without beveling the ground and with fractal slots cut in the patch

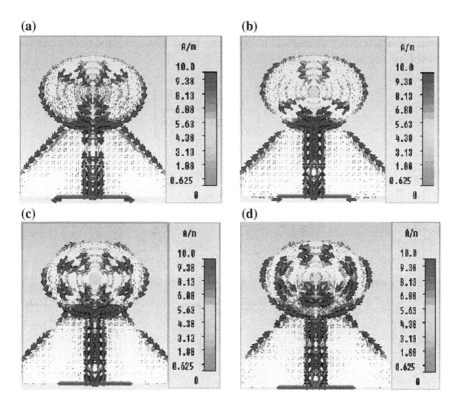

Fig. 5 Simulated current distribution of proposed antenna at **a** 3.0 GHz, **b** 5.5 GHz, **c** 7.5 GHz, and **d** 10.0 GHz

3.3 Radiation Pattern

The radiation patterns of this proposed antenna are simulated at selective frequencies 2.0 to 16 GHz in E-plane and H-plane. H-plane radiation patterns are depicted in Fig. 6 which shows that it is nearly good omnidirection and E-plane is bidirectional. The simulated radiation patterns at 3.0, 5.0, and 10 GHz plotted are shown below.

3.4 Peak Gain and Group Delay

The peak gain of the proposed antenna is simulated 4.9 dBi as shown in Fig. 7. The peak gain increase as the frequency increases. Beyond a certain higher frequency, it almost constant. Figure 8 shows group delay of the proposed antenna which is within 1 ns, confirming the proposed antenna to be non-dispersive. The proposed antenna shows a nearly flat feedback in 3.1–10.6 GHz ultrawide band, where the group delays make large outing.

This ensures that this antenna is distortion free and exhibits satisfactory time domain characteristics throughout its operating band. The correlation coefficient is defined by

$$\rho = \max_{\tau} \left[\frac{\int s_1(t)s_2(t-\tau)}{\sqrt{s_1^2(t)dt}\sqrt{s_2^2(t)dt}} \right]$$

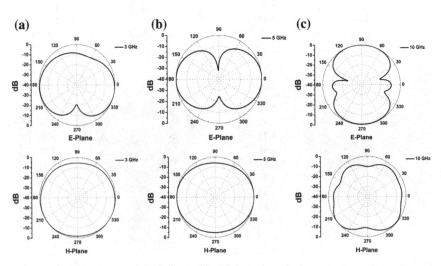

Fig. 6 Simulated radiation patterns in H-plane and E-plane at **a** 3.0 GHz, **b** 5.0 GHz, and **c** 10 GHz

Fig. 7 Simulated peak gain
of this proposed antenna

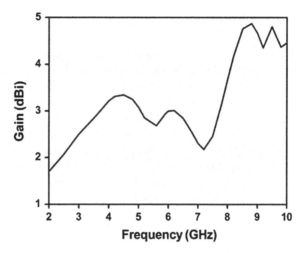

Fig. 8 Simulated group
delay

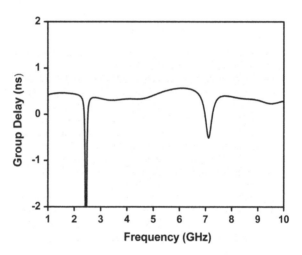

τ is the delay which is changed to make the numerator in the equation maximum.
It obtains the correlation between the electric field signals s_1 (t) and s_2 (t). The
input signal is fifth derivative Gaussian pulse and its fifth derivative. The excited
pulses are chosen as the signal s_1 (t), while the received pulse as signal s_2 (t).
Indeed, it reflects the similarity between the source pulse and the received pulse.
When the two signal waveforms are identical, this means that the antenna system
does not distort the input signal at all. The correlation coefficient found from the
slotted fractal antenna is 0.83 and from the unslotted fractal antenna is 0.89.

4 Conclusion

A novel CPW-fed elliptical fractal antenna is proposed. The monopole antenna with elliptical fractal slots in the radiating patch having UWB characteristics. The simulated radiation pattern of this antenna is very close to bidirectional in E-plane and omnidirectional in H-plane. The gain of the antenna varies from 2 to 4.9 dBi. The impedance bandwidth of the antenna ranges from 2 GHz to 16 GHz which corresponds to 140 % impedance bandwidth. The simulated group delay exhibits within 1 ns over the desired frequency. The total antenna dimension is 48 × 45 mm. This specifies the proposed antenna potential for use in military application. The antenna is simple to design, size of the antenna is compact and easy to fabricate, and it is suitable for MIC/MMIC circuits.

References

1. Report of the spectrum efficiency working group.: FCC spectrum policy task force, Tec Rep (2002)
2. Elsadek, H., Nashaat, D.M.: Band compact size trapezoidal PIFA antenna. J. Electromagn. waves Appl. **21**(7), 865–876 (2007)
3. Liu, W.C., Liu, H.J.: Miniaturized asymmetrical CPW-FED meandered strip antenna for triple-band Operation. J. Electromagn. Waves Appl. **21**(8), 1089–1097 (2007)
4. Deng, C., Xle, Y.J., Li, P.: CPW fed planer printed monopole antenna with impedance bandwidth, enhanced. IEEE Antennas Wirel. Propag. Lett. **8**, 1394–1397 (2009)
5. Zhu, F., Gao, S., Li, J.Z., Xu, J.D.: Planer asymmetrical ultra wide band antenna with improved multiple band-notched characteristics. Electron. Lett. **48**(11) (2012)
6. Bazaz, R., Koul, S.K., Kumar, M., Basu, A.: An ultra wide band antenna with band reject capability and its characterization in time domain. Prog. Electromagnet. Res. C. **19**, 223–234 (2011)
7. Pei, Q.Q., Qiu, C.W., Yuan, T., Zouhdi, S.: Hybrid shaped ultra wideband antenna. Microwave Opt. Technol. Lett. **49**(10), 2412–2415 (2009)
8. Das, S., Gorai, A., Ghatak, R.: An enhanced bandwidth CPW-fed fractal antenna for UWB communication, National Conference on Materials, Devices and Circuits in Communication Technology, pp. 59–62 (2014)
9. Werner, D.H., Ganguly, S.: An overview of fractal antenna Engineering research. IEEE Antennas Propag. Mag. **45**(1) (2003)
10. Saleem, R., Brown, A.K.: Empirical miniaturization analysis of an inverse parabolic step sequence based UWB antennas. Prog. Electromagn. Res. **114**, 369–381 (2011)
11. Karmakar, A., Verma, S., Pal, M., Ghatak, R.: An ultra wide band monopole antenna with multiple fractal slots with dual band rejection characteristic. Prog. Electromagn. Res. C **31**, 185–197 (2012)
12. Naghshvarian-Jahromi, M.: Novel wideband planar fractal monopole antenna. IEEE Trans. Antennas Propag. **56**(12), 3844–3849 (2008)

Design of Microstrip Branch Line Coupler Phase Shifter in L-Band

Thiyam Romila Devi, Satyabrata Maiti, Abhishek Jena and Amlan Datta

Abstract The goal of this paper is to analyze, simulate, and design a microstrip quadrature hybrid coupler phase shifter operating at resonant frequency of 1.5 GHz with input impedance of 50 Ω. This works as a phase shifter to provide reflections which cancel at the input port and sum to a phase shifted version of the input on the fourth port. The microstrip line dimensions such as length, width, substrate dielectric constant, effective dielectric constant, and parameters like return loss, insertion loss, isolation and its desired phase shift are calculated using HFSS (version 13.0). The perfect 3 dB power division is observed at output port 2 and port 3 of around −3.6 and −3.4 dB at 1.6 GHz. The return loss and isolation are observed at port 1 and port 4 of around −36 and −22 dB at operating frequency of 1.6 GHz.

Keywords Branch-line coupler · Microstrip line · Insertion loss · Isolation · Return loss · HFSS

1 Introduction

Generally, quadrature hybrid coupler is four port devices that split the incident input power signal port (port 1) into two output ports named as transmitted port (port 2) and coupled port (port 3). The two output signals are attenuated by three

T.R. Devi (✉) · S. Maiti · A. Jena · A. Datta
School of Electronics Engineering, KIIT University, Bhubaneswar, Odisha, India
e-mail: romilath@gmail.com

S. Maiti
e-mail: satyabratamaiti12@gmail.com

A. Jena
e-mail: abskjena@gmail.com

A. Datta
e-mail: amlandatta01@gmail.com

© Springer India 2015
L.C. Jain et al. (eds.), *Intelligent Computing, Communication and Devices*,
Advances in Intelligent Systems and Computing 309,
DOI 10.1007/978-81-322-2009-1_28

237

decibels (3-dB) directional couplers and have a 90^0 phase difference between them. Three decibel attenuation means that 50 % of the incident input power is lost [1]. The fourth port on the device is isolated from the first port. This hybrid is created often in microstrip form and additionally called as a branch-line hybrid coupler. As the name implies that the power is equally divided between the outputs ports and therefore, the structure is electrically and mechanically symmetrical. The analysis of this hybrid is done using a technique of even-odd mode decomposition. The device exhibits reciprocal behavior and as a result any port can act as an input and operate in the same manner [2]. The coupler is made of 4 sections of quarter-wavelength transmission line. Two sections have a characteristic impedance of Z_0 and other two have an impedance of $Z_0/\sqrt{2}$. Each port is fed with a transmission line of impedance Z_0 [2]. The characteristic impedance Z_0 was decided to be set to 50Ω because this value is the most commonly used in industry. The 3-dB branch-line coupler was optimized to have frequency band of 1–2 GHz in L-band. The path of coupled arm (port 3) is ¼ of wavelength long in order to provide a phase shift of 90° between incident input signal at port 1 and isolated port 4 and also in between output transmitted signal at port 2 and output coupled signal at port 3. Primary application of hybrid couplers is in phase shifters. Other applications are in microwave systems or subsystems as attenuators, balanced amplifiers, balanced mixers, modulators, and discriminators [3].

2 Methodology

The designing of the microstrip hybrid coupler and its corresponding character-istics are simulated by using HFSS (version 13.0). The Fig. 1 shows the 3-dB hybrid coupler geometry in microstrip form. Table 1 shows the definitions of four ports of branch-line coupler.

3 Parameters of Hybrid Coupler

A coupler can be categorized by following main parameters:

1. Bandwidth: The frequency range where the device provides a phase shift within $\pm 10^0$ of the desired phase shift.
2. Insertion loss: The additional loss within the device above the splitting loss. This is due to reflection of signals, dielectric losses, and conductor losses.
3. Coupling ratio: The ratio of the lower of the two output powers i.e., at port 2 and port 3 to the input power at port 1.
4. Isolation: The ratio between the input power and the leakage power at the isolated port [4]. Over bandwidth of 13:1 maximum, the couplers will typically

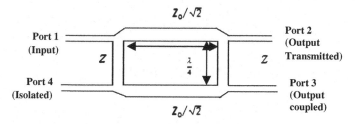

Fig. 1 3 dB hybrid coupler geometry [2]

Table 1 Definitions of four ports of hybrid coupler

Name	Description
Input port 1 (incident port)	Main system input signal, $P_{InputPort}$
Output port 2 (transmitted port)	Main system output signal, $P_{OutputPort}$
Output port 3(forward coupled port)	Power sampled from input port 1, $P_{CoupledPort}$
Isolated port 4(reverse coupled port)	Power sampled from output port 2, $P_{IsolatedPort}$

exhibit about 3 dB of insertion loss, a coupling ratio of 3 dB and isolation of 20 dB [4].

$$\text{Isolation (dB)} = 10 * \log\left(\frac{P_{\text{isolatedport}}(W)}{P_{\text{inputport}}(W)}\right)$$

5. Coupling Factor: The power which is transferred to the output coupled port 3 with respect to input port 1.

$$\text{Coupling Factor(dB)} = 10 * \log\left(\frac{P_{\text{coupledport}}(W)}{P_{\text{inputport}}(W)}\right)$$

6. Coupling Loss: The amount of power lost from the input port 1 to the output coupled port 3 and the power lost from input power port 1 to the isolated port 4. Therefore, coupling loss is given as,

$$\text{Coupling Loss(dB)} = 10 * \log\left(\frac{P_{\text{inputport}}(W) - P_{\text{coupledport}}(W) - P_{\text{isolatedport}}(W)}{P_{\text{inputport}}(W)}\right).$$

4 Design Specification of Hybrid Coupler

The 3 dB hybrid coupler phase shifter is classified by many parameters like phase response over bandwidth, insertion loss, isolation and its desired phase shift [5]. A phase shifters specification inclined to have about 2–3 dB of insertion loss, 90^0 or 180^0 of phase shift. The basic hybrid coupler phase shifter structure comprises of ground plane, dielectric substrate, and microstrip line. Generally, the length of the branch-line coupler and series stub microstrip line of the coupler is taken as one-fourth of the design wavelength by considering the dimensions of the hybrid coupler [6].

$$L = \frac{v_p}{f} = \frac{c}{f\sqrt{\varepsilon_r}} \tag{1}$$

4.1 Microstrip Line

The most popular types of planar transmission lines is microstrip line as it can easily be integrated with passive or active microwave devices and can also be fabricated by photolithographic processes. The effective dielectric constant of a microstrip line is given by [2]

$$\varepsilon_e = \frac{\varepsilon_r + 1}{2} + \frac{\varepsilon_r - 1}{2}\frac{1}{\sqrt{1 + \frac{12d}{W}}} \tag{2}$$

Which satisfies the relation $1 < \varepsilon_e < \varepsilon_r$ and is dependent on the substrate thickness, d and conductor width, W.

4.2 Given ε_r and Z_0

Given the dimensions of the microstrip line, the characteristic impedance can be found out by using the formula below:

$$Z_0 = \frac{60}{\sqrt{\varepsilon_r}} \ln \left(\frac{8d}{W} + \frac{W}{4d} \right) \quad \text{for } \frac{w}{d} \leq 1$$

$$= \frac{120\pi}{\sqrt{\varepsilon_r}[\frac{W}{d} + 1.393 + 0.667 \ln(\frac{W}{d} + 1.444)]} \quad \text{for } \frac{w}{d} \geq 1$$

(3)

The W/d ratio can be found as,

$$\frac{W}{d} = \frac{8e^A}{e^{2A} - 2} \quad \text{for } \frac{w}{d} < 2$$

$$= \frac{2}{\pi} \left[B - 1 - \ln(2B - 1) + \frac{\varepsilon_r - 1}{2\varepsilon_r} \left\{ \ln(B - 1) + 0.39 - \frac{0.61}{\varepsilon_r} \right\} \right] \quad \text{for } > 2$$

(4)

where $A = \frac{z_0}{60} \sqrt{\frac{\varepsilon_r + 1}{2}} + \frac{\varepsilon_r - 1}{\varepsilon_r + 2} \left(0.23 + \frac{0.11}{\varepsilon_r} \right)$ and $B = \frac{377\pi}{2Z_0\sqrt{\varepsilon_r}}$.

A conductor loss is much more important than dielectric loss for most of the microstrip substrates, exceptions may occur with semiconductor substrates [7]

$$\emptyset = 90° = \beta l = \sqrt{\varepsilon_e} k_0 l$$

(5)

where, $\beta = \sqrt{\varepsilon_e} k_0.s$ the propagation constant; $k_0 = \frac{2\pi f}{c}$, where f is the frequency (GHz).

The length "1" of the microstrip line is given as

$$l = \frac{90(\frac{\pi}{180})}{\sqrt{\varepsilon_e} k_0}$$

(6)

5 Simulation of Hybrid Coupler

The hybrid coupler for this design will be matched to ports with 50Ω impedance and thus the quarter-wavelength sections of the device have 50Ω and 35.35Ω impedances. The coupler is built out of microstrip on a substrate of FR4 Epoxy (Tables 2, 3, and 4).

The following design in Fig. 2 is the layout of quadrature hybrid phase shifter which is built by using the parameters shown above.

6 Simulated Results

The results have been evaluated for return loss, isolation, and phase difference for the design through HFSS (version 13.0).

Table 2 Parameters and values of hybrid coupler

Parameter	Dimensions
Dielectric constant (ε_r)	4.4 (FR4 Epoxy)
Length of the substrate (L)	42.045 mm
Width of the substrate (W)	72.33 mm
Height of the dielectric substrate (d)	1.6 mm

Table 3 Parameters of $Z_0 = 50$ ohm of input impedance of microstrip line

Parameters	Dimensions
Length (l)	24.11 mm
Width (W)	3.1 mm
Effective dielectric constant (ε_e)	4.3946

Table 4 Parameters of $Z_0/\sqrt{2} = 35.35$ ohm of input impedance of microstrip line

Parameters	Dimensions
Length (l)	23.845 mm
Width (W)	5.2224 mm
Effective dielectric constant (ε_e)	4.3968

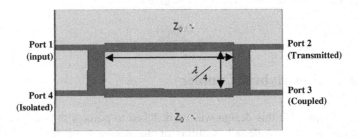

Fig. 2 Layout of Quadrature (3-dB) hybrid coupler

6.1 Return Loss, Insertion, Isolation in dB of all Four Ports with Respect to Port 1

The result shown in Fig. 3 is focus reasonable value of S-Parameters of four ports with respect to input port. The perfect 3 dB power division is obtained in port 2 and port 3; Perfect return loss and isolation are obtained in port 1 and port 4. The Return loss ($S_{1,1}$) and isolation ($S_{4,1}$) are found at 1.6 GHz of about -36 and -22 dB. The insertion loss ($S_{2,1}$) is found of -3.61 dB.

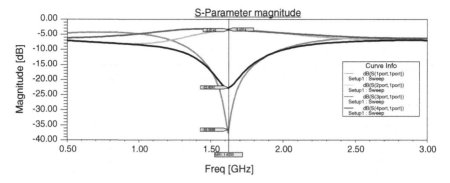

Fig. 3 *S*-parameters magnitudes versus frequency of all ports relative to input port 1

6.2 Phase Difference of Output Ports i.e. Transmitted Port 2 and Coupled Port 3 Relative to Input Port 1

The result shown in Fig. 4 is phase difference between the two output ports 2 and 3 with respect to input port 1where there is a phase difference of about 91^0 which is very close to 90^0 at frequency 1.6 GHz.

6.3 Coupling Ratio of Two Output Ports with Respect to Input Port

Figure 5 shows the coupling ratio between the outputs ports 2 and port 3 relative to input port 1 where the output coupled port 3 with respect to input port 1 is lower than output transmitted port 2 with respect to input port 1. The lower coupling loss is found of about −3.42 dB at frequency 1.6 GHz.

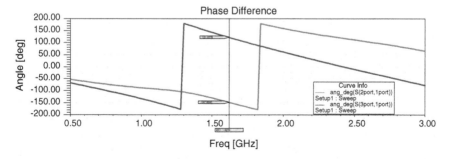

Fig. 4 Phase difference between two outputs ports with respect to input port

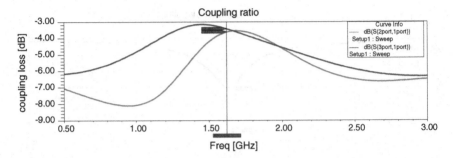

Fig. 5 Coupling ratio between output ports i.e., at port 2 and port 3 relative to input port 1

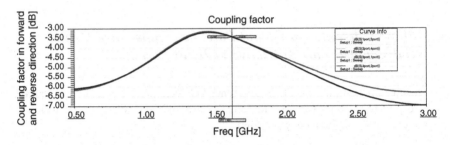

Fig. 6 Coupling factor in forward and reverse direction of all four ports

6.4 Coupling Factor in Forward and Reverse Direction of Four Ports

Figure 6 shows the coupling factor in forward and reverse direction of all four ports where all the coupling loss is obtained at frequency 1.6 GHz of about -3.4 dB.

7 Conclusion

The proposed branch-line hybrid coupler implemented as a phase shifter is designed by using HFSS (version 13.0). The investigation confirmed that the hybrid coupler described here may be used in filters, phase shifters, and resonators which are operating in the 1.5 GHz (L-band). It has a significant effect on the bandwidth, return loss, insertion loss, isolation, and coupling loss. It is made in order to provide the phase shift of 90^0. However, the simulation result for the proposed hybrid coupler phase shifter gives 91^0 which are very close to 90^0. The main goal of this research work is to reduce insertion loss and to increase isolation. The simulated result of insertion loss and isolation at 1.6 GHz is -3.61 dB which is below 3 and −22 dB which is nearer to 20 dB. It was assumed that a currently used hybrid coupler can also be used as reflection type phase shifter by adding two reflective loads at the two output ports of hybrid coupler. The studies are currently being carried out in the KIIT University, Bhubaneswar, India.

References

1. The Decibel Defined. [Online]. Available: http://sss-mag.com/db.html, 5 Aug 2008
2. Pozar, D.M.: Microwave engineering, 3rd edn. Wiley, Hoboken (2005)
3. Liao, S.Y.: Microwave devices and circuits, Eastern Economy Edition, Third Edition (2004)
4. Microwave power divider and coupler tutorial, http://www.markimicrowave.com/menus/appnotes/Microwave_Power_Dividers_and_Couplers_Primer.pdf, Marki Microwave. Accessed Jan (2012)
5. Shiban, K., Bhat, B.: Microwave and millimetre wave phase shifters, vol I and II. Artech House, Norwood (1991)
6. Dawar, P., Tegh, G.: Bahadur Institute of Technology, New Delhi India, Analysis of Microstrip Branch-line Coupler using Sonnetlite, IJECT Vol. 3, Issue1, ISSN: 2230-7109 | ISSN: 2230-9543 (2012)
7. Bahl, I.J., Trivedi, D.K.: Designer's guide to microstrip line, microwaves, pp. 174–182 (1977)

Capturing Node Resource Status and Classifying Workload for Map Reduce Resource Aware Scheduler

Ravi G. Mude, Annappa Betta and Akashdeep Debbarma

Abstract There has been an enormous growth in the amount of digital data, and numerous software frameworks have been made to process the same. Hadoop MapReduce is one such popular software framework which processes large data on commodity hardware. Job scheduler is a key component of Hadoop for assigning tasks to node. Existing MapReduce scheduler assigns tasks to node without considering node heterogeneity, workload type, and the amount of available resources. This leads to overburdening of node by one type of job and reduces the overall throughput. In this paper, we propose a new scheduler which capture the node resource status after every heartbeat, classifies jobs into two types, CPU bound and IO bound, and assigns task to the node which is having less CPU/IO utilization. The experimental result shows an improvement of 15–20 % on heterogeneous and around 10 % of homogeneous cluster with respect to Hadoop native scheduler.

Keywords MapReduce · Homogeneous cluster · Heteregeneous cluster · Hadoop · Scheduler

1 Introduction

In the recent years, internet growth due to social networking sites and other means of uses has produced an enormous amount of data. As the amount of data increases, time to process the data and the amount of resources required to process

R.G. Mude (✉) · A. Betta · A. Debbarma
Department of Computer Science and Engineering, National Institute of Technology
Karnataka, Surathkal 575025, Karnataka, India
e-mail: ravimude20@gmail.com

A. Betta
e-mail: annappa@ieee.org

A. Debbarma
e-mail: deepakashz02@gmail.com

© Springer India 2015
L.C. Jain et al. (eds.), *Intelligent Computing, Communication and Devices*,
Advances in Intelligent Systems and Computing 309,
DOI 10.1007/978-81-322-2009-1_29

also increases. Google came up with an approach called as MapReduce [1], which is designed for building large cluster made up of commodity hardware. A popular open source implementation of MapReduce and distributed file system [2] is Apache Hadoop [3], which is used for processing petabytes of data with the use of thousands of machines in a reliable and fault-tolerant manner. Cloud providers such as Amazon offer a Hadoop platform as an on demand service termed as Elastic MapReduce [4]. Users have to pay for this service, so they need to use resources optimally.

The scheduler is one of the key parts of Hadoop, which is used for assigning tasks to node. Each job running on the Hadoop system can be categorized into the CPU bound and I/O bound which will help in scheduling the tasks to node. Hadoop native scheduler does not consider heterogeneity in workload and cluster.

But nowadays, due to a system failure or requirement of the larger cluster than previous, we need to add systems to the available cluster set up which makes it heterogeneous.

In this paper, we are proposing a scheduler for Hadoop that takes into account node resource status and the type of workload. Our scheduler learns node status and job type online which will classify node into two types: CPU, I/O busy and job to CPU, I/O intensive. After classifying the job and capturing node resource status, scheduler will perform the task assignment to proper node. In this work, we concentrate on hardware utilization for different kinds of job run on the cluster.

The rest of the paper was organized as follows. Section 2 describes different types of proposed and available scheduler for Hadoop. Section 3 describes the structure of our scheduler in details. Section 4 gives an evaluation of our proto-type. Finally, we conclude and discuss future work in Sect. 5.

2 Related Work

Currently, the focus of research in the field of Hadoop scheduler has moved from homogeneous to heterogeneous cluster [3, 15]. Speculative [5] execution a feature of Hadoop leads to performance degradation on heterogeneous clusters because the task that takes longer than normal to complete the job gets re-executed pre-emptively on a second node, assuming the first may fail. This concept works for homogeneous cluster but in the case of heterogeneous cluster a node with low resources will consume more time to execute.

The Hadoop default FIFO scheduler [1] uses whole cluster for each job by which all other jobs wait for their turn in FIFO order. Yahoo and Facebook also designed Hadoop scheduler called as Fair scheduler [6] and the Capacity scheduler [7], respectively. These schedulers use multiple queues to allocate the resources and are mostly used for fair allocation of resources. They are not considering heterogeneous workload and heterogeneous clusters.

The LATE [5] and SAMR [8] scheduler tries to improve Hadoop performance in a heterogeneous environment. LATE scheduler attempts to find real slow tasks by computing remaining time of all tasks without taking into account data locality for launching speculative map tasks. SAMR scheduler calculates progress of task dynamically and finds out which tasks need backup.

The context aware scheduler for Hadoop (CASH) [9] learns resource capabilities, resource requirements, and schedule the job on node that are more efficient to satisfy the requirements. However, learning approach used by CASH is offline, whereas our scheduler offers online learning resource requirements and node status.

A classification and optimization-based scheduler (COSHH) [10] assigns the job to nodes by considering heterogeneity at both application and cluster levels. It uses K-means clustering approach for classifying jobs. It makes different classes of the job, according to their execution times. COSHH uses prediction mechanism/ approach to estimate mean time execution time of the incoming job on all resources.

In [11], author proposed workload characteristic-oriented scheduler. It characterizes workload by estimator with the help of static task selection strategies. Dynamic MapReduce scheduler [12] and Multiple-job optimization [13] assume homogeneous cluster, learn job resource profile on the fly, and schedule job by dividing the workload into two types I/O, CPU intensive/bound.

The aforementioned scheduler proposed by us is similar with CASH and dynamic MapReduce scheduler. However, our scheduler learns the node status, job resource requirements on the fly, classifies the workload into two types, and then schedules job without overloading the node.

3 Proposed Algorithm

In our proposed approach, we capture node resource status and find job resource requirements by running sample map/reduce task of a job. JobTracker after receiving every heartbeat from the TaskTracker notes down node resource status (including CPU and I/O utilization). After knowing job resource requirements and node resource status, the scheduler running in JobTracker assigns tasks to node in such a way that, the node will not be overloaded by one type of job (i.e., CPU and I/O).

3.1 Capturing Node Resource Status

Communication in master and slave node of Hadoop happen using heartbeat protocol. TaskTracker (slave) sends a heartbeat message to master (JobTracker) after some particular time indicating that the node is alive. We have added some metrics to capture the CPU usage and IO usage of TaskTracker. JobTracker obtains resource utilization information of that node after receiving a heartbeat message from TaskTracker. Formula 1 gives the information about the percentage of CPU usage in TaskTracker, while formula 2 gives IO rate of TaskTracker.

$$\text{CPU Usage} = \frac{(\text{Cumulative CPU Time} - \text{Last Cumulative CPU Time})}{(\text{Sample Time} - \text{Last Sample Time}) * \text{Num\_of\_processors}} \quad (1)$$

Cumulative CPU Time   Actual CPU cycles used to execute the program
Sample Time   Current CPU Time
Last Sample Time   Old CPU Time/Last Heartbeat CPU Time

$$\text{I/O Rate} = \frac{\begin{array}{c}(\text{Cumulative Reads} - \text{Last Cumulative Reads}) \\ +(\text{Cumulative Writes} - \text{Last Cumulative Writes})\end{array}}{(\text{Sample Time} - \text{Last Sample Time})} \quad (2)$$

3.2 Capturing Job Type and Requirement

Every job in Hadoop is divided into several maps and reduces tasks. If we get to know the resource usage of some map/reduce task, then we can anticipate the resource usage of the rest. The same rule applies to the reduce task. On arrival of new job, our scheduler will allow execution of some of the map and reduce tasks called as sample task. After execution of sample map/reduce task, JobTracker [14] will get the resource required to execute that task, as JobTracker knows the node resource status before and after execution of task.

After execution of sample task, JobTracker will check the job summary logs. It contains all the information about task like remaining time required to execute the task and many others. JobTracker chooses M_{rtm}, R_{rtm}, D_{map}, and D_{reduce} from the available information which will help in calculating the IO rate of map/reduce task.

M_{rtm} run time for map task
R_{rtm} run time for reduce task
D_{map} data read/written for map task
D_{reduce} data read/written for reduce task

$$\text{Map IO rate, MI/O rate} = \frac{D_{map}}{M_{rtm}} \quad (3)$$

$$\text{Reduce IO rate, RI/O rate} = \frac{D_{reduce}}{R_{rtm}} \quad (4)$$

The ratio of the amount of input and output data generated by map/reduce task depends on the type of workload. If the workload is CPU bound then it will generate less data, while IO bound task generate more data than the disk bandwidth. So to find the task CPU/IO bound, we need to compare MI/O rate and RI/O rate with Disk IO rate. We use formula 5 to calculate map IO rate and compare with a Disk IO rate. If the Map IO rate is smaller than Disk IO rate the map task is CPU bound otherwise the task is IO bound.

$$\text{MI/O Rate} = \frac{D_{\text{map}}}{M_{\text{rtm}}} < \text{Disk IO Rate} \tag{5}$$

Formula 6 is used for finding the reduce task as CPU or IO bound. As we discuss before, JobTracker first calculate reduce IO rate and compares it with the Disk IO rate. If reduce IO rate is greater than the Disk IO rate then the task is IO bound, otherwise the task is CPU bound.

$$\text{RI/O Rate} = \frac{D_{\text{reduce}}}{R_{\text{rtm}}} < \text{Disk IO Rate} \tag{6}$$

3.3 Capturing Resource Aware Scheduler

Figure 1 shows that after every heartbeat from the TaskTracker, JobTracker will get the node resource usage information. Whenever new jobs are submitted by the user to the JobTracker, it will be added to the queue and the sample of that job will be executed on the TaskTracker. JobTracker will get resource requirement of job by execution of sample map/reduce task on TaskTracker. Now master is having all node resource status, job type/requirement which includes CPU and I/O requirement. So scheduler classifies the node into two types: CPU- and I/O busy node. Then it starts placing the nodes to CPU-busy queue and I/O busy queue according to their percentage of uses. The CPU-busy node queue contains the node in increasing order of percentage of their CPU usage, while I/O busy node queue contains the node in increasing order of their I/O rate. Therefore, scheduler is having two separate queues of nodes and type of job i.e., CPU or I/O bound. Now scheduler will assign tasks to TaskTracker according to node resource to job matching.

Algorithm 1 is used for matching job with node having valuable amount of resources. Most important priority for scheduler is locality and requirement match. After knowing that map task is CPU/IO bound, the scheduler will select the node from the CPU/IO bound queue having less CPU/IO utilization to assign the task. Before assigning the task, first it will check for local map slot and node having less CPU/IO utilization. Secondly, if the slot is not local then it will be allocated to the node having rack local map slot with less CPU/IO utilization. Finally, it will select node containing nonrack local map slot having less CPU/IO utilization. The scheduler will apply the same algorithm to reduce task for assigning task to TaskTracker. In algorithm *MT-CB* represents the map task CPU bound while *RT-CB* is reduce task CPU bound (Fig. 1).

Algorithm 1: Task Placement Algorithm

```
1.    For every heartbeat received from node n
2.    Sort TaskTracker, TT in order of CPU, IO utilization
3.    for Job in J do
4.    run sample map, reduce of Job for IO,CPU requirement
5.    if a slot is free then
6.    ------------Place Mappers---------
7.    if MT-CB
8.        then
9.            //place task on node from CPU node queue with less CPU utilization
10.           if(free slot in local node)
11.               then
12.                   run task t on n;
13.               else if(free slot in Rack-local node)
14.               then
15.                   run t on n;
16.               else
17.                   run t on n(non rack local)
18.    else
19.            //place task on node from IO node queue with less IO rate
20.           if(free slot in local node)
21.               then
22.                   run task t on n;
23.               else if(free slot in Rack-local node)
24.               then
25.                   run t on n;
26.               else
27.                   run t on n(non rack local)
28.    ----------Place Reducers-----------
29.    if RT-CB
30.        then
31.            //place task on node from CPU node queue with less CPU Utilization
32.           if(free slot in local node)
33.               then
34.                   run task t on n;
35.               else if(free slot in Rack-local node)
36.               then
37.                   run t on n;
38.               else
39.                   run t on n(non rack local)
40.    else
41.            //place task on node from IO node queue with less IO rate
42.           if(free slot in local node)
43.               then
44.                   run task t on n;
45.               else if(free slot in Rack-local node)
46.               then
47.                   run t on n;
48.               else
49.                   run t on n(non rack local)
```

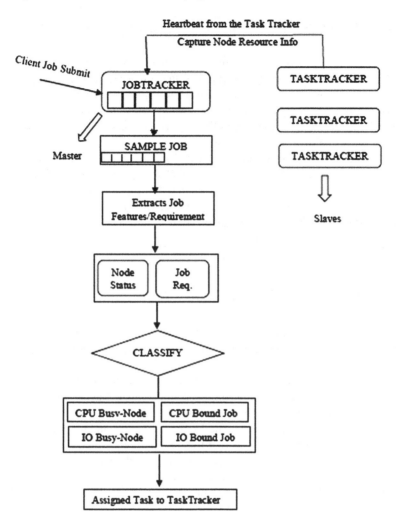

Fig. 1 Capturing resource status for resource aware scheduler

4 Evaluation and Results

In this section, we test the performance of our scheduler and compare the result with Hadoop native FIFO and Fair scheduler. We used two different scenarios to test the performance of our scheduler. First scenario will check the performance with heterogeneous cluster while in second case it considers homogeneous cluster. We have used Hadoop 0.20.203 version for implementation of our capturing resource aware scheduler. The workload used for testing the performance is described first then the executions of the benchmarks on heterogeneous and homogeneous cluster are shown.

4.1 Description of Workload

Apache Hadoop contains inbuilt benchmarking and testing tools which are used for testing the performance of Hadoop. From that we have chosen Teragen, Terasort, wordcount, Grep, and Pi benchmark to test the performance. Terasort is used for sorting the large collection of data while wordcount count the number of words present in the file. Grep search the keyword in huge file and Pi benchmark calculate the value of Pi for some random points.

We have two types of workload CPU and I/O bound. Terasort and Grep are I/O bound while wordcount and Pi are CPU bound workload. First, we will execute single job and evaluate the performance. Then we will mix the CPU and I/O bound jobs and compare the performance with respect to heterogeneous and homogeneous cluster. Table 1 shows the combination of job that was executed on the homogeneous and heterogeneous cluster.

4.2 Performance on Heterogeneous Cluster

First we test the performance of our scheduler with heterogeneous cluster for that we have configured 7 node cluster. Table 2 gives the details about nodes in the cluster.

Job Completion Time. The time required to execute job implies job completion time. First, we will execute single job and compare the performance of our resource aware scheduler with FIFO, a Fair scheduler in overall job completion time. We used two different benchmarks Teragen and random-writer for data generation. Teragen is used to generate the numeric data which will be used for performing Terasort while random-writer generates random text data which will be used for performing wordcount and Grep benchmark. A second technique checks the performance of our scheduler by running the mix of CPU, IO bound job on

Table 1 Combination of job

Comb1	Wordcount (10 GB) + Pi (1,500) samples
Comb2	Terasort (10 GB) + Pi (1,500) samples
Comb3	Pi (1,500) samples + Grep (10 GB)
Comb4	Terasort(10 GB) + Grep (10 GB)

Table 2 Cluster node details

No. of system	Role	CPU	Ram (GB)	Disk (GB)
1	Master	Core i7	4	250
4	Slave	Core i7	8	250
1	Slave	Core 2 duo	2	100
1	Slave	Core i3	4	150

Table 3 Average execution time for single job

Job	Size (GB)	FIFO	FAIR	RAWare
Terasort	1	2 min 45 s	3 min 14 s	2 min 15 s
	10	28 min 20 s	34 min 0 s	25 min 19 s
Wordcount	5	24 min 10 s	23 min 15 s	21 min 20 s
	10	44 min 47 s	42 min 3 s	41 min 18 s

heterogeneous cluster. And it compares their job completion time with respect to other scheduler. Table 1 gives the details about combination of job that we ran on different cluster.

Table 3 shows the result obtained after execution of single job having CPU or IO bound nature. The result shows that our resource aware scheduler improves the overall job completion time. Fair scheduler took more time to execute for Terasort. On execution of Terasort benchmark, Fair scheduler assigns the reduce task to the node which is having the less amount of resources. The results we have considered here are the average execution time taken by the scheduler to execute the job. The improvement increases with the size of data because time for classifying workload for the small and large job is almost same. As scheduler needs to run the same number of sample tasks for both jobs. Figure 2 shows the result obtained after execution of combo of jobs. From the result, we can say that our scheduler improves the performance with respect to FIFO and Fair Scheduler. For combo 2, Fair scheduler takes more time than FIFO scheduler. As we have discussed, it is due to execution of reduce tasks on the node which is having the less amount of resources. The result shows that the average improvement of our scheduler with respect to FIFO and the Fair is around 15–20 %.

4.3 Performance on Homogeneous Cluster

The second scenario checks the performance of our resource aware scheduler with respect to homogeneous cluster for that we have constructed a cluster containing 5 nodes, each having a configuration of core i7 processor, 8 GB of Ram, and 250 GB disk space. Among the 5 nodes, one system acts as the master and other 4 nodes are slaves. Similar to heterogeneous cluster, we have validated the performance of our scheduler in overall job completion time with single and combination of Job. Table 1 gives the details of combination of job.

Results from the Table 4 shows that our scheduler improves the performance over other native scheduler while the enhancement is minimal as compared to heterogeneous clusters. Figure 3 shows the improvement with respect to FIFO and Fair scheduler. X axis represents combo of a job and the Y axis shows, job completion time (in second). The improvement for our scheduler with respect to homogeneous cluster is around 10 %.

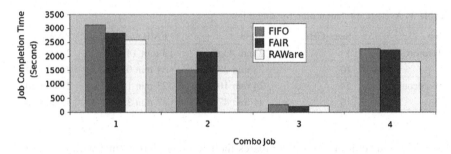

Fig. 2 Job completion time in second (*y* axis) with respect to combination of job (*x* axis)

Table 4 Average execution time for single job

Job	Size (GB)	FIFO	FAIR	RAWare
Terasort	1	1 min 57 s	1 min 57 s	1 min 53 s
	10	27 min 32 s	28 min 8 s	24 min 1 s
Wordcount	5	19 min 45 s	20 min 28 s	19 min 32 s
	10	37 min 30 s	36 min 56 s	36 min 20 s

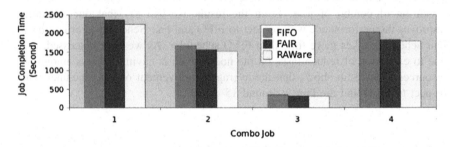

Fig. 3 Job completion time in second (*y* axis) with respect to combination of job

5 Conclusion and Future Work

In this paper, we have proposed a resource aware scheduler which takes into account node resource status (CPU, IO), finds job requirement type, and then allocates task to the TaskTracker. It classifies the job into two types: CPU and IO bound and allocate jobs to the node having less CPU/IO utilization, respectively. We have designed job placement algorithm which assigns tasks to local node first and if the local node is busy then it will go for nonlocal node. The proposed algorithm has been implemented on Hadoop 0.20.203 version and tested with a benchmark on heterogeneous and homogeneous cluster. It is compared with FIFO, Fair Scheduler and the results show that our scheduler improves in overall execution time of MapReduce job. Also, it enhances Hadoop performance by 15–20 % with respect to heterogeneous cluster while around 10 % in homogeneous cluster.

We intend to enhance our work to consider the abstract requirement that will be needed for the job which will help us to assign tasks to the node having a valuable amount of resources. Moreover, we aim to take into account resources such as usage of network bandwidth and the amount of power present in the nodes.

References

1. Dean, J., Ghemawat, S.: MapReduce: simplified data processing on large clusters. Technical Report, Google (2004)
2. Hadoop Distributed File System. http://hadoop.apache.org/common/docs/current/hdfs_design.html
3. Hadoop MapReduce. http://hadoop.apache.org/MapReduce
4. Amazon Elastic Map Reduce, http://aws.amazon.com/elasticmapreduce/
5. Joseph, A.D., Katz, R., Zaharia, M., Konwinski, A., Stoica, I.: (2008) Improving MapReduce performance in heterogeneous environments. In: OSDI'08. USENIX Association, Berkeley, pp. 29–42 (2008)
6. Fair Scheduler. http://hadoop.apache.org/MapReduce/docs/r0.21.0/fairscheduler.html
7. Yahoo! Inc. Capacity Scheduler. http://developer.yahoo.com/blogs/hadoop/posts/2011/02/capacity-scheduler/
8. Chen, Q., Zhang, D., Guo, M., Deng, Q., Guo, S.: SAMR: a self-adaptive MapReduce scheduling algorithm in heterogeneous environment. In: 10th IEEE International Conference on Computer and Information Technology (CIT 2010), pp. 2376–2743 (2010)
9. Arun kumar, K., Konishetty, V.K., Voruganti, K., Prabhakara Rao, G.V.: CASH: context aware scheduler for Hadoop. In: Proceedings of the International Conference on Advances in Computing, Communications and Informatics, New York, 2012, ICACCI '12. ACM, pp. 52–61
10. Rasooli, A., Down, D.G.: COSHH: a classification and optimization based scheduler for heterogeneous Hadoop systems. High Performance Computing, Networking Storage and Analysis, SC Companion. IEEE, pp. 1284–1291 (2013)
11. Lu, P., Lee, Y.C., Wang, C., Zhou, B.B., Chen, J., Zomaya, A.Y.: Workload characteristic oriented scheduler for MapReduce. In: 2012 IEEE 18th International Conference on Parallel and Distributed Systems, pp. 156–163 (2012)
12. He, Y., Tian, C., Zhou, H., Zha, L.: A dynamic MapReduce scheduler for heterogeneous workloads. In: Eighth International Conference on Grid and Cooperative Computing, IEEE 2009, pp. 218–224
13. Hu, W., Tian, C., Liu, X., Qi, H., Zha, L., Liao, H., Zhang, Y., Zhang, J.: Mutiple-job optimization in MapReduce for heterogeneous workloads. In: 2010 Sixth International Conference on Semantics, Knowledge and Grids, IEEE 2010, pp. 135–140
14. JobTracker Architecture. http://hadoop.apache.org/common/docs/current/mapred_tutorial.html
15. Murthy, A.: Next Generation Hadoop [Online]. Available: http://developer.yahoo.com/blogs/hadoop/posts/2011/03/MapReduce-nextgen-scheduler/

Grid Computing-Based Performance Analysis of Power System: A Graph Theoretic Approach

Himansu Das, A.K. Jena, P.K. Rath, B. Muduli and S.R. Das

Abstract Electrical power grid is a complex network infrastructure. It is necessary to design one infrastructure which can communicate and control the different grid stations. Topological analysis provides the static properties of power grid, which does not meet the real-time requirement of power grid. It is focused on only physical significance of the power grid. By taking resistance as electrical parameters to generate, one weighted graph which provides the different parameters of power grid. This paper compares the topological and electrical characteristics of power grid. Topological analysis focuses on geographic distance rather than electrical distance. It may mislead that topological analysis provides the solution to the electrical power grid.

Keywords Grid computing · Power grid · Complex network · Topological analysis · Electrical analysis

H. Das (✉) · A.K. Jena
KIIT University, Odisha, India
e-mail: das.himansu2007@gmail.com

A.K. Jena
e-mail: ajay.bbs.in@gmail.com

P.K. Rath · B. Muduli · S.R. Das
Roland Institute of Technology, Odisha, India
e-mail: pradeep.ratha@gmail.com

B. Muduli
e-mail: bhagaban.muduli@gmail.com

S.R. Das
e-mail: srdas1984@gmail.com

© Springer India 2015
L.C. Jain et al. (eds.), *Intelligent Computing, Communication and Devices*,
Advances in Intelligent Systems and Computing 309,
DOI 10.1007/978-81-322-2009-1_30

259

1 Introduction

High-performance computing (HPC) is used to solve computational problems which are not flexible to solve by using conventional computers owing to the huge amount of processing power, network, and memory space requirements. Most of the complex problems [1] such as World Wide Web, Internet, social interacting species, neural network, chemical systems, and coupled biological systems are solved by using topological analysis of graph. Though power system is a complex network [2], it is necessary to go for topological analysis of power grid. Topological analysis is performed based on the physical significance like geographical distance of the network structure of the graph. But it will ignore the electrical properties of power grid such as resistance and impedance of the network. To study the electrical connectivity of power grid, it is necessary to use electrical distances rather than geographical distances. In topological analysis, the power network is treated as the undirected graphs [3], where each node represents a bus and connection between the different buses are represented by a dedicated connection called edges of the grid network. It is important to note that in the physical grid, these buses can have different electrical properties, like nodes are assumed to be homogeneous. Physical length and electrical impedance are ignored in case of topological analysis of the undirected graph representation. But in electrical analysis, power grid network is treated as weighted graph. The weighs of the corresponding edge represents the equivalent resistance of the connecting paths between the different buses. This weighed graph is again converted to its equivalent un-weighted graph by using reduced distance matrix principle. The goal of this paper is to characterize the topological and electrical structure of the power grid and evaluate the performance of electricity infrastructures.

In this paper, a framework for electric power systems has been presented that employs the topological and electrical structure of power grid. A middleware called GridGain [4] which is java-based has been employed in this paper. The rest part of this paper is planned as follows, Sect. 2 reflects the graph theoretic model of power grid. Section 3 provides the topological features of power network. Section 4 provides the electrical analysis of power grid. Section 5 provides the different parameters of topological and electrical analysis result of IEEE 14 bus system. It also provides the comparative study between topological analysis and electrical analysis of power grid. Section 6 concludes the paper.

2 Power System Model

We can represent power grid as a complex network, we build an un-weighted and undirected graph [3] composed of nodes and edges. The number of node of the graph is specified by N, the number of edges is m, the average nodal degree is $\langle k \rangle$, the average shortest path length in hops is $\langle l \rangle$, Pearson degree correlation coefficient is p, the fraction of nodal degrees which is larger than the average nodal degree seen at the end of the randomly selected link is $r\{k_i > k\}$. The incidence matrix A having dimension $m \times N$ of the graph [3] contains N nodes and m links that can be represented as

$$A = \begin{cases} 1 & \text{If there is a direct path between } i \text{ to } j \\ -1 & \text{if there is a path between } j \text{ to } i \\ 0 & \text{Otherwise} \end{cases} \tag{1}$$

The Laplacian matrix L can be obtained as $L = A^T A$ with

$$L(i,j) = \begin{cases} -1, & \text{if there exists link } i \text{ to } j, \text{ for } i \neq j \\ k, & \text{with } k = -\sum_{j \neq i} L(i,j), \text{ for } j = i \\ 0, & \text{otherwise,} \end{cases} \tag{2}$$

with $i, j = 1, 2, ..., N$.

3 Topological Properties of Power Grid

To represent the power grid as a complex network, we build an un-weighted and undirected graph [3] composed of nodes and edges. Using metrics from graph theory and modern complex networks analysis, the results provide insight into the properties of power grids, considering the topological and electrical information. The goal is to characterize the topological structure of the power grids and highlight implications for the performance of electricity infrastructures [5–7].

3.1 Complex Network Model to Analyze the Topology of Power Grid

To study the power grids as a complex network, some simplifications are necessary. There are numerous useful numerical measures [8] for graphs. These measures provide a useful set of statistics for comparing power grids with other graph structures. The total number of links of any graph is $m = \frac{1}{2}\sum_i L(i,i)$. The average nodal degree can be represented as the average number of edges is connected to a node. The average nodal degree is $\langle k \rangle = \frac{1}{N}\sum_i L(i,i)$. The degree of a node [8] specifies the number of nodes adjacent to that node. The nodal degree vector is determined by $\underline{K} = \{k_1, k_2, k_3, ..., k_N\} = \text{diagonal}(L)$. The average node degree can found at the end of a randomly selected edge is:

$$\bar{k} = (2m)^{-1} \sum_{(i,j)} (k_i + k_j) = (2m)^{-1} \sum_{(i)} (k_i^2) = \frac{\langle k^2 \rangle}{\langle k \rangle} \tag{3}$$

Then, the ratio $r\{k > \bar{k}\}$ can be obtained as

$$r\{k > \bar{k}\} = \frac{\|\{k_i; k_i > \bar{k}\}\|_\infty}{N} \tag{4}$$

3.2 Degree Distribution

The degree of a node indicates the number of nodes adjacent to that node. In degree distribution [3, 9], we can represent the global connectivity of the network. Degree distribution describes the diversity of connectedness in a graph. These networks tend to have extremely connected hubs, which can make the network vulnerable to directed attack. The degree of node i in a graph with adjacency matrix A is:

$$k_i = \sum_{j=1}^{N} a_{ij} \tag{5}$$

3.3 Clustering Coefficient

The clustering coefficient [10], C, is a common metric that provides information about the transitivity of a network; i.e., if two pairs of nodes, $\{x, y\}$ and $\{y, z\}$, are clustered, then there also exists an edge between nodes x and z. In that case, they would form a cluster. C is defined as follows in terms of the coefficient C_i or the individual clustering coefficient for each node.

$$C(G) = \frac{1}{N} \sum_{i=1}^{N} C_i \tag{6}$$

where the clustering of node i (C_i) is $C_i = \frac{\lambda_G(i)}{\tau_G(i)}$, $\lambda_G(i)$ is the number of edges between the neighbors of node i, $\tau_G(i)$ the total number of edges that could possibly exist among the neighbors of node. For undirected graphs, $\tau_G(i) = k_i(k_i - 1)/2$ is the node degree. The clustering coefficient for a random graph network theoretically equals the probability of randomly selecting links from all possible links is represented as $C(R) = \frac{2m}{N(N-1)} = \frac{\langle k \rangle}{N-1}$.

4 Electrical Properties of Power Grid

The topology of electrical power grids is similar to the complex networks. However, the topological analysis ignores the electrical properties of power grids and we need to represent metrics that capture the electrical structure of the power grid. Here, we measure electrical properties of power grid in order to understand the structure of power grids.

4.1 Resistance Distance Matrix

The connections between different components of the power grid depend on the physical properties that govern current and voltage. The 'Resistance Distance' matrix provides an additional method for computing electrical distance [11] of the power grid. Resistance distance provides the effective resistance between different points in a network. The distance metric represents the active power transfers and nodal phase angle. One way to analyze connectivity between components in an electrical system is to look at the properties of sensitivity matrices. A power grid network can be transformed into its corresponding impedance matrix. This matrix forms the basis of power flow or load flow analysis and short circuit analysis. Here, we will focus on the formulation of bus admittance matrix known as Y_{bus} matrix and bus impedance matrix called as Z_{bus} matrix. The relationship between these two matrices is $Z_{bus} = Y_{bus}^{-1}$.

4.2 Formation of Bus Admittance Matrix

By using Norton's theorem in any electrical circuit, which can be formulated by a current source I_S with an equivalent admittance of Y_S. The relations between the original system and the Norton equivalent system are represented as $I_S = V_S/Z_S$ and $Y_S = 1/Z_S$. Norton's theorem is used for the formulation of the Y_{bus} matrix. The Y_{bus} matrix is a sum of each elements of the k_{th} column is Y_{kk}. The electrical resistance distance matrix [13] E is a simple way of measuring the electrical connectedness of nodes within a power system. The matrix E can be defined as the absolute value of the inverse of the system admittance matrix, i.e. $E = |Y^{-1}|$. To obtain a node degree, equivalent measure of each node can be obtained by E_a. This represents the sensitivity between voltage and current changes for any node with respect to every other node in the network.

$$E_a = \sum_{\substack{b=1 \\ b \neq a}}^{n} \frac{Eab}{n-1} \tag{7}$$

4.3 *Un-weighted Reduced Resistance Distance Matrix*

The resistance distance matrix E describes the amount of connectivity between all pair of nodes in the system. Kirchhoff's and Ohm's laws provides the connectivity among all nodes in the system, the graph is fully connected and weighted. The reduced resistance distance matrix [12] R is a way of adapting E in such a way that the resulting graph is equivalent in size to its un-weighted topological structure representation. The m edges are replaced with the m smallest entries in the upper (or lower) triangle of E. The result of graph G (n, m) and edges represents the strong electrical connections rather than direct physical connections. The adjacency matrix of new graph (R) can be obtained as

$$R = \begin{cases} Rab = 1 & \forall Eab < t \\ Rab = 0 & \forall Eab \geq t \end{cases} \tag{8}$$

where t is a threshold adjusted to produce exactly m links in the network.

5 Case Study and Results

This section presents the detailed methodology involved to present the topological and electrical properties of IEEE 14 bus system and how a grid service is deployed subsequently. The service efficacy has been demonstrated by means of a case study

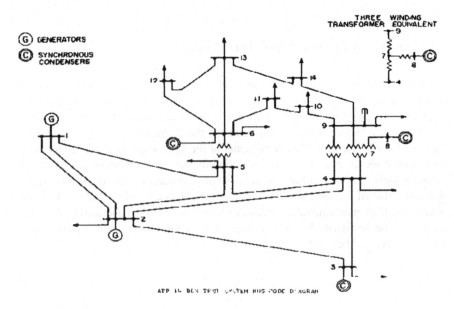

Fig. 1 IEEE 14 bus system

Table 1 Comparison of topological and electrical structure of power grid

Comparison of topological and electrical parameters of IEEE 14 power grid		
Parameters of power network	Topological properties	Electrical properties
Nodes (N)	14	14
Edges (m)	18	18
Average nodal degree $\langle K \rangle$	2.5714	2.5714
Average shortest path $\langle l \rangle$	3.030612	1.367347
Diameter (D)	7	3
$r\{k > k$ bar$\}$	0.42857	0.1428
$C(G)$	0.35714	0.16224
$C(R)$	0.1978	0.1978

that considers the IEEE 14 power system as a test case. The IEEE 14 bus test case [13] represents a portion of the American Electric Power System. The IEEE 14 bus system is shown in Fig. 1. Table 1 shows the topological and electrical properties of IEEE 14 bus system power grid as well as compares the each properties of power grid. IEEE 14 power grid follows the exponential degree distribution.

6 Conclusion

In this paper, we have presented a number of results that characterize the topological and electrical structure of the power grid. Topological analysis focuses on physical significance of the power grid-like geographical distance rather than electrical distance. But in electrical analysis, different parameters such as resistance and impedance are considered to analyze the different properties of power grid network. Though different electrical parameters are considered in electrical analysis, it will provide the better result than topological analysis of power grid.

References

1. Boccaletti, S., Latora, V., Moreno, Y., Chavez, M., Hwang, D-U.: Complex networks: structure and dynamics. Phys. Rep. **424**(4), 175–308 (2006)
2. Pagani, G.A., Aiello, M.: The power grid as a complex network: a survey. arXiv preprint arXiv:1105.3338 (2011)
3. Wang, Z., Scaglione, A., Thomas, R.J.: Generating statistically correct random topologies for testing smart grid communication and control networks. IEEE Trans. Smart Grid **1**(1), 28–39 (2010)
4. GridGain: www.gridgain.com last accessed on 21 Feb 2014
5. Das, H., Roy, D.S.: A grid computing service for power system monitoring. Int. J. Comp. Appl. 62 (2013)
6. Das, H., Panda, G.S., Muduli, B., Rath, P.K.: The complex network analysis of power grid: a case study of the West Bengal power network. In: Intelligent Computing, Networking, and Informatics, pp. 17–29. Springer India (2014)

7. Das, H., Mishra, S.K., Roy, D.S.: The topological structure of the Odisha power grid: a complex network analysis. IJMCA 1(1), 012–016 (2013)
8. Wang, Z., Scaglione, A., Thomas, R.J.: On modeling random topology power grids for testing decentralized network control strategies. In: 1st IFAC Workshop Estimation and Control Network System (NecSys' 09), Venice, Italy (2009)
9. Whitney, D.E., Alderson, D.: Are technological and social networks really different? In: Unifying Themes in Complex Systems, pp. 74–81. Springer, Berlin (2008)
10. Wang, Z., Robert J. T., Scaglione, A.: Generating random topology power grids. In Hawaii International Conference on System Sciences, Proceedings of the 41st Annual, pp. 183–183. IEEE, 2008
11. Klein, D.J., Randić.: Resistance distance. J. Math. Chem. 12(1), 81–95 (1993)
12. Hines, P., Blumsack, S., Cotilla Sanchez, Barrows, C.: The topological and electrical structure of power grids. In: 43rd Hawaii International Conference on System Sciences (HICSS), 2010, pp. 1–10. IEEE, 2010
13. Power System Test Case Archive, http://www.ee.washington.edu/research/pstca/pf14/pg_tca14bus.html last accessed on 01 Mar 2014

Investigation on Power Saving Mechanism for IEEE 802.16m Networks in Mobile Internet Traffic

A. Rajesh and R. Nakkeeran

Abstract In IEEE 802.16m networks, the existing power saving models (PSMs) had been developed by assuming traffic as Poisson arrival with exponential interarrival time or generalized traffic process. However, the mobile Internet traffic presumes Pareto/Weibull arrival with Pareto interarrival time. The existing PSM with this traffic causes frequent battery drain out at the mobile stations (MSs) with inappropriateness of the sleep parameters. Hence, in this paper, a PSM is derived for MS with mobile Internet traffic by considering the traffic parameters, namely scaling, shaping and location or threshold. A combined power saving class (PSC) is also suggested to improve the system performance with the developed PSM. The effectiveness of the developed model is validated by means of numerical results.

Keywords Power saving class · IEEE 802.16m · Power saving model

1 Introduction

Power saving mechanisms by means of power saving classes (PSC) play a vital role in the field of wireless communication. Although many power saving models (PSMs) have been suggested for IEEE 802.16m networks [1], they suffer from precise modeling of unique traffic pattern [2]. Besides, the traffic patterns vary for real-time and non-real-time services. In particular, the power saving model using Poisson arrival process is no longer compelling when the mobile Internet traffic

A. Rajesh (✉)
School of Electronics Engineering, VIT University, Vellore, India
e-mail: rajesha@vit.ac.in

R. Nakkeeran
Department of Electronics Engineering, School of Engineering and Technology,
Pondicherry University, Puducherry, India
e-mail: nakkeeranpu@gmail.com

© Springer India 2015
L.C. Jain et al. (eds.), *Intelligent Computing, Communication and Devices*,
Advances in Intelligent Systems and Computing 309,
DOI 10.1007/978-81-322-2009-1_31

patterns divulge long tail or heavy tail in their probability distributions [3]. In this paper, an attempt is made to model the power saving mechanism for mobile stations (MSs) with mobile Internet traffic. The need for such model is due to frequent battery drain out by MSs with high-speed Internet connectivity.

With sleep-mode mechanism, the IEEE 82.16m defines three PSCs, namely PSC-I for best effort and non-real-time services, PSC-II for real-time and unsolicited grant services, and PSC-III for management-related services. The importance of determining the suitable values of sleep parameters such as minimum sleep interval, T_{min} and maximum sleep interval, T_{max}, in PSC-I has been studied in [4, 5] for Poisson-distributed traffic. Improper values of T_{min} and T_{max} will reduce the performance of the system in terms of energy consumption and/or medium access control (MAC) response delay. For a chosen set of T_{min} and T_{max}, to optimize the sleep cycle of PSC-I, many techniques on combined PSC have been proposed in the literature [6].

The main drawback of the existing combined PSC is that the number of listen intervals is fixed (in terms of duration and position) between the sleep intervals. Recently, modeling of inactivity timer with variable listen interval has been proposed in [7] to determine the state transition conditions. Although many works have been attempted with IEEE 802.16m power management, the data traffic analysis over a decade reveals that the self-similar traffic with long-tailed patterns are different from Poisson-distributed traffic. Therefore, the conventional power saving mechanism with Poisson arrival and exponential interarrival has to be revisited [4–8, 12, 13].

Although optimal power saving policies amid various non-Poisson distributions such as lognormal distribution, gamma distribution, half-normal distribution, generalized Pareto distribution, and Weibull distribution have been detailed in [9], it does not suggest a power saving model for mobile Internet traffic. The rest of the paper is organized as follows: The development of the proposed power saving scheme is presented in Sect. 2. Results are discussed in Sect. 3, and concluding remarks are given in Sect. 4.

2 Proposed Power Management for Mobile Internet Traffic

2.1 Model Description

The power saving model is derived by formulating interarrival time of the mobile Internet traffic model. The traffic flow interarrival time is defined as the difference between two consecutive traffic arrivals. In case of mobile Internet traffic [10], if the traffic arrival is assumed to trail Pareto or Weibull distribution, then the mean interarrival time (λ_T) follows Pareto distribution and is obtained as follows:

$$\lambda_T = \frac{\alpha\beta}{\alpha - 1} \quad \text{for} \quad \alpha > 1 \tag{1}$$

$$\lambda_T = \frac{\beta}{\alpha - 1} \quad \text{for} \quad \alpha > \gamma \tag{2}$$

where α is the shaping parameter, β is the scaling parameter, and γ is the location or threshold parameter. The location parameter can be given as the ratio of scaling parameter to the shaping parameter. From [10], the probability density function (pdf) for mobile Internet traffic is represented by

$$f(t = T_{ON}) = \alpha_1 \beta_1 t^{\beta_1 - 1} e^{-\alpha_1 t^{\beta_1}} \quad \begin{matrix} \alpha_1 > 0 \\ \beta_1 \geq 0 \end{matrix} \tag{3}$$

$$f(t = T_{OFF}) = \frac{\alpha_2 \beta_2^{\alpha_2}}{(\beta_2 + t)^{\alpha_2 + 1}} \tag{4}$$

where α_1 and α_2 denote the shaping parameter during the ON period and OFF period, respectively. Also, β_1 and β_2 refer to the scaling parameter during the ON period and OFF period, respectively. The cumulative distribution function (CDF), over a given interarrival time (t_{ia}), can be defined in terms of pdf, denoted by $F_{t_{ia}}(t) = Pr[t_{ia} \leq t]$, which corresponds to $\int_0^t f_{t_{ia}}(t)\, dt$. The expected value of the number of sleep cycles or energy consumption or response delay over the interarrival time is given by $\sum_{t=-\infty}^{+\infty} t\, Pr[t_{ia} = t]$. Consequently, $F_{t_{ia}}(t)$ is derived as

$$F_{t_{ia}}(t) = \begin{cases} 1 - \left\{ 1 + \left(\left(\frac{\beta - \gamma(\alpha - 1)}{\alpha - 1} \right) / \beta \right) \right\}^{-\alpha} & \alpha > \gamma \\ 0 & \text{otherwise} \end{cases} \tag{5}$$

Let N_S denote the number of sleep intervals before the MS enters into the wake mode, i represent the monitor period, T_i correspond to the duration of the ith sleep interval, L denote the duration of the listen interval, and $T_i + L$ stand for the duration of the sleep cycle. Then, the length of the sleep interval given in [5] can also be represented as follows:

$$T_i = \begin{cases} \min(2^{i-1}T_{\min}, T_{\max}), & \text{if} \quad 2^{i-1}T_{\min} < T_{\max} \\ T_{\max}, & \text{if} \quad 2^{i-1}T_{\min} \geq T_{\max} \end{cases} \tag{6}$$

The awakening from sleep mode can be initiated either by MS or by BS and therefore known as mobile station initiation of awakening (MIA) and base station initiation of awakening (BIA). The probability that there is only one BIA can be calculated as follows:

$$Pr(\text{BIA} = \text{True}) = Pr\left[0 < \frac{N_S}{T_i + L} \le 1\right]$$
$$= Pr[0 < N_S \le T_i + L]$$
$$= 1 - \{1 + (((T_i + L) - \gamma)/\beta)\}^{-\alpha}$$

(7)

Therefore, the probability that there is at least one initiation of awakening in the ith sleep cycle can be obtained through

$$Pr(N_S = i) = Pr\left[(i-1) < \frac{N_S}{T_i + L} \le i\right]$$
$$= Pr[(i-1)(T_i + L) < N_S \le i(T_i + L)]$$
$$= \left(1 + \left(\sum_{j=1}^{i-1}((T_j + L) - \gamma)/\beta\right)\right)^{-\alpha}(1 - \{1 + (((T_i + L) - \gamma)/\beta)\}^{-\alpha})$$

(8)

where the probability of arrival follows type II Pareto distribution and the probability that there is no BIA is described as follows:

$$P(\text{BIA} = \text{false}) = \{1 + (((T_i + L) - \gamma)/\beta)\}^{-\alpha}$$

(9)

The mean or expected number of sleep cycles ($E[N_S]$) can be given as the product of number of monitor intervals and the initiation of awakening. From Eqs. (6) and (8), the $E[N_S]$ for mobile Internet traffic model would be

$$E[N_S] = \sum_{i=1}^{\infty} iPr(N_S = i)$$
$$= Pr(N_S = 1) + \sum_{i=2}^{\infty} iPr(N_S = i)$$
$$= (1 - \{1 + (((T_i + L) - \gamma)/\beta)\}^{-\alpha}) + \sum_{i=2}^{\infty} i\left(1 + \left(\sum_{j=1}^{i-1}((T_j + L) - \gamma)/\beta\right)\right)^{-\alpha}$$
$$(1 - \{1 + (((T_i + L) - \gamma)/\beta)\}^{-\alpha})$$

(10)

Since the power management with downlink traffic is alone considered in this paper, only BIA exists and there is no MIA from sleep mode to wake mode. Hence, the mean energy consumption with BIA can be derived as follows:

$$E[c] = \sum_{i=1}^{\infty} \left(\sum_{m=1}^{i} (T_m E_S + L E_L) \right)$$
$$\left(1 + \left(\sum_{j=1}^{i-1} ((T_j + L) - \gamma)/\beta \right) \right)^{-\alpha} (1 - \{1 + (((T_i + L) - \gamma)/\beta)\}^{-\alpha}) \tag{11}$$

where E_S and E_L denote the energy consumption in sleep interval and energy consumption in listen interval, respectively. The mean response delay ($E[r]$) with BIA can be estimated as follows:

$$E[r] = \sum_{i=1}^{\infty} ((T_i + L)/2)$$
$$\left(1 + \left(\sum_{j=1}^{i-1} ((T_j + L) - \gamma)/\beta \right) \right)^{-\alpha} (1 - \{1 + (((T_i + L) - \gamma)/\beta)\}^{-\alpha})$$
$$\tag{12}$$

By comparing the developed $E[N_S]$, $E[c]$, and $E[r]$ with reference to that in [5], one could infer that the duration of the sleep cycle is not an integer multiple of arrival rate. Nevertheless, it is inversely proportional to the scale parameter (β) and to the power of the shape parameter (α).

2.2 Combined PSC

The motivation toward a heuristic combined PSC is to apply the amendments of IEEE 802.16m standard which specifies that the duration of the listen interval of PSC-I may be varied and extended as long as the BS has packets buffered for any of its MS. However, assigning the minimum and maximum duration of the listen intervals over a given sleep cycle remains critical as the power saving specification updates from the BS may not impart the current energy state and traffic condition of individual MS. Hence, in order to reduce the response delay with low energy consumption, a heuristic power saving mechanism is suggested by combining the sleep cycles of PSC-I and PSC-II [11].

In addition, the combined PSC can be updated by the MS at any point of time without terminating the normal sleep-mode operation. The duration of sleep interval is made cyclic after reaching 512 ms (1,024 ms) with binary exponential increment. This process ensures the occurrence of more number of listen intervals and thus reduces the response delay than the sleep cycle of existing PSC algorithm. However, the illustration shown in Fig. 1a repeats after 8 ms due to space constraints. To reduce the energy consumption with reduced response delay, the integration of PSC-I and PSC-II is attempted. The duration of sleep interval in

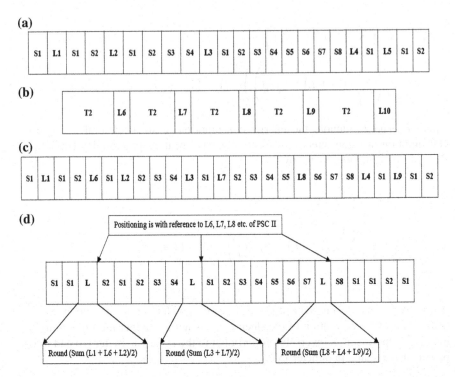

Fig. 1 Illustration of the proposed combined *PSC* (*S* timeslot—represents the duration of sleep interval, *L* timeslot—represents the duration of listen interval, and *T2* timeslot—represents the duration of sleep interval in PSC-II). **a** PSC I using cyclic binary exponential (CBE) with $T_{min} = 1$ ms, $T_{max} = 8$ ms and $L = 1$ ms.**b** PSC II with $T2 = 4$ ms and $L = 1$ ms. **c** Conventional combined PSC I. **d** Proposed power saving mechanism

PSC-II is chosen as 4 ms (shown in Fig. 1b) for illustration. Amalgamation of PSC-I and PSC-II is accomplished by retaining the listen intervals of both PSCs and maintaining the sleep interval of PSC-I. After merging, it is observed that the number of listen intervals increases with the resultant PSC as illustrated in Fig. 1c.

```
Algorithm 1: Pseudo-code of the combined PSC scheme
Initialize: Tmin, Tmax, L, T2
Sleep_Mode (Tmin, Tmax, L, T2) {

    TPSC-II = generate_fixed (T2, L);
    TPSC-I = cyclic_binary_exp (Tmin, Tmax, L);
    Tapp = append (TPSC-I, TPSC-II);
    Tsca = scale_listen (Tapp);
    TPSC-CB = position_listen (Tsca); }

Update: Tcurrent = TPSC-CB;
```

Therefore, the proposed PSC system is derived by summing the number of listen intervals, which is then downscaled by a factor of 2. The number of listen intervals between PSC-I and PSC-II is summed by referring the listen interval of PSC-I which in turn depends on the listen interval position of PSC-II. Once the final value of sleep interval is derived, the next challenging task is to position the listen interval. The listen interval is positioned by referring the listen interval of PSC-II (refer Fig. 1d). The pseudo-code for the combined PSC is shown in Algorithm 1. The T_{app} is the resultant sleep cycle of PSC-I and PSC-II. The sleep cycle after scaling is denoted as T_{sca}, and the absolute combined sleep cycle is represented as T_{PSC-CB}. Since the position of listen interval in the proposed PSC is placed with reference to PSC-II, it is robust to delay sensitive traffic with increased interarrival time.

3 Performance Evaluation

To evaluate the performance of the IEEE 802.16m system with proposed power saving mechanism, numerical studies are conducted using MATLAB 2010a simulator. The results are obtained by averaging over 15 runs, and each run is repeated for 10^5 iterations. The MAC parameters are configured in accordance with the standard discussed in [1]. Figure 2 illustrates the average number of sleep intervals at the MS for different values of T_{min} and T_{max}. The performance is compared between using existing PSC-I algorithm and proposed combined PSC algorithm. The number of sleep intervals of the developed PSM with existing PSC algorithm

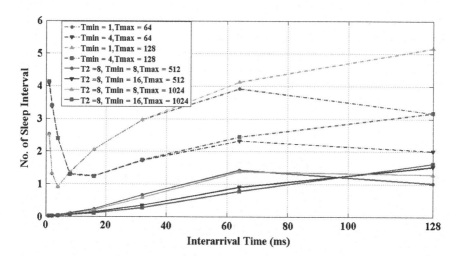

Fig. 2 Average sleep intervals as a function of interarrival time with PSC using existing PSC-I (*dotted lines*) and proposed combined PSC (*solid lines*)

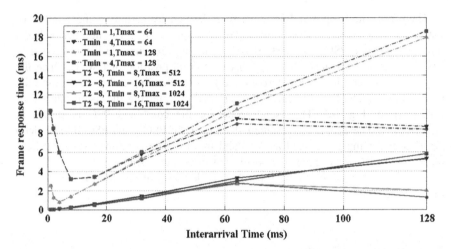

Fig. 3 Average frame response time (delay) as a function of interarrival time

keeps on decreasing until an interarrival time of 16 ms and then starts to increase. It is observed that the number of sleep intervals with the combined PSC increases with the increase in the interarrival time.

However, this response is reduced beyond interarrival time of 64 ms for $T_{\min}$ and $T_{\max}$ of 8 and 512 ms, respectively. Considering the reduced number of sleep intervals observed for the chosen $T_{\min}$ and $T_{\max}$ in both algorithms, it is noted that the combined PSC shows an improvement of 80.85 % compared to the existing PSC algorithm. This reduction in the number of sleep intervals with combined PSC would have resulted because of appropriate scaling and positioning of listen intervals between the sleep intervals. Although a heuristic means of scaling the sleep and listening intervals is attempted, the combined PSC is effective only if it guarantees the response delay with reduced energy consumption. Figure 3 reveals the response delay for various values of $T_{\min}$ and $T_{\max}$. It is observed that the response delay with existing PSC-I algorithm decreases as the duration of inter-arrival time increases from 1 to 8 ms. However, the response delay drastically increases beyond 8 ms.

Despite the response delay increases linearly with combined PSC, the mean response delay is less compared to existing PSC-I algorithm. From the grid, it is understood that the existing PSC with the set of value of $T_{\min}$ and $T_{\max}$ (1 and 64 ms) exhibits a mean delay of 3.8829 ms. On the other hand, the combined PSC shows a response delay of 0.8006 ms for $T_{\min} = 8$ ms and $T_{\max} = 512$ ms. Thus, the combined PSC exhibits an improvement of 79.38 % than the existing PSC-I algorithm. The energy consumption of the MS is shown in Fig. 4 for different values of $T_{\min}$ and $T_{\max}$. It is observed that the trend of energy consumption follows the profile of response delay.

A mean energy consumption of 4.3832 mW is noticed in existing PSC-I algorithm for $T_{\min} = 1$ ms and $T_{\max} = 64$ ms. On the other hand, in the case of

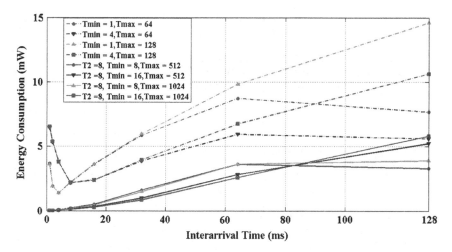

Fig. 4 Average energy consumption as a function of interarrival time

combined PSC for $T_{min} = 8$ ms and $T_{max} = 512$ ms, the mean energy consumption is reduced to 1.1718 mW. This leads to an improvement of 73.26 % compared to the system with existing PSC-I algorithm. Though the minimum and maximum sleep intervals with proposed combined PSC are higher than existing PSC-I, the former performs better than the latter due to heuristic scaling of listen intervals that limit frequent retransmission attempt by the MS.

4 Conclusion

In this paper, a power saving model is developed for mobile Internet traffic that exhibits varied interarrival time. The model is evaluated with existing PSC-I and combined PSC algorithm. The proposed model along with the combined PSC allows the MS to conserve a considerable amount of energy with reduced delay due to appropriate scaling of listen and sleep intervals.

References

1. IEEE Standard for Local and Metropolitan Area Networks: Part 16: Air Interface for Fixed and Mobile Broadband Wireless Access Systems. IEEE P802.16m/D4 (2010)
2. Ghani, S., Iradat, F.: Loss probability in networks with Pareto distributed traffic. In: Proceedings of IEEE 2nd International Conference on Intelligent Systems, Modelling and Simulation, pp. 355–360 (2011)

3. Singhai, R., Joshi, S.D., Bhatt, R.K.P.: Offered-load model for Pareto inter-arrival network traffic. In: Proceedings of IEEE 34th Conference on Local Computer Networks, pp. 364–367 (2009)
4. Zhang, Y., Xiao, Y., Leung, V.C.M.: Energy management analysis and enhancement in IEEE 802.16e WirelessMAN. IEEE Trans. Veh. Technol. **58**(7), 3738–3752 (2009)
5. Xiao, Y.: Energy saving mechanism in the IEEE 802.16e wireless MAN. IEEE Commun. Lett. **9**(7), 595–597 (2005)
6. Kwon, S.-W., Cho, D.-H: Enhanced power saving through increasing unavailability interval in the IEEE 802.16e systems. IEEE Commun. Lett. **14**(1), 24–26 (2010)
7. Tseng, Y.-C., Chen, J.-J., Yang, Y.-C.: Managing power saving classes in IEEE 802.16 wireless MANs: a fold-and-demultiplex method. IEEE Trans. Mob. Comput. **10**(9), 1237–1247 (2011)
8. Jin, S., Chen, X., Qiao, D.: Analytical modeling of inactivity timer in IEEE 802.16m sleep mode. IEEE Commun. Lett. **16**(5), 650–653 (2012)
9. Almhana, J., Liu, Z., McGorman, R.: Nearly optimal power saving policies for mobile stations in wireless networks. Elsevier Comput. Commun. **33**(1), 595–602 (2010)
10. Wong, D.T.C., Kong, P.-Y., Liang, Y.-C., Chua, K.C., Mark, J.W.: Wireless broadband networks. Wiley (2009)
11. Rajesh, A., Nakkeeran, R.: Performance analysis of unified power saving mechanism for IEEE 802.16m networks. In: Proceedings of IEEE 2nd World Congress on Information and Communication Technologies (WICT'12), pp. 1–5. IIITM, Kerala (2012)
12. Lin, Y.-W., Wang, J.-S.: An adaptive QoS power saving scheme for mobile WiMAX. Springer Wireless Pers. Commun. **69**(4), 1435–1462 (2013)
13. Ferng, H.-W., Li, H.-Y.: Design of predictive and dynamic energy-efficient mechanisms for IEEE 802.16e. Springer Wireless Pers. Commun. **68**(4), 1807–1835 (2013)

Hybrid Edge Detection-Based Image Steganography Technique for Color Images

Deepali Singla and Mamta Juneja

Abstract Steganography, Greek meaning covered writing, is a branch of information security which hides the existence of important information in order to prevent any unauthorized access. The number of image steganography techniques has been proposed so far to achieve the goals of steganography, i.e., high payload, less imperceptibility, and more robustness. In this paper, we are proposing a new steganography technique for colored image (i.e., RGB images). The proposed scheme has taken into consideration the property that more information can be hidden into the contrast areas rather than in the bright areas. This scheme makes the use of hybrid edge detector, i.e., combination of fuzzy and canny edge detector. After detecting edges, embedding is done accordingly. The proposed scheme achieves all the three goals of steganography appropriately.

Keywords Adaptive LSB · AES · Fuzzy edge detector · Stego image

1 Introduction

Steganography, Greek meaning covered writing, is a branch of information security which hides the existence of important information in order to prevent any unauthorized access [1]. There are number of innocent media used by the steganography, e.g., text, signals, videos, and images [1]. Basic mechanism of image steganography is as shown in Fig. 1.

D. Singla (✉) · M. Juneja
Computer Science and Engineering, Panjab University, Chandigarh, India
e-mail: deepysingla91@gmail.com

M. Juneja
e-mail: er_mamta@yahoo.com

© Springer India 2015
L.C. Jain et al. (eds.), *Intelligent Computing, Communication and Devices*,
Advances in Intelligent Systems and Computing 309,
DOI 10.1007/978-81-322-2009-1_32

277

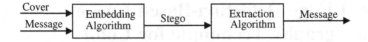

Fig. 1 Basic mechanism of steganography [1]

This paper is organized as follows: Sect. 2 describes the newly proposed image steganography technique. Section 3 describes the results and discussion of the newly proposed technique. Section 4 presents the conclusion and future scope.

2 The Proposed Method

The literature survey conducted from [2–12] we have concluded that more data can be hidden in the edge areas rather than in the smooth areas. In the proposed method, we are using a hybrid edge detector which is a combination of canny edge detector [12] and the fuzzy edge detector [13]. Now, for edge pixels, 4 bits of green channel and 8 bits of blue channel are replaced with secret message bits. For smooth area pixels, adaptive LSB-based technique is used which is based on Kekre 2009 [14] algorithm. Embedding is done on the basis of Table 1.

2.1 Steganography Algorithm at Sender End

At sender end, cover image is given as the input and stego image is taken as output. The stego image will be obtained by following below given steps (Fig. 2).

2.2 Extraction at Receiver End

At receiver end, stego image is received and it is considered that cover image is present at both ends. For extracting the message at receiver end, reverse procedure is followed to obtain the secret message.

Adaptive LSB technique used in embedding is as given below:

Table 1 Adaptive embedding in smooth areas

Pi_red	Pi_blue	Pi_green	Utilized bits
255–240	255–240	255–240	Green 4 bits blue 8 bits
239–224	239–224	239–224	Green 3 bits blue 7 bits
223–192	223–192	223–192	Green 2 bits blue 6 bits
191–0	191–0	191–0	Green 1 bits blue 5 bits

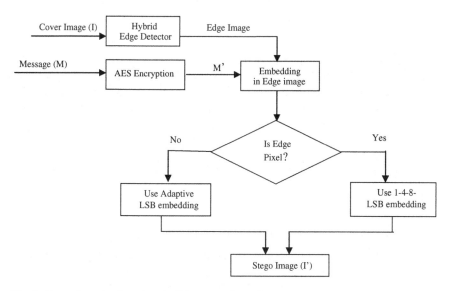

Fig. 2 General mechanism of embedding at sender end

3 Experimental Results

The above-proposed technique based on hybrid edge detector is analyzed in this section. The following table shows the results of applying different edge detectors used in these techniques. For analyzing the performance of the different steganography techniques, we use two parameters, i.e., peak signal-to-noise ratio (PSNR) and mean square error (MSE). On the basis of these parameters, we are comparing our method with the hybrid-edged approach explained in [8] and with the performance of other basic edge detector-based steganography using 6 bits of edge pixel and 2 bits of smooth pixel. The following table shows the performance of these different techniques and our proposed method (Tables 2 and 3).

Table 2 Comparison of the different methods with proposed technique for Lena image

Methods parameter	Sobel	Prewitt	Robert	Gaussian	Canny	Hybrid	Proposed method
MSE	9.855	9.279	8.514	14.659	17.999	32.751	12.455
PSNR	38.19	38.45	38.82	36.46	35.57	32.97	46.68

Table 3 Comparison of the different methods with proposed technique for temple image

Methods parameter	Sobel	Prewitt	Robert	Gaussian	Canny	Hybrid	Proposed method
MSE	14.726	14.658	10.252	22.061	25.601	45.365	11.344
PSNR	36.44	36.47	38.02	34.69	34.04	31.56	48.24

4 Conclusion

The proposed method is a hybrid approach using hybrid edge detector, 1–4–8 LSB technique, and adaptive LSB technique. The proposed method is working on the principal that more number of bits can be embedded in contrast area than in brighter areas. The proposed method has achieved desirable quality with high capacity. The proposed method is prone to many attacks as it has achieved a high-quality stego image.

References

1. Artz, D.: Digital steganography: hiding data within data. IEEE Internet Comput. J. **5**(3), 75–80 (2001)
2. Thien, C.C., Lin, J.C.: A simple and high-hiding capacity method for hiding digit-by-digit data in images based on modulus function. Pattern Recogn. **36**, 2875–2881 (2003)
3. Maroney, C.: Hide and seek 5 for windows 95, computer software and documentation, originally released in Finland and the UK
4. Hempstalk, K.: hiding behind corners: using edges in images for better steganography. In: Proceedings of the Computing Women's Congress, 2006
5. Singh, M., Singh, B., Singh, S.S.: Hiding encrypted message in the features of images." IJCSNS **7**, 302 (2007)
6. Hossain, M., Al Haque, S., Sharmin, F.: Variable rate steganography in gray scale digital images using neighborhood pixel. In: 12th international conference Dhaka, information computers and information technology, ICCIT '09, 2009
7. Chen, W.J., Chang, C.C., Le, T.H.N.: High payload steganography mechanism using hybrid edge detector. Expert Syst. Appl. **37**, 3292–3301 (2010)
8. Amirtharajan, R., Bose, B., Imadabathuni, S., Bosco, J.: Security building at the line of control for image stego. Int. J. Comp. Appl. **12**(5), 46–53 (2010)
9. Hussain, M., Hussain.: Embedding data in edge boundaries with high PSNR. In: Proceedings of 7th international conference on emerging technologies (ICET 2011), pp. 1–6, 2011
10. Juneja, M., Sandhu, P.S.: A new approach for information security using an improved steganography technique. J. Inf. Process Syst. **9**, 405–424 (2013)
11. Specification for the Advanced Encryption Standard (AES), Federal Information Processing Standards Publication 197, 2001
12. Canny, J.: A computational approach to edge detection. IEEE Trans. Pattern Anal. Mach. Intell. **8**, 679–687 (1986)
13. Talai Z., Talai A.: A fast edge detection using fuzzy rules. In: International conference on communication and control application, pp. 1–5, 2011
14. Kekre, H.B., Athawale, A., Halarnkar, P.N.: Performance evaluation of pixel value differencing and Kekre's modified algorithm for information hiding in images. In: ACM international conference on advances in computing, communication and control, 2009

Haze Removal: An Approach Based on Saturation Component

Khitish Kumar Gadnayak, Pankajini Panda and Niranjan Panda

Abstract Outdoor images those are taken under bad weather conditions are basically degraded by the various atmospheric particles such as smoke, fog, and haze. Due to the atmospheric absorption and scattering phenomena while capturing the images, the irradiance received by the camera from the scene point is attenuated along the line of sight. The incoming light flux is attenuated with the light from all other directions called the airlight. Due to this reason, there is a resultant decay in the color and the contrast of the captured image. Haze removal from an input image or dehazing of an image is highly required so as to increase the visibility of the input image. Removing the haze layer from the input hazy image can significantly increase the visibility of the scene. The haze-free image is basically visually pleasing in nature. The paper focuses on the haze removal process by considering the HSI color model of an image instead of RGB color space. In the HSI color model, the saturation component describes the contrast of an image. From the saturation component, it is possible to estimate the transmission coefficient or the alpha map, and from this, a haze-free image can be recovered which has the better visibility than that of the captured hazy image.

Keywords Scattering · Airlight · Attenuation · Haze · Saturation · Image modeling · Transmission coefficient

K.K. Gadnayak (✉)
Computer Science and Engineering, C. V. Raman College of Engineering,
Bhubaneswar, Odisha, India
e-mail: khitish05071983@gmail.com

P. Panda
Information Technology, C. V. Raman College of Engineering,
Bhubaneswar, Odisha, India
e-mail: mrs.pankajini.panda@gmail.com

N. Panda
Institute of Technical Education and Research, SOA University,
Bhubaneswar, Odisha, India
e-mail: niranjanpanda@soauniversity.ac.in

© Springer India 2015   281
L.C. Jain et al. (eds.), *Intelligent Computing, Communication and Devices*,
Advances in Intelligent Systems and Computing 309,
DOI 10.1007/978-81-322-2009-1_33

1 Introduction

Images of outdoor scenes are basically degraded due to the presence of atmospheric particles. Due to the atmospheric absorption and scattering of these particles, irradiance received by the camera from the scene point is attenuated along the line of sight. The incoming light flux is blended with the light from all other directions called the airlight [1]. The degradation of the image is variant in nature, and due to this degradation, there is a resultant decay in the color and the contrast of the captured image. Haze removal is one of the challenging tasks because the haze is dependent on the unknown depth, and this problem is under constrained if the input is a single image. Therefore, many algorithms have been demonstrated using multiple images or some additional information. Polarization-related methods [1] remove the haze effect by considering two or more images with different degree of polarization. In Narasimhan and Nayar [2–4], more constraints are obtained from multiple images of the same scene or image under different weather conditions (Fig. 1).

In a computer vision, the widely used model for the formation of hazy images is given as:

$$I(x) = t(x) * J(x) + (1 - t(x)) * A \tag{1}$$

where x indicates the position of the pixel, I is the observed hazy image, J is the scene radiance which is the haze-free image that is to be restored, A is the global atmospheric light, and t is the medium of transmission describing the portion of the light that is not scattered and reaches the camera. The transmission has a scalar value ranges from 0 to 1 for each pixel, and the value indicates the depth of the information of the scene objects directly.

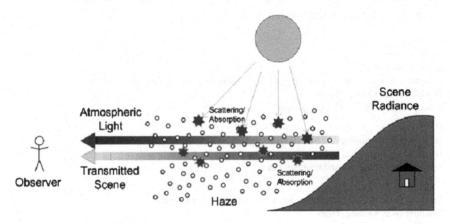

Fig. 1 Haze formation model

2 Literature Survey

Schechner et al. [1] paper describes the image formation process considering the polarization effect of atmospheric scattering and inverting the process is required to get a haze-free image. The input image is basically composed of two unknown components; one is the scene radiance in the absence of the haze and the airlight.

Tan [5] proposed single image dehazing method which is based on the optical model and is given as

$$I(x) = L_\infty \rho(x) e^{-\beta d(x)} + L_\infty (1 - e^{-\beta d(x)})$$ (2)

This proposed approach is based on the assumption that the clear day images have high contrast as compared to the images those are affected by the bad weather. Relying upon this assumption, Tan removed the haze by maximizing the local contrast of the restored image.

Fattal's [6] single image dehazing approach based on refined image formation model that relates to the surface shading and the transmission function. Fattal grouped the pixel belonging to the same surface having the same reflectance and the same constant surface albedo, and he proposed independent component analysis method to determine the surface shading and the transmission.

He et al. [7] dark channel prior is a single image dehazing approach which is based on the statistic approach of the outdoor haze-free image. This approach describes most local regions that do not cover the sky in a haze-free image; some of the pixels have very low intensity in at least one color (RGB) channel and such pixels are known as dark pixels. In the captured hazy images, these dark pixels are used to estimate the haze transmission. After estimating the transmission map for each pixel, and then by combining the haze imaging model and soft matting technique it is possible to recover a high quality haze-free image.

Ancuti et al. [8] describes the dehazing approach which is based on the fusion strategy that is derived from the original hazy image inputs by applying a white balance and contrast-enhancing procedure. The fusion enhancement technique estimates perceptual-based qualities known as the weight maps for each pixel in the image. These weight maps control the contribution of each input to the final obtained result.

Chu et al. [9] paper is based on the concept that the degradation level affected by the atmospheric haze is basically dependent on the depth of the scene. Pixel in each of the part of the image tends to have the similar depth. Based on these assumptions, the given is segmented into different regions. After segmentation, the transmission map is estimated for each region and the transmission map is refined using soft matting [10].

Xie et al. [11] paper describes the implementation of multiscale retinex algorithm on the luminance component in YCbCr space of input image to get the pseudo-transmission map. The input hazy image has been transformed from RGB color space to YCbCr space, and then, by using the multiscale retinex algorithm, better haze-free image is recovered.

In our survey work, we have gone through various haze removal algorithms [12] which focus on the information regarding to get a haze-free image which improves the performance of the scene visibility in the hazy images.

3 Haze Removal Approach

Both the RGB and HSI color models are widely used in image processing applications. The HSI color model describes the hue, saturation, and intensity components where the hue describes the pure color, saturation component gives the measure of white light added to the pure color, and the intensity describes the brightness of the pixels in an image. The equivalent relationship between the HSI color model and the RGB color model is described as follows:

$$\begin{cases} H = \cos^{-1}\left\{ \frac{\frac{[(R-G)+(R-B)]}{2}}{\sqrt{(R-G)^2+(R-B)(G-B)}} \right\} \\ S = 1 - \frac{3}{(R+G+B)}[\min(R,G,B)] \\ I = \frac{(R+G+B)}{3} \end{cases} \tag{3}$$

According to the equivalent relationship of HSI and RGB model and from the equation of haze formation model, it can be stated as follows:

$$\begin{cases} H_I(x) = H_J(x) \\ S_I(x) = t(x)\frac{I_J(x)}{I_I(x)}S_J(x) \\ I_I(x) = t(x)I_J(x) + (1-t(x))A \end{cases} \tag{4}$$

In this, H_I, S_I, and I_I are the hue, saturation, and the intensity components of the observed hazy image and H_J, S_J, and I_J are the hue, saturation, and the intensity components of the haze-free image. The hue component of the original haze-free image is not affected by the haze and remains unchanged. But the saturation component is affected according to this above equation. The haze degrades the saturation component of the image, and the intensity component is also affected by the haze as defined in the Eq. (4). For the saturation layer,

$$\frac{t(x)I_J}{I_I} = \frac{t(x)I_J}{t(x)I_J + (1-t(x))A} \leq 1 \tag{5}$$

From this, it can be observed that $S_I \leq S_J$. Hence for a real scenario, the haze does not affect the hue of the image, but it degrades the saturation and the intensity channel of the original image.

4 Proposed Formulation

Markov random field theory is a probabilistic theory and is used in labeling to establish probabilistic distribution of interacting labels. Let $F = \{F_1, F_2, \ldots, F_m\}$ be a family of random variables defined on the set S in which each random variable F_i takes a value f_i in label set L. The family F is called a random field.

The optimization algorithms are used to either maximize or minimize the given objective function. The two optimization algorithms iterated conditional mode (ICM) and simulated annealing (SA) are basically used for image restoration as a basic objective functions. The basic objective function to be minimized is given as:

$$E(f) = \sum_{i,j} \left(Y_{i,j} - \widehat{Y}_{i,j}\right)^2 + \lambda \sum_{i,j} \sum_{i',j' \in N_{i,j}} \left(f_{i,j} - f_{i',j'}\right)^2 \tag{6}$$

where Y is the observed RGB image and $\widehat{Y}$ is the RGB image formed as:

$$\widehat{Y} = \alpha * X + (1 - \alpha) * A \tag{7}$$

where $X = \text{HSI\_to\_RGB}\ (H_{\text{init}} S_{\text{init}} I_{\text{init}})$ and A is airlight vector.

Approaching Method 1: (Estimation of saturation by assuming original intensity to be known) For this formulated approach, as there is one unknown, i.e., saturation of the original image is to be estimated, local minimization algorithm ICM is used for one unknown.

The basic objective function to be minimized is:

$$E(f) = \sum_{i,j} \left(Y_{i,j} - \widehat{Y}_{i,j}\right)^2 + \lambda \sum_{i,j} \sum_{i',j' \in N_{i,j}} \left(f_{i,j} - f_{i',j'}\right)^2 \tag{8}$$

Approaching Method 2: (Estimation of both saturation and intensity) For this formulated approach, as there are two unknowns are to be estimated, global minimization algorithm SA is used for two unknowns (i.e., double simulated annealing).

The basic objective function to be minimized is:

$$E(f) = \sum_{i,j} \left(Y_{i,j} - \widehat{Y}_{i,j}\right)^2 + \lambda_1 \sum_{i,j} \sum_{i',j' \in N_{i,j}} \left(f_{i,j} - f_{i',j'}\right)^2$$
$$+ \lambda_2 \sum_{i,j} \sum_{i',j' \in N_{i,j}} \left(f_{i,j} - f_{i',j'}\right)^2 \tag{9}$$

Fig. 2 Haze removal. **a** Input Lena image. **b** Hazy image. **c** Reconstructed images using ICM. **d** Using single SA

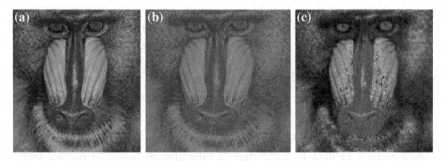

Fig. 3 Haze removal. **a** Input mandrill image. **b** Hazy mandrill image. **c** Reconstructed image using double simulated annealing

5 Experimental Result

Single SA is basically used for single unknown problem. In the experimental work from the original Lena color image, synthetic hazy image has been obtained. By assuming the intensity of the original image, the saturation has been estimated, and from this, a haze-free image has constructed (Figs. 2, and 3).

Double SA is basically used for two unknown problems. In this, from the input mandrill image, synthetically a hazy image has been formed. Using double SA, both the saturation and intensity components have been estimated, and after correct estimation of both saturation and intensity, the original image has been reconstructed.

6 Discussion

In this paper work, we have proposed an approach for a single haze removal process. We have observed that for a real case scenario, the hazy image captured under a bad weather condition has less contrast as compared to the haze-free image. Therefore, in this paper, we have considered the RGB color space as well as the HSI color space in which the saturation component gives the measure of contrast and intensity component gives the color description. In order to estimate the transmission coefficient, it is necessary to estimate the correct saturation and the intensity values. After estimating the correct alpha map, we have able to reconstruct the recovered haze-free image from a synthetic hazy image where we have assumed that the airlight is to be known.

References

1. Schechner, Y.Y., Narasimhan, S.G., Nayar, S.K.: Instant dehazing of images using polarization. In: Proceedings IEEE Conference Computer Vision and Pattern Recognition, vol. 1, pp. 325–332 (2001)
2. Narasimhan, S., Nayar, S.: Vision in bad weather. In: Proceedings IEEE International Conference on Computer Vision, pp. 820–827 (1999)
3. Narasimhan, S.G., Nayar, S.K.: Chromatic framework for vision in bad weather. In: IEEE Conference on Computer Vision and Pattern Recognition, vol. 1, pp. 598–605 (2000)
4. Narasimhan, S.G., Nayar, S.K.: Contrast restoration of weather degraded images. IEEE Trans. Pattern Anal. Mach. Intell. 25(6), 713–724 (2003)
5. Tan, R.: Visibility in bad weather from a single image. In: Proceedings IEEE Conference Computer Vision and Pattern Recognition (2008)
6. Fattal, R.: Single image dehazing. ACM Trans. Graph. SIGGRAPH 27(3), 72 (2008)
7. He, K., Sun, J., Tang, X.: Single image haze removal using dark channel prior. In: IEEE Conference on Computer Vision and Pattern Recognition, pp. 1957–1963 (2009)
8. Ancuti, C.O., Ancuti, C., Bekaert, P.: Effective single image dehazing by fusion. In: 17th IEEE International Conference on IEEE Image Processing (ICIP) (2010)
9. Chu, C.-T., Lee, M.-S.: A content-adaptive method for single image dehazing. Advances in Multimedia Information Processing, pp. 350–361 (2011)
10. Levin, A., Lischinski, D., Weiss, Y.: A closed form solution to natural image matting. In: Proceedings IEEE Conference Computer Vision and Pattern Recognition, vol. 1, pp. 61–68 (2006)
11. Xie, B., Guo, F., Cai, Z.: Improved single image dehazing using dark channel prior and multi-scale Retinex. Intelligent System Design and Engineering Application (2010)
12. Gadnayak, K.K., Panda, P., Panda, N.: A survey on image dehazing. Int. J. Eng. Res. Technol. 462–466 (2013)

Biomedical Image Registration Using Genetic Algorithm

Suraj Panda, Shubhendu Kumar Sarangi and Archana Sarangi

Abstract This paper focuses on the state-of-the-art technology which is useful for medical diagnosis and proper treatment planning. Using this scheme, different data formats such as MRI (magnetic resonance image), CT (computed tomography), PET (positron emission tomography), and SPECT (specialized positron emission tomography) of the same patient can be registered. These medical images provide complementary information which is conflicting occasionally due to nonalignment problem. However, the registered image provides more information for medical personals. In the registration process, images are aligned with each other and the size of the object is made equal. So in this process, the nonaligned image is transformed with respect to the reference image. Here, we have registered the biomedical images by maximizing the mutual information. Genetic algorithm (GA) is used to optimize rotation, scaling and translation parameters. Results presented reveal the suitability of the proposed method for biomedical image registration.

Keywords Biomedical image registration · Mutual information · Genetic algorithm

S. Panda (✉) · S.K. Sarangi
Department of Electronics and Instrumentation Engineering, ITER, SOA University, Bhubaneswar, India
e-mail: suraj_panda05@yahoo.co.in

S.K. Sarangi
e-mail: shubhendu1977@gmail.com

A. Sarangi
Department of Electronics and Communication Engineering, ITER, SOA University, Bhubaneswar, India
e-mail: archanasarangi24@gmail.com

© Springer India 2015
L.C. Jain et al. (eds.), *Intelligent Computing, Communication and Devices*,
Advances in Intelligent Systems and Computing 309,
DOI 10.1007/978-81-322-2009-1_34

1 Introduction

Medical images are captured from different sensors at different time and at different view point. Nonalignment of images are seen due to various reasons like calibration and setup errors of different scanning machines, respiratory, and cardiac motion of the patient. Different scanners have different scan geometry such as slice position, orientation, magnification, and thickness. Thus, images taken from different scanners are different and needs registration. By image registration [1, 2], one dataset is mapped onto another which involves establishment of a coordinate transformation relating the native coordinate system of the two datasets. One image is taken as the reference image, and the other image is transformed with respect to the reference image.

Registration can be achieved by maximizing the mutual information. Anatomical regions may appear with different relative intensities in different modalities, but these intensities are highly co-related. When images are correctly registered, its joint entropy is highly peaked and when misregistered, it spreads out. Peaked vs. spread out nature is expressed formally in terms of entropy of the distribution. Mutual information is defined in terms of entropy, and expresses the degree to which one image's intensities can be predicted given knowledge of the other, similar to correlation but more informative [3, 4]. Recently, mutual information-based approaches for biomedical image registration are gaining popularity among its clan [5, 6]. This has motivated us to investigate an efficient method for biomedical image registration using GA [7].

Genetic Algorithms are search algorithms which are inspired by evolution. They combine survival of the fittest among string structures with a structured yet randomized information exchange. GA is a computational model which optimizes functions. GA works with coded variables which discretizes the search space even if the function is continuous [7]. As GA uses randomized operators, it improves the search space in an adaptive manner. The programmer provides the function to be evaluated and the individual score determines how efficiently the task is performed by them. Based on their fitness values, the individuals are selected. Higher is the fitness value, greater is the chances of being selected. GAs are different from traditional methods in the sense that (i) GA works with a coding parameter set, not the parameter themselves; (ii) GA search from a population of points, not a single point; (iii) GA uses objective function information to maximize or minimize; and (iv) GA uses probabilistic transition rules, not deterministic rules.

In this paper, GA is used to optimize all three parameters rotation, scaling, and translation required for registration. The presented idea is simple yet interesting. Biomedical image registration results are produced to reveal the suitability of this proposed algorithm.

2 Methodology

2.1 Mutual Information

It carries the statistical dependence, which is the information that one image carries about the other [4, 6]. Let A and B are two images, then the mutual information $I(A, B)$ is defined as:

$$I(A, B) = \sum_{a,b} p_{AB}(a, b) \ \log \frac{p_{AB}(a, b)}{p_A(a)p_B(b)}$$

$$I(A, B) = H(A) + H(B) - H(A, B)$$

where $P_A(a)$ and $P_B(b)$ are the marginal probability distribution, and $P_{AB}(a, b)$ is the joint probability distribution.

$H(A)$ and $H(B)$ are individual entropies, and $H(A, B)$ is Joint entropy.

The joint probability distribution can be calculated as $p_{AB}(a, b) = \frac{h(a,b)}{\sum_{a,b} h(a,b)} p_A(a)$ $= \sum_b p_{AB}(a, b)$ and $p_B(b) = \sum_a p_{AB}(a, b)$ where $h(a, b)$ is the joint histogram.

Joint histogram is calculated by using following steps:

- Calculate the $h = $ individual histogram h_a and h_b
- Initialize the joint histogram to zero that is $h(i, j) = 0$
- Calculate the intensity partition numbers a and b to A and B, increment $h(a, b)$ Then calculate $\sum_{i,j} h(i, j)$

During registration procedure, it is needed to coincide the grid points of the nonaligned image with the reference image. Thus, maximization of mutual information or minimization of the mutual entropy is carried out in this work.

2.2 Optimization Using GA

Here we present the proposed method for optimization of different parameters using GA.

In Fig. 1, the structure of a chromosome is represented. The chromosome contains 28 bits out of which 8-bits are reserved for rotation 'R,' 8-bits for scaling 'S,' 12-bits for translation 'T.' Again, out of 12-bits for translation, 6-bits are reserved for translation along x-axis 'T_x' and other 6-bits for translation along y-axis 'T_y' (Fig. 2).

R	S	T	
		T_x	T_y
8-bits	8-bits	6-bits	6-bits

Fig. 1 Chromosome structure

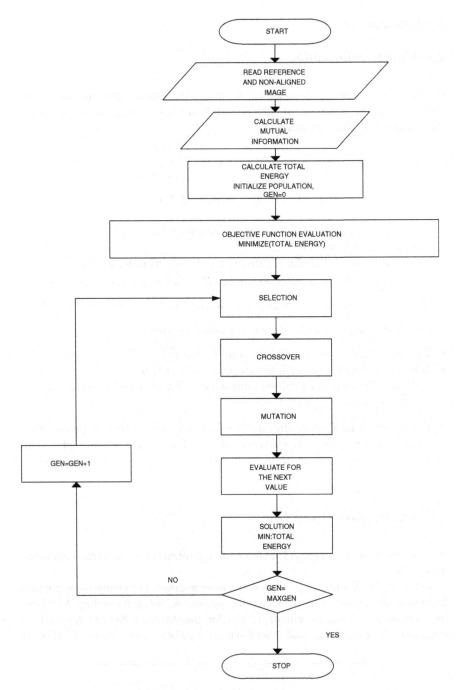

Fig. 2 Flow chart for biomedical image registration using genetic algorithm

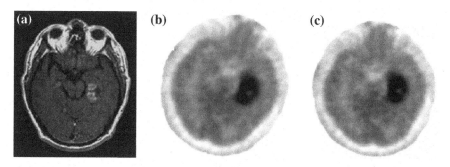

Fig. 3 **a** MRI brain images (Reference images), **b** PET brain images (Nonaligned images θ = −15°), and **c** registered images

Parameter setting:

- Population size = 100
- Maximum generations = 30
- Length of chromosome = 28 bits (Binary encoded)
- Crossover = 80 %
- Mutation = 2 %

Different parameters used for implementing GA are shown below.

3 Results and Discussions

The new algorithm is implemented in MATLAB. MRI images are considered for this experiment [8–12]. Note that size of the aligned and nonaligned images is same.

As shown in Figs. 3, 4, 5, 6, and 7, the nonaligned and reference images of same size are taken and registered using genetic algorithm. Actual values of '*R*,' '*S*,' and '*T*' are displayed in Table 1. The optimum values of '*R*,' '*S*,' and '*T*' obtained after 30 runs are displayed in Table 2. Note that '+' sign refers to an anticlockwise rotation while '−' sign refers to a clockwise rotation.

The use of GA for biomedical image registration is quite justified from the experimental results. This method may be well suited for clinical applications. Proper diagnosis and treatment planning may be possible for head CT, MRI, and PET images [11, 12].

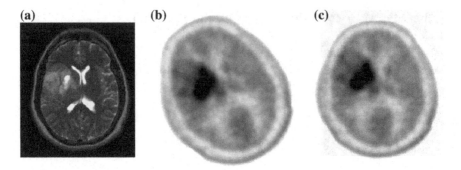

Fig. 4 a MRI brain images (Reference images), **b** PET brain images (Nonaligned images $\theta = 22°$), and **c** registered images

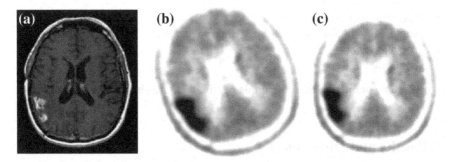

Fig. 5 a MRI brain images (Reference images), **b** PET brain images (Nonaligned images $\theta = 18°$), and **c** registered images

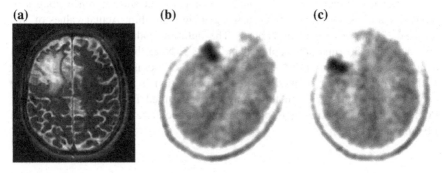

Fig. 6 a MRI brain images (Reference images), **b** PET brain images (Nonaligned images $\theta = -33°$), and **c** registered images

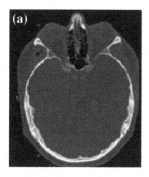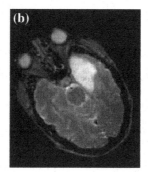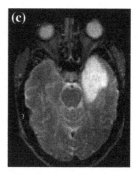

Fig. 7 **a** CT brain image (Reference image), **b** MRI brain image (Nonaligned image θ = 40°), and **c** registered image

Table 1 Actual values of '*R*,' '*S*,' and '*T*'

R	S	T_x	T_y
−15.00	1.10	3.00	1.00
22.00	1.20	2.00	0.00
18.00	1.10	3.00	2.00
−33.00	1.25	2.00	1.00
40.00	1.30	1.00	0.00

Table 2 Optimum values of '*R*,' '*S*,' and '*T*'

R	S	T_x	T_y
−15.15	1.11	3.02	1.02
22.14	1.23	2.01	0.10
18.11	1.12	3.04	2.30
−33.05	1.23	2.03	1.01
40.12	1.31	1.01	0.30

4 Conclusion

We have proposed an effective evolutionary algorithm for biomedical image registration. Here, mutual information is maximized effectively using GA. In this work, we also handle low-texture information very wisely. The initial parameters have been chosen efficiently to get the accurate result. The reference image and the registered image are now ready for fusion. One can extract more information from the fused images, which will be investigated in the future work. The optimization of parameters using other evolutionary computation techniques such as bacteria foraging optimization and Cuckoo search algorithm may be investigated in the future work.

References

1. Gonzalez, R.C., Woods, R.E., Eddins, S.L.: Digital image processing using MATLAB. Pearson, 2010
2. Gonzalez, R.C., Woods, R.E.: Digital image processing, 2nd edn. Pearson (2010)
3. Yamamura, Y., Kim, H., Yamamoto, A.: A method for image registration by maximization of mutual information. In: International Joint Conference, pp. 124–128, 18–21 Oct 2006
4. Lin, J., Gao, Z., Xu, B., Cao, Y., Yingjian Z.: The affection of gray levels on mutual information based medical image registration. In: Proceeding of 26th Annual International Conference of the IEEE EMFS, pp. 27–32, Sept 2004
5. Azzawi, N.A., Abdulla, W.: MRI monomodal feature based registration based on the efficiency of multiresolution representation and mutual information. Am. J. Biomed. Eng. 2(3), 98–104 (2012)
6. Suganya, R., Rajaram, S.: Mutual information and genetic algorithm based registration of MRI brain images. Can. J. Biomed. Eng. Technol. 2(2) (2011)
7. Goldberg, D.E.: Genetic algorithms in search, optimization, and machine learning. Pearson Education Asia (2000)
8. http://www.medicalencyclopedia.com for bio-medical images
9. http://www.Umdnj.edu for medical images
10. http:// www.Jankharia.com for medical images
11. pacs.carestreamhealth.com
12. www.kubtec.com

Bi-dimensional Statistical Empirical Mode Decomposition-Based Video Analysis for Detecting Colon Polyps Using Composite Similarity Measure

Mainak Biswas and Debangshu Dey

Abstract The third leading cause of all deaths from cancer is colorectal cancer (10 % of the total for men and 9.2 % of the total in case of women) (Globocan in Cancer Incidence and Morality Worldwide (2008) [1]. The only prevention is to detect and remove the cancerous adenomatous polyps during optical colonoscopy (OC). This paper proposes bi-dimensional statistical empirical mode decomposition (BSEMD)-based colon polyp detection strategy, wherein a composite similarity measure (CSM) has been used. In this work, separate sets of training and testing samples are opted. Only few samples are randomly chosen for training database, remaining samples make the testing database. The proposed method is implemented on sequences of sample images from an OC video database (Park et al in IEEE Trans. Biomed. Eng 59:1408–1418) [2] provided by American College of Gastroenterology (American College of Gastroenterology, http://gi.org/) [3]. PCA-based feature extraction is used in this work, as it reduces the dimensions efficiently from the main object. The obtained results demonstrate the achieved improvement in the recognition rates, in comparison with other detection procedures.

Keywords Bi-dimensional statistical empirical mode decomposition (BSEMD) · Composite similarity measure (CSM) · Colon polyps (CP) · Optical colonoscopy (OC)

M. Biswas (✉) · D. Dey
Electrical Engineering Department, Jadavpur University, Kolkata, India
e-mail: mainakbiswas041@gmail.com

D. Dey
e-mail: debangshudey80@gmail.com

© Springer India 2015
L.C. Jain et al. (eds.), *Intelligent Computing, Communication and Devices*,
Advances in Intelligent Systems and Computing 309,
DOI 10.1007/978-81-322-2009-1_35

1 Introduction

The fourth most common type of cancer is *Colon Cancer*, which produce almost a million cases per annum. These cancerous tissues are called colon polyps (CP). Generally, there are two types of polyps i.e., *Hyperplastic polyps* and *Adenomatous polyps*. It is not necessary to remove hyperplastic polyps, because they are not premalignant tumor. But adenomatous polyps confer high clinical risk because they are cancerous. The dangerous CP can only be prevented by early detection and removal of cancerous adenomatous polyps during optical colonoscopy (OC) [4]. An OC is a technique by which an endoscopist examines the entire colon with the help of colonoscope, which is very thin, long and at the end of the colonoscope there is a light along with a tiny fiber optic video camera [5]. This widely accepted procedure has some drawbacks, such as high cost, risky to some extent, bad patient-endoscopist ratio, and visualization problem.

Extensive use of bi-dimensional empirical mode decomposition (BEMD) has been found in recent research works in a wide spectrum of state-of-the art applications, especially in the field of image fusion, image compression [6], image de noising [7], and texture analysis [8]. A massive success has been achieved when MR images and CT images are fused using BEMD and dual channel PCNN [9]. BEMD-based fault detection for rotating machines using relevance vector machine (RVM) along with generalized discriminant analysis (GDA) feature extraction, resulting in a 100 % recognition rate [10]. Iris detection can be done using this novel process [11].

In this work, two colonoscopy videos are taken, one is from normal colon video sets and another is the colon polyp video. Each video is divided into number of image samples.

In this work, an adaptive image processing algorithm for detecting colon polyp called bi-dimensional statistical empirical mode decomposition (BSEMD) which is the extension of BEMD, for computing bi-dimensional intrinsic mode functions (IMFs) [12]. While constructing the upper and lower envelopes, BSEMD replaces the 2-D interpolation and follows 2-D smoothing technique.

It is so time consuming to check all the pixel values of an image with another, it needs a feature extraction, by which a new dataset is obtained containing maximum information about that data but in dimensionally reduced form. In this work, PCA-based feature extraction is employed on these processed images.

During video observation, it is found that whenever the probe light falls on the wall of colon, it creates a bright inner circle and a darker outer circle [13]. So, it can be stated that the video frames are captured under varying light intensity [14]. To overcome this problem, *Euclidean-Based Similarity Measure* is not the optimal one. Therefore, to address this difficulty, a composite similarity measure (CSM) is employed.

Some approaches for detecting colon polyp have been proposed earlier. In the year of 2003, Maroulis et al. [15] proposed this detection using wavelet transform-based artificial neural network (ANN) classifier whose detection rate is 95 %.

Using color wavelet covariance texture analysis along with support vector machine (SVM) as the classifier and linear discriminant analysis (LDA) results 97 % specificity and 90 % sensitivity [16].

Ninety-four per cent detection accuracy is achieved when Iakovidis et al. [17] propose gray level with color texture features using SVM classification. By using this proposed method, the best recognition rate of detecting CP from OC image sample is 100 % and the mean (average) is 97.72 %.

The paper is organized as follows: Sect. 2 presents the analysis of empirical mode decomposition (EMD), BEMD, and BSEMD. The mathematical expression of composite similarity analysis and entire proposed method are stated in Sect. 3. Section 4 provides the detection result. Finally, Sect. 5 concludes the paper.

2 Proposed Image Decomposition Algorithm

In this section, at first the conventional 1 dimensional EMD, BEMD algorithm is described and then BSEMD algorithm is used not only for fast decomposition but also for constructing the 2-D smooth upper and lower envelops.

2.1 Empirical Mode Decomposition

The main part of Hilbert–Huang Transform (HHT) is EMD [18]. A nonstationary and nonlinear signal can be decomposed into a number of frequency components adaptively by EMD [6, 18]. These frequency components are called IMFs. Each IMF satisfies these two properties [19]:

(1) The difference between number of extrema and zero-crossings should be zero or one.
(2) At each point, the value of mean created by the upper and lower envelopes defined by the local extrema is zero.

Applying EMD on a given signal $y(t)$ can be decomposed

$$y(t) = \sum_{k=1}^{n} \text{IMF}_k(t) + \text{RES}_k(t)$$

where, $\text{IMF}_k(t)$ is kth IMF and $\text{RES}_k(t)$ is the kth residue (RESs) of that given signal $y(t)$ [20]. RESs are the low frequency trend of the given signal. From these IMFs and RESs, the main signal can be reconstructed again. Here are the basic steps for computing the adaptive signal processing process (1-D EMD) for a given signal $y(t)$ is given in Table 1.

Table 1 EMD

EMD algorithm
1. Identify all the local extrema (maxima, minima) of the given signal $y(t)$
2. Connect all the maxima and create the upper envelope $e_{up}(t)$
3. Connect all the maxima and create the upper envelope $e_{low}(t)$
4. Calculate the mean envelope $e_{mean}(t) = (e_{up}(t) + e_{low}(t))/2$
5. Discard the mean from the main signal to obtain the detail signal, $d(t) = y(t) - e_{mean}(t)$
6. Set $y(t) = d(t)$ and repeat the iterative processes from Step 1 until the resulting detail signal $d(t)$ obeys the stopping criterion
7. Stopping criterion $= \sum_{t=0}^{T} \left\| \frac{d_{1(k-1)}(t) - d_{1k}(t)}{d_{1(k-1)}^2(t)} \right\|^2 \in [0.2, 0.3]$

2.2 Bi-dimensional Empirical Mode Decomposition

BEMD is nothing but the 2-D form of EMD. BEMD decomposes an input image into two parts. First is bi-dimensional Intrinsic Mode Functions (BIMFs) and the second one is RESs [19]. For constructing the envelopes, BEMD follows 2-D interpolation scheme [6]. The name of the procedure is shifting process by which BIMFs are obtained. The steps of that shifting process for a given image $I(x,y)$, $x = 1,\ldots,M; y = 1,\ldots,N$, are given in Table 2.

Table 2 BEMD

BEMD algorithm
1. Initialize the residue image for k = 1, $RES_{(k-1)}(x,y) = RES_0(x,y) = I(x,y)$
2. For j = 1, $h_{j-1}(x,y) = h_0(x,y) = RES_{k-1}(x,y)$
3. Identify all the minima and maxima (extrema) of $h_{j-1}(x,y)$
4. Compute the 2-D upper and lower envelopes, $Z_{up}(x,y)$ and $Z_{low}(x,y)$ using 2-D interpolation respectively
5. Evaluate the mean $Z_{mean}(x,y) = (Z_{up}(x,y) + Z_{low}(x,y))/2)$
6. Discard the mean $Z_{mean}(x,y)$ from residue $h_{k-1}(x,y)$ to obtain the detail image $h_j(x,y) = h_{j-1}(x,y) - Z_{mean}(x,y)$
7. Compute stopping criterion
8. Update j = j + 1, repeat steps 3–step 7 until stopping criterion $< \epsilon, \epsilon \in [0.2, 0.3]$
9. Update $RES_k = RES_{k-1} - IMF_k$, where, IMF_k is the kth IMF
10. Repeat steps 2–step 9 with $k = k + 1$
11. Stopping criterion $= \sum_{t=0}^{T} \left\| \frac{d_{1(j-1)}(t) - d_{1j}(t)}{d_{1(j-1)}^2(t)} \right\|^2$

2.3 Bi-dimensional Statistical Mode Decomposition [12]

BSEMD is the extended version of BEMD. While constructing the upper and lower envelopes, BEMD adopts 2-D interpolation. In case of BSEMD, it obeys 2-D smoothing while constructing the upper and lower envelopes instead of 2-D interpolation [12]. The proposed algorithm is as follows: (Table 3).

The performance of smoothing shifting depends on the right selection of smoothing parameter(λ). For the low value such as $\lambda = 0$, these two envelopes are created by interpolation. Choosing $\lambda = \infty$ the result becomes a straight plane which is not desirable for representing an image. So need for an optimal choice of λ (Fig. 1).

3 Proposed Colon Polyp Detection Scheme

This section presents the full overview of the entire detection procedure. The overall scheme is given below in the Fig. 2.

3.1 Dataset

The two sets of database contain total 453 images of colon polyp (data01) and 168 normal colon images (data02), provided by American College of Gastroenterology [3]. Separate training and testing samples are created. Only few 'training samples'

Table 3 BSEMD [12]

BSEMD Algorithm
1. Initialize the residue image for k = 1, $RES_{(k-1)}(x, y) = RES_0(x, y) = I(x, y)$
2. For j = 1, $h_{j-1}(x, y) = h_0(x, y) = RES_{k-1}(x, y)$
3. Identify all the minima and maxima (extrema) of $h_{j-1}(x, y)$
4. For a given smoothing parameter λ, compute the 2-D upper and lower envelopes, $Z_{\lambda(up)}(x, y)$ and $Z_{\lambda(low)}(x, y)$ using 2-D interpolation, respectively
5. Evaluate the mean $Z_{\lambda(mean)}(x, y) = (Z_{\lambda(up)}(x, y) + Z_{\lambda(low)}(x, y))/2)$
6. Discard the mean $Z_{\lambda(mean)}(x, y)$ from residue $h_{k-1}(x, y)$ to obtain the detail image $h_j(x, y) = h_{j-1}(x, y) - Z_{\lambda(mean)}(x, y)$
7. Check for the stopping criterion
8. Update j = j + 1, repeat steps 3–step 7 until stopping criterion $< \epsilon. \in [0.2, 0.3]$.
9. Update $RES_k = RES_{k-1} - IMF_k$, where, IMF_k is the kth IMF
10. Repeat steps 2–step 9 with $k = k + 1$
11. Stopping criterion $= \sum_{t=0}^{T} \left\| \frac{d_{1(j-1)}(t) - d_{1j}(t)}{d_{1(j-1)}^2(t)} \right\|^2$

Fig. 1 Decomposition technique **a** source image, **b** computed 1st–3rd IMFs and 1st–3rd Residues using BEMD algorithm, **c** computed 3-IMFs and 3-RESs using BSEMD algorithm [19]

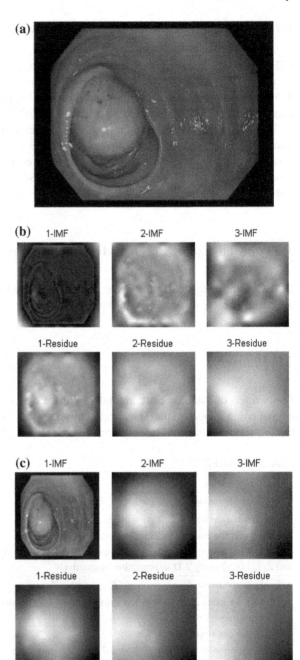

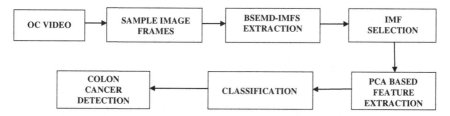

Fig. 2 Proposed colon polyp detection scheme

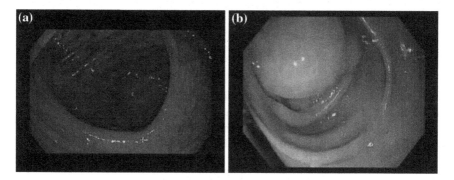

Fig. 3 a Normal colon, b cancerous polyp in colon

are randomly selected from data01 as well as from data02. And remaining images are considered as 'testing samples.' The size of each image is 32 × 32. Figure 3 shows the normal and cancerous polyp colon.

3.2 IMF Selection

The selection of IMF is a challenging task. Figure 4 shows the result of detection rate at various IMFs. Hence, first IMF is chosen for feature extraction.

3.3 Feature Extraction

Feature selection means when an input data is too long to be expressed, it needs for a sub data instead of main object which contain enough information about that object. Extract a subset from an object for constructing model, which will be used further instead of that object is called feature extraction. PCA is performed by solving Eigenvalue problem; here PCA is used for extracting the feature from the first IMF of image samples. This orthogonal linear transform, transforms data to

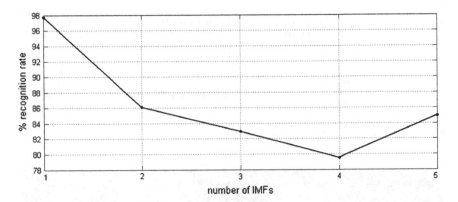

Fig. 4 Recognition rate at various IMFs

another new coordinate system. In the first coordinate, the greatest variance of object lies and the second greatest variance lies in the second coordinate. First coordinate is called first principal component and so on. Covariance matrix (C) of data (D_k) is diagonalizable and is defined as [21]

$$C = \frac{1}{m} \sum_{i=1}^{m} D_i D_i^T$$

where, $D_k \in \Re^n$, $k = [1, 2, \ldots\ldots, m]$ and $\sum_{i=1}^{m} D_k = 0$

3.4 Composite Similarity Measure

During video observation, it is found that whenever the probe light falls on the wall of colon it creates a bright inner circle and a darker outer circle [13]. So, it can be stated that the video frames are captured under varying light intensity [14]. To overcome this problem, *Euclidean-Based Similarity Measure* is not the optimal one. Therefore, to address this difficulty, a CSM is employed. The proposed algorithm is as follows:

$$\text{CSM} = \frac{\log\left(\sum_{i=1}^{n} |p_i - q_i|\right)}{\frac{\sum_{i=1}^{n} (p_i - \bar{p})(q_i - \bar{q})}{\left\{\sum_{i=1}^{n} (p_i - \bar{p})^2\right\}^{1/2} \left\{\sum_{i=1}^{n} (q_i - \bar{q})^2\right\}^{1/2}}}$$

Intensities of two images are $P = \{p_i : i = 1, \ldots, n\}$ and $Q = \{q_i : i = 1, \ldots, n\}$ where, $\bar{p} = \frac{1}{n} \sum_{i=1}^{n} p_i$ and $\bar{q} = \frac{1}{n} \sum_{i=1}^{n} q_i$

For best possible similarity measure, Pearson [22] modified the denominator part of CSM.

$$\text{CSM} = \frac{\log\left(\sum_{i=1}^{n} |p_i - q_i|\right)}{\frac{\frac{1}{n}\sum_{i=1}^{n}(p_i-\bar{p})(q_i-\bar{q})}{\left\{\frac{1}{n}\sum_{i=1}^{n}(p_i-\bar{p})^2\right\}^{1/2}\left\{\frac{1}{n}\sum_{i=1}^{n}(q_i-\bar{q})^2\right\}^{1/2}}}$$

$$= \frac{\log\left(\sum_{i=1}^{n} |p_i - q_i|\right)}{\frac{1}{n}\sum_{i=1}^{n}\left(\frac{p_i-\bar{p}}{\sigma_p}\right)\left(\frac{q_i-\bar{q}}{\sigma_q}\right)}$$

where, $\left\{(p_i - \bar{p})^2\right\}^{\frac{1}{2}} = \left(\frac{(p_i-\bar{p})}{\sigma_p}\right)$ and $\left\{(q_i - \bar{q})^2\right\}^{\frac{1}{2}} = \left(\frac{(q_i-\bar{q})}{\sigma_q}\right)$

So, CSM can simply be written as:

$$\text{CSM} = \frac{\log\left(\sum_{i=1}^{n} |p_i - q_i|\right)}{\frac{1}{n}\overline{P^tQ}}$$

4 Experimental Results

This proposed method is implemented in Matlab (version R2013a) using (Core 2 Duo CPU, 2.66 GHz, 3 GB RAM). To make the experiment harder, the chance of presence of training samples in the testing dataset is fully eliminated. For this datasets, the BSEMD–CSM algorithm is run 20 times, and the mean detection rate along with standard deviation is recorded.

4.1 Time Taken by BSEMD and BEMD

Table 4 illustrates the time taken by BEMD and BSEMD when a source image of size 128×128 is decomposed in 3 IMFs and 3 RESs.

It is clear from Table 4 that time taken for decomposing an image using BSEMD is less compared with BEMD.

Table 4 Time taken by BEMD and BSEMD

Methods	Smoothing parameter (λ)	Time taken (in sec)
BSEMD	20	3.99
BEMD	1	7.38

Table 5 PCA-based feature extraction is employed among lot of benchmark process

Methods	% Recognition rate
PCA	97.72 % ± 0.90

Table 6 Comparative study between CSM and some benchmark classifiers

Classifiers	% Recognition rate
CSM	97.72 % ± 0.90
Manhattan norm	88.89 ± 1.60
Euclidean distance	72.22 ± 6.07
Correlation norm	74.26 ± 8.48
Mahalanobis distance	71.00 ± 6.25

4.2 Recognition Rate at Various Feature Extraction Methods

In this paper, PCA-based feature extraction is employed among lot of benchmark process. Here is a comparative study of that features with PCA is shown in Table 5. PCA-based feature extraction gives the best result among some well-known feature extraction processes.

4.3 Classification Result for Common Type of Classifiers

CSM achieved a massive success in this work. Table 6 illustrates the comparative study between CSM and some benchmark classifiers. In this work, this classifier gives the best recognition rate among others.

4.4 Performance Comparison Based on Best Recognition Rate

Eventually, a comparative study of recognition rates of BSEMD–CSM with respect of other benchmark and well-known methods is given in the Table 7, which shows that the proposed BSEMD–CSM algorithm is a clear winner.

Table 7 Comparative study of recognition rates of BSEMD_CSM

Method	% Recognition rate
BSEMD-CSM (453 polyps)	97.72
Method in [15]	95
Method in [17]	94

5 Conclusion

In this proposed work, a new similarity measure for computing IMFs and RESs in a quick succession for colon polyp detection using BSEMD. The proposed method is implemented on sequences of sample images from an OC video database [2] provided by American College of Gastroenterology. PCA-based feature extraction is used in this work, as it reduces the dimensions efficiently from the main object. Detail experimental results show the improvement achieved by using BSEMD–CSM. However, automatic selections of video frames have not been implemented yet, which is well within the scope of future work.

Acknowledgements The work is supported by University Grants Commission, India (UGC) under the University with Potential for Excellence (UPE), phase II scheme awarded to Jadavpur University, Kolkata, India.

References

1. Globocan: Cancer Incidence and Morality Worldwide, (2008) http://www.iarc.fr/en/media-centre/iarcnews/2010/globocan2008.php
2. Park, S.Y., Sargent, D., Spofford, I., Vosburgh, K.G., Rahim, Y.A.: A colon video analysis framework for polyp detection. IEEE Trans. Biomed. Eng. **59**(5), 1408–1418 (2012)
3. American College of Gastroenterology, http://gi.org/
4. Ganz, M., Yang, X., Slabaugh, G.: Automatic segmentation of polyps in colonoscopic narrow-band imaging data. IEEE Trans. Biomed. Eng. **59**(8), 2144–2151 (2012)
5. Medical Supplies & Equipment Co., http://www.medical-supplies-equipment-company.com/
6. Tian, Y., Zhao, K., Yiping, X., Peng, F.: An image compression method based on the multi-resolution characteristics of BEMD. Comput. Math. Appl. 61(8):2142–2147, Springer (2011)
7. Arfia, F.B.,Sabri, A., Messaoud, M.B., Abid, M.: The bidimensional empirical mode decomposition with 2D-DWT for gaussian image denoising. In: Proceedings of 17th International Conference on Digital Signal Processing, pp. 1–5. Corfu (2011)
8. Pan, J., Zhang, D., Tang, Y.: A Fractal-based BEMD method for image texture analysis. In: IEEE International Conference on Systems Man and Cybernetics (SMC), pp. 3817–3820. Istanbul (2010)
9. Zhang, B., Zhang, C., Wu, J., Liu, H.: A medical image fusion method based on energy classification of BEMD components. Optik **125**(1), 146–153 (2014)
10. Tran, V.T., Yang, B.S., Gu, F., Ball, A.: Thermal image enhancement using bi-dimensional empirical mode decomposition in combination with relevance vector machine for rotating machinery fault diagosis. Mech. Syst. Signal Process. **38**(2), 601–614 (2013). Springer
11. Chen, W.K., Lee, J.C., Han, W.Y., Shis, C.K., Chang, K.C.: Iris recognition based on bi-dimensional empirical mode decomposition and fractal dimension. Inf. Sci. **221**, 439–451 (2013). Springer
12. Kim, D., Park, M., Oh, H.S.: Bidimensional statistical empirical mode decomposition. IEEE Signal Process. Lett. **19**(4), 191–194 (2012)
13. Jolliffe I.T.: Principal component analysis. Series: Springer Series in Statistics, 2nd edn., pp. 487. Springer, New York (2002)
14. Goshtasby, A.A.: Similarity and dissimilarity measures. Image Registration Principles, Tools and Methods, pp. 7–66. Springer London (2012)

15. Maroulis, D.E., Iakovidis, D.K., Karkanis, S.A., Karras, D.A.: CoLD: a versatile detection system for colorectal lesions in endoscopy video-frames. Comput. Methods Programs Biomed, vol. 70, (2), pp. 151–166. Springer-Verlog (2003)
16. Karkanis, S.A., Iakovidis, D.K., Maroulis, D.E., Karras, D.A., Tzivras, M.: Computer-aided tumor detection in endoscopic video using color wavelet features. IEEE Trans. Inf. Tech. Biomed. 7(3), 141–152 (2003)
17. Iakovidis, D.K., Maroulis, D.E., Karkanis, S.A.: An intelligent system for automatic detection of gastrointestinal edennomas in video endoscopy. Comput. Biol. Med. 36(10), 1084–11003 (2006). Springer
18. Huang, N.E., Wu, M.L., Qu, W., Long, S.R., Shen, S.S.: Applications of hilbert-huang transform to non-stationary financial time series analysis. Appl. Stochast. Mod. Bus. Ind. 19(3), 245–268 (2003)
19. Linderhed, A.: 2D empirical mode decompositions-in the spirit of image compression. In: Proceedings SPIE 4738, Wavelet and Independent Component Analysis applications IX, vol. 4738, pp. 1–8 (2002)
20. Ge, G., Yu, L.: Extrema points coding based on empirical mode decomposition: an improved image sub-band coding method. Comput. Elect. Eng. 39(3), 882–892 (2013)
21. Spearman, C.: The proof and measurement of association between two things. J. Psychol. 15(1), 72–101 (1904)
22. Pearson, K.: Contributions to the mathematical theory of evaluation, III, regression, heredity, and panmixia. Philos. Trans. R. Soc. Lond. Ser. A 187, 253–318 (1896)

A Wavelet-Based Text-Hiding Method Using Novel Mapping Technique

Tanusree Podder, Abhishek Majumdar, Lalita Kumari
and Arpita Biswas

Abstract Steganography is a technique of invisible communication through multimedia objects. This paper provides a technique to hide secret messages using a unique mapping technique in discrete wavelet domain (DWT) of an image. In the proposed method, red, green, and blue channels of cover image are taken to apply DWT method. Secret message are hidden only in the vertical and diagonal coefficients of cover channel because these coefficients are less sensitive to human eyes. The reason of using frequency domain technique is to provide higher level of security. Visual difference is null between the cover and stego images. Experimental result shows that the proposed method gives good peak signal-to-noise ratio (PSNR) and low MSE.

Keywords Steganography · Spatial domain · Frequency domain · Discrete wavelet transform · PSNR · MSE

1 Introduction

Secure data transfer become essential with the advancement of internet technology. To provide security, cryptography and steganography [1, 2] are the most widely used techniques. Cryptographic technique concerns with transforming data from one

T. Podder (✉) · L. Kumari
Computer Science and Engineering Department, National Institute of Technology, Agartala, Tripura, India
e-mail: tanusreepodder29@gmail.com

L. Kumari
e-mail: kumari2003@yahoo.co.in

A. Majumdar · A. Biswas
Computer Science and Engineering Department, SSCET, Badhani, Punjab, India
e-mail: abhishekmajumdar91@gmail.com

A. Biswas
e-mail: arpita.abiswas.biswas8@gmail.com

© Springer India 2015
L.C. Jain et al. (eds.), *Intelligent Computing, Communication and Devices*,
Advances in Intelligent Systems and Computing 309,
DOI 10.1007/978-81-322-2009-1_36

form to another, known as encryption whereas the decryption can be done only by intended receiver. On the other hand, steganography is the art of convert communication, means it makes the existence of the data invisible. It uses multimedia object like audios, videos, or image for transferring the secret data. The multimedia object is known as cover object. In the proposed method, we will use image as cover object for secretly sending confidential information. Spatial and frequency domain methods are two categories where steganographic techniques can be applied. Least significant bit technique is the first steganographic method which was in spatial domain. We directly deal with pixel values in spatial domain which increases the chances of attack.

Performance of various steganographic methods can be rated by three parameters: capacity, security, and imperceptibility. So, to offer more security, we are proposing a frequency domain technique, which helps to spread the secret data in the entire image and makes steganalysis more difficult.

2 Literature Review

LSB is one of the widely popular spatial domain [3, 4] steganographic technique. In LSB steganography, the least significant bits of the cover media's digital data are used to conceal the message. The simplest of the LSB techniques is LSB replacement. LSB replacement steganography flips the last bit of each of the data values to reflect the message that needs to be hidden. To embed message in frequency domain, many methods have been proposed based on JPEG, which works on discrete cosine transform such as Jsteg and OutGuess. Present study is going on discrete wavelet transform. Lai and Chang [5] proposed an adaptive data-hiding technique to hide data in horizontal, vertical, and diagonal details based on the coefficient value, but the method needs the original cover image to extract the secret message. Tolba et al. [6] proposed a method which hides data in the LSB's of all coefficient after applying IWT on the cover image. Ghasemi et al. [9] proposed a DWT-Based approach to hide the secret message in 4×4 sub-block of the cover image.

3 Integer-to-Integer Wavelet Transform

Discrete wavelet transform is widely used for steganographic work in frequency domain. One-dimensional Discrete wavelet (DWT) is nothing but a processed work of high- and low-pass filtering on an object, which helps to produce a smoother version of the object by convolution. In 2D-DWT, first 1D-DWT is applied on rows followed by columns, results four level of details as approximation, horizontal, vertical, and diagonal details.

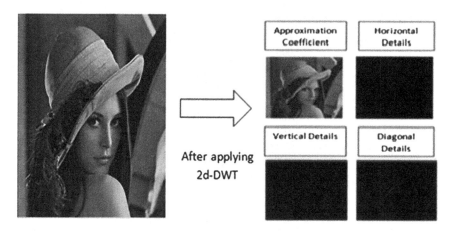

Fig. 1 Integer-to-integer wavelet transform

Normally, image contains Integer value but after applying DWT, the resulting output is a floating point value, which makes reconstruction difficult. So, for our proposed algorithm, we are using integer-to-integer wavelet transform (IWT) [7, 8]. Let the original image I is of size $M \times N$, and red, green, and blue channels of I are IR, IG, and IB, respectively. Now, IWT on IR gives us

$$IRA_{i,j} = floor(IR_{2i,2j} + IR_{2i+1,2j/2})$$
$$IRH_{i,j} = IR_{2i+1,2j} - IR_{2i,2j}$$
$$IRV_{i,j} = IR_{2i,2j+1} - IR_{2i,2j}$$
$$IRD_{i,j} = IR_{2i+1,2j+1} - IR_{2i,2j}$$

where IRA, IRH, IRV, and IRD are nothing but the approximation, horizontal, vertical and diagonal sub-image of the red plane of cover image as shown in Fig. 1.

4 Proposed Method

In the proposed algorithm, sender needs to input secret message and a cover image. A new mapping technique is applied on secret message. Mapped secret message gets embedded into the IWT coefficient of cover image with the help of embedding algorithm. For extraction of the secret message, user needs to input the stego image. Figure 2 shows the embedding and extraction process of proposed algorithm.

We are converting each character of secret message into 8-bit binary and constructing a binary message array of secret message. Let the secret message which need to be embedded is of k-bit as $b_0, b_1, \ldots, b_{k-2}, b_{k-1}$. Now, we will take 4

Table 1 Predefined mapping table

4-bit secret message	Mapped ASCII value	8-bit binary value
0000	35	00100011
0001	38	00100110
0010	64	01000000
0011	87	01010111
0100	125	01111101
0101	91	01011011
0110	184	10111000
0111	198	11000110
1000	224	11100000
1001	181	10110101
1010	223	11011111
1011	246	11110110
1100	174	10101110
1101	190	10111110
1110	200	11001000

bits from the binary array (e.g., b_0, b_1, b_2, and b_3) and map it into a predefined ASCII value. In this way, there can be 16 combinations of mapping possible from 0000 to 1111. We will map each 4 bit of secret message into an ASCII value means we will map each 4 bit into 8 bit value. After mapping select coefficients from vertical and diagonal sub-image of red (IRV, IRD), green (IGV, IGD), blue (IBV, IBD) hide 1 bit at a time in the 2nd LSB of all coefficient. If the first bit of the message is matching with the 2nd-lsb of the coefficient, then we will not change the coefficient value else, and if the message bit is 0, then we will make 2nd lsb as 0 and 1st as 1. If the message bit is 1, then we will make 2nd lsb as 1 and 1st as 0 (Table 1).

To find the data close to the original, minimum-error replacement method is used. In our proposed algorithm, replacing 1-bit data in the ith LSB will create maximum error of $\pm 2i - 1$ and minimum 0.

4.1 Embedding Process

1. Let the secret message is "I will come to meet you on 21st dec, time and place will be same as previous."
2. Converting a short part of secret message "I will" in binary, we will get "01001001 0010000011101110110100101101100011011100."
3. Take the first 4-bit b_0, b_1, b_2, b_3, and map it according to the predefined table. Repeat the same process until whole secret message is mapped (Fig. 3).

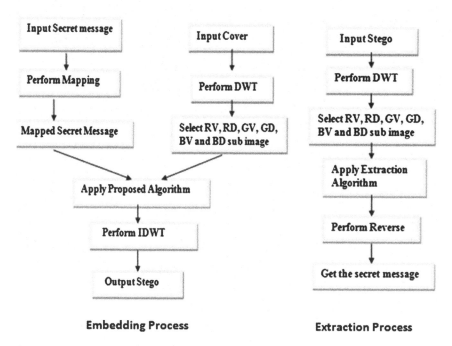

Embedding Process   **Extraction Process**

Fig. 2 Block diagram of embedding and extraction process

Fig. 3 Mapping technique of secret message

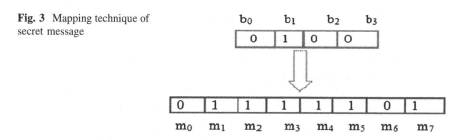

4. After mapping, we will get the mapped secret message as "0111110110110 1010100000001000111100011011000110101110001011010101101110001010 11101011100010101110."

5. Now, apply IWT on red channel of cover image to get RA, RH, RV, RD sub-image. Similarly, apply IWT on green and blue channel. Take a coefficient P from the sub-image RV and select the value at 2nd LSB if it matches with the 1st bit m_0 of the mapped secret message then keep the coefficient value same. The secret message is hidden based on different cases described below.

Case 1:   while $m_0 = 0$ and coefficient $P_1 = 01010100$, 2nd LSB is 0, and it is matching with m_0. So, we would not change the coefficient value (Fig. 4).

Fig. 4 Case 1 of proposed method

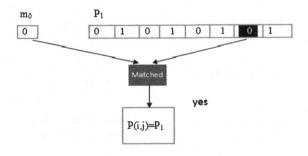

Fig. 5 Case 2 of proposed method

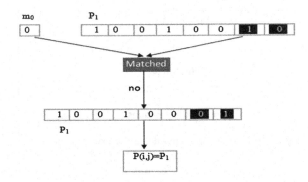

Case 2: while $m_0 = 0$ and coefficient $P_1 = 10010010$, 2nd LSB is 1 and it is not matching with m_0. So, we will change the 2nd LSB as 0, and if the 1st LSB is 0 then make it 1 (Fig. 5).

Case 3: while $m_0 = 0$ and coefficient $P_1 = 10010011$, 2nd LSB is 1 and it is not matching with m_0. So, we will change the 2nd LSB as 0, and if the 1st LSB is 1 then no change (Fig. 6).

Case 4: while $m_0 = 1$ and coefficient $P_1 = 10010001$, 2nd LSB is 1 and it is not matching with m_0. So, we will change the 2nd LSB as 0, and if the 1st LSB is 0, then make it 1 (Fig. 7).

Case 5: while $m_0 = 1$ and coefficient $P_1 = 10010000$, 2nd LSB is 0 and it is not matching with m_0. So, we will change the 2nd LSB as 1, and if the 1st LSB is 0, then make no change (Fig. 8).

6. Repeat step-5 until all mapped secret message $m_0, m_1, \ldots, m_{2l-1}$ are embedded in RV, RD, GV, GD, BV, BD sub-images of red, green and blue channel of cover image.

7. The above process gives stego image as output.

Extraction process is reverse of embedding process.

Fig. 6 Case 3 of proposed method

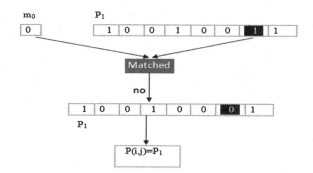

Fig. 7 Case 4 of proposed method

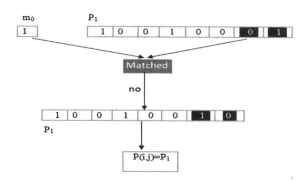

Fig. 8 Case 5 of proposed method

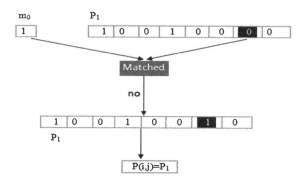

5 Result and Analysis

For analysis of the proposed algorithm, we took "peppers.png" as cover image and 1, 2, 3 KB data as secret message. After applying the proposed algorithm, we got good-quality image. Mean square error (MSE) and peak signal-to-noise ratio

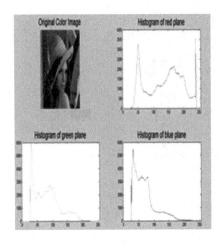 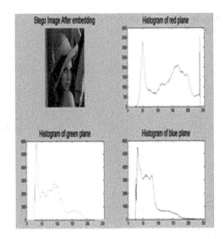

Fig. 9 Histogram of cover and stego images

(PSNR) is used to measure the image quality after embedding the secret message. Histogram of the Cover image is shown in Fig. 9 and stego image is in Fig. 9 after hiding 3 KB as secret data.

5.1 Mean Square Error

Higher poor-quality stego image is produced after embedding the secret message if MSE gives higher value (Table 2).

5.2 Peak Signal-to-Noise Ratio

Higher PSNR value indicates the proposed algorithm produced good-quality stego image (Table 3).

Table 2 Result of MSE after applying on cover and stego images

Size of secret message (KB)	MSE (red channel)	MSE (green channel)	MSE (blue channel)
1	0.0087	0.0089	0.0087
2	0.0175	0.0176	0.0175
3	0.0267	0.0261	0.0261

Table 3 Result of PSNR after applying on cover and stego images

Size of secret message (KB)	PSNR (red channel)	PSNR (red channel)	PSNR (blue channel)
1	68.7880	68.7880	68.7842
2	65.7422	65.7422	65.7444
3	63.9036	63.9036	64.0060

5.3 Histogram Analysis

It is clear from Fig. 9 that there is no visual difference between cover and stego images.

6 Conclusion

Frequency domain is mainly used to give better security. With the help of integer wavelet transform, we are spreading the secret data over the entire image. Proposed method does not directly store the secret message still the result shows it gives better PSNR ratio implies low error rate. The error rate is minimized by changing 8th LSB according to the secret message and cover image value which gives maximum of $\pm 2i - 1$ and minimum of 0 as error. Histogram of both cover and stego images is identical. So, we can say histogram analysis also gives better result.

References

1. Li, B., He, J., Huang, J., She, Y.Q.: A survey on image steganography and steganalysis. J. Inf. Hiding Multimedia Signal Process. **2**(2), 142–172 (2011)
2. Kumar, A., Pooja, K.: Steganography–a data hiding technique. Int. J. Comput. Appl. **9**(7), 19–23 (2010)
3. Swain, G., Lenka, S.K.: LSB Array Based Image Steganography Technique by Exploring the Four Least Significant Bits, pp. 479–488. Springer, Berlin Heidelberg (2012)
4. Roque, J.J., Minguet, J.M.: SLSB: Improving the Steganographic Algorithm LSB. WOSIS, INSTICC Press, pp. 57–66 (2009)
5. Lai, B.-L., Chang, L.-W.: Adaptive Data Hiding for Images Based on Haar Discrete Wavelet Transform, pp. 1085–1093. Springer, Berlin Heidelberg (2006)
6. Tolba, M.F., Ghonemy, M.A., Taha, I.A., Khalifa, A.S.: Using Integer Wavelet Transforms in Colored Image-Steganography. IJICIS, vol. 4(2) (2004)

7. Calderbank, A.R., Daubechies, I., Sweldens, W., Yeo, B.-L.: Lossless image compression using integer to integer wavelet transform. In: International Conference on Image Processing. IEEE Press, vol. 1, pp. 596–599 (1997)

8. Bilgin, A., Sementilli, P.J., Sheng, F., Marcellin, M.W.: Scalable image coding using reversible integer wavelet transform. IEEE Trans. Image Process. **9**(11), 1972–1977 (2000)

9. Ghasemi, E., Shanbehzadeh, J., Fassihi, N.: High capacity image steganography based on genetic algorithm and wavelet transform. Intelligent Control and Innovative Computing. Springer Science and Business Media, Berlin, pp. 395–404 (2012)

Unsupervised Segmentation of Satellite Images Based on Neural Network and Genetic Algorithm

P. Ganesan, V. Rajini, B.S. Sathish and V. Kalist

Abstract Segmentation is one of the important processes in image analysis to extract the necessary but hidden information in the image. The success of the image analysis is based on the outcome of the segmentation process. There are number of methods proposed for the segmentation of satellite images. Soft computing approaches such as fuzzy logic, neural networks, and genetic algorithm are most widely used for the segmentation of satellite images. But, every method has its own advantages and disadvantages. In this paper, a new approach based on the combination of genetic algorithm and feed forward neural network is proposed for the segmentation of satellite images. In this process, the genetic algorithm selects and feeds the fittest individual to the neural network. This feed forward network performs the segmentation. The computational cost is drastically reduced due to this cooperative and parallel approach. Experimental result illustrates the efficiency of the proposed approach.

Keywords Segmentation · Clustering · Image enhancement · Genetic algorithm · Neural network

P. Ganesan (✉) · B.S. Sathish · V. Kalist
Department of Electronics and Control Engineering, Sathyabama University, Chennai, India
e-mail: gganeshnathan@gmail.com

B.S. Sathish
e-mail: subramanyamsathish@yahoo.co.in

V. Kalist
e-mail: kalist.v@gmail.com

V. Rajini
Department of Electrical and Electronics Engineering, SSN College of Engineering, Chennai, India
e-mail: rajiniv@ssn.edu.in

© Springer India 2015
L.C. Jain et al. (eds.), *Intelligent Computing, Communication and Devices*,
Advances in Intelligent Systems and Computing 309,
DOI 10.1007/978-81-322-2009-1_37

1 Introduction

Image segmentation is the process in which an image is split into number of sub-images, called clusters, based on any one of the image characteristics. This low level but important process plays important role in the success of the image analysis [1, 2]. Satellite images find applications in monitoring resources, military, and commercial applications. The images collected from satellite are useful and have enormous amount of data. It is very difficult to segregate the necessary information from the image [3]. In this paper, the segmentation of satellite images based on the combination of genetic algorithm and neural network is proposed. Both GA and NN have been individually applied for the image segmentation but none of them are successful [4]. Even though, GA is the best suited for optimization problems, this algorithm takes more time for convergence, i.e., suffers from slowness [5]. Similarly, the selection of best individual is very difficult for neural network. The genetic algorithm works on a population (for example, image) using various operators (crossover, mutation, etc.) that applied to the population [6]. In GA, population is a collection or set of points (pixels in image) in the search space. The GA solver generates the initial population by default. Based on the fitness of the individuals in the current generation, the next generation population is computed. The selection of fittest individuals for the next generation is based on the outcome from the genetic operations: crossover and mutation. The best individuals after some generation (iteration) is given to the feed forward two-layer neural network that is trained by Levenberg–Marquardt algorithm. The segmented image as output produced by the network is reasonable, if the following conditions are satisfied: (i) a smaller value of mean square error (MSE), (ii) the validation and test set error have the similar characteristics, and (iii) the best validation performance. The remainder of the paper is organized as follows. The genetic algorithm and feed forward network are explained in the Sect. 2. The Sect. 2 explained the methodology used for the segmentation and detection of water resources in satellite images. The experimental results are discussed in detail in Sect. 3. The Sect. 4 concluded the paper.

2 Methodology

The genetic algorithm (GA) is an approach based on Darwin's theory of natural selection that is applied for solving optimization problems [5, 7]. GA searches for a minimum of a function, i.e., minimize the fitness function [8, 9]. The GA solver generates the initial population by default. Based on the fitness of the individuals in the current generation, the next generation population is computed. The initial population is generated by using the population size and the range of the initial population. In GA solver, the default population size is 20, but this is not sufficient

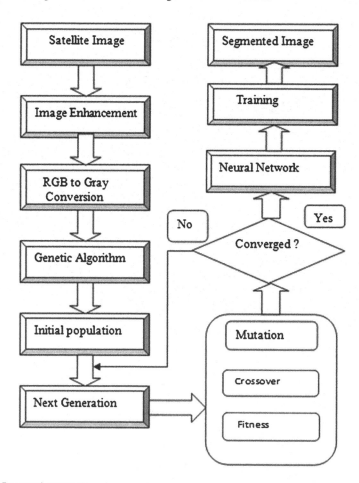

Fig. 1 Proposed approach

for real-life problems with very large number of variables. For a smaller problems (less number of variables), a smaller population size of 8 or 10 is enough. The default initial population is created using a uniform random number generator in a default range of [0, 1].

Feed forward neural network can be used for multi-dimensional mapping problems such as input–output fitting problems [4, 10]. In this work, the network is trained by the default Levenberg–Marquardt algorithm. For a small problem with few hundred weights, this algorithm will have fastest convergence as compared to other training algorithms. The accurate training and obtaining lower MSE than other training algorithms are the significant advantages of this Levenberg–Marquardt

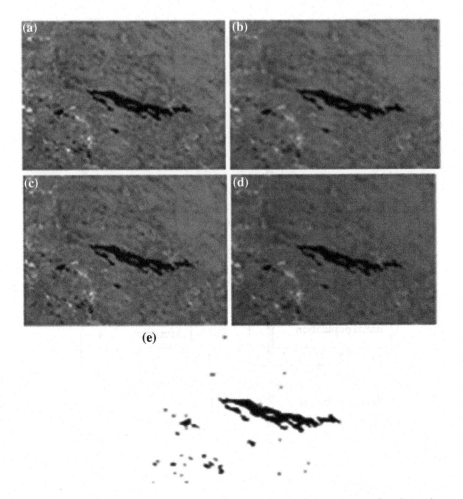

Fig. 2 Segmentation using proposed approach **a** input image, **b** input image after sharpening, **c** input image after smoothening, **d** intensity image, and **e** segmented image

algorithm. In this feed forward network, the hidden layer has tan-sigmoid transfer function and the output layer is with linear transfer function.

Figure 1 shows the proposed approach for the segmentation of satellite images. The images received from satellite are stored in a database after initial processing. The test image is enhanced by two image enhancement process: sharpening and smoothening. The test image is sharpened using unsharp masking in spatial domain. For this sharpening process, the lower limit is kept as 10 and the upper limit is 100. Then, the sharpened image is smoothened by mean filter of mask size 3. The intensity image is applied to genetic algorithm (GA) solver to find the best fittest point (pixels in image). The selection of fittest individuals for the next generation is based on the outcome from the genetic operations: crossover and

(a)   **(b)**

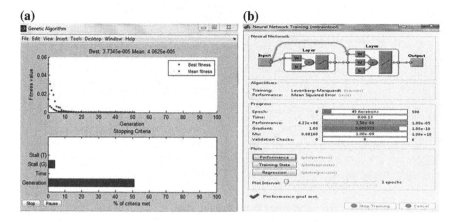

Fig. 3 **a** Genetic algorithm for the selection, and **b** neural network training

mutation. The best individuals after some generation (iteration) are given to the
feed forward two-layer neural network, which is trained by Levenberg–Marquardt
algorithm. After some iteration, the segmented image as output produced by the
network is reasonable, if the following conditions are satisfied: (i) a smaller value
of MSE, (ii) the validation and test set error have the similar characteristics, and
(iii) the best validation performance.

3 Experimental Result and Discussion

The image, Salmon River reservoir, shown in Fig. 2 is collected from the NASA's
landsat imagery and taken by ETM + multispectral sensor on July 21, 1985. This
image is taken in bands 1, 2, 5, 7 with spectral range of 0.450–2.35 μm and pixel
resolution of 30 m. In 1914, the Salmon River reservoir was created in New York.
This multipurpose reservoir is a major source for electrical power generation, flood
control, and recreation. Landsat satellite continuously monitors this reservoir since
1985 and sends the data. Figure 3 shows the GA and neural network processes for
the selection and training of fittest individuals.

GA stops after 50 generations, and the best fittest pixels computed from the GA
solver are given to neural network. The best training performance is achieved at
epoch (iteration) 49. If the validation error enhanced for six iterations, the network
will be stopped. The output produced by the network is reasonable, if the following
conditions are satisfied: (i) a smaller value of MSE, (ii) the validation and test set
error have the similar characteristics, and (iii) the best validation performance (no
significant over fitting). The performance curve of training and training state is
depicted in Fig. 4.

In experiment 2, Lake Ayakkum, which is located in central china, is segmented using proposed method as shown in Fig. 5.

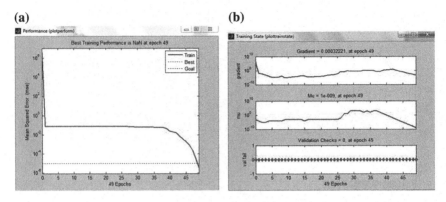

Fig. 4 **a** Performance curve of training, and **b** training state

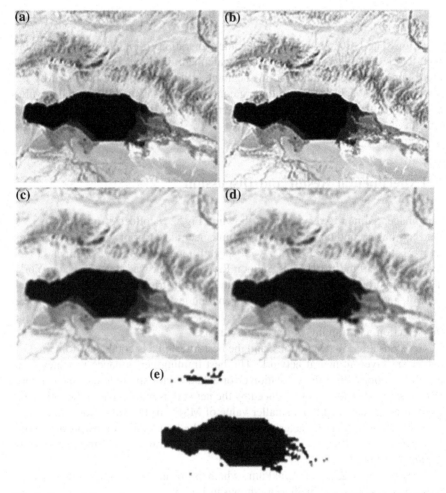

Fig. 5 Segmentation using proposed approach **a** input image, **b** input image after sharpening, **c** input image after smoothening, **d** intensity image, and **e** segmented image

Table 1 Comparison of segmentation result

Image	Best fitness	Iteration	MSE	Execution time (s)
Salmon River	3.7345e−005	49	3.50e−06	13
Lake Ayakkum	2.6093e−005	30	3.72e−06	05

Table 2 Comparison of proposed method with GA and NN based on execution time

Image	Proposed method	GA	NN
Salmon River	13	62	31
Lake Ayakkum	05	47	24

4 Conclusion

The segmentation of satellite images based on the combination of genetic algorithm and feed forward neural network is proposed. In this approach, the genetic algorithm finds the fittest individuals and presents to the neural network. This feed forward neural network performs the segmentation. The computational cost is drastically reduced due to this cooperative approach. The experimental result clearly shows the efficiency of the proposed approach. Table 1 shows that MSE is minimum in both experiments. The smaller value of MSE results in good segmentation. From Table 2, we can easily conclude that the computational cost (execution time) for the proposed method is very less as compared to the feed forward neural network or genetic algorithm. The performance of the network can be improved by any one of the following approaches: (i) either increasing number of training vectors or hidden neurons, (ii) reinitialize the network with new values of weights and biases, (iii) increasing the number of input values, and (iv) use different training algorithms.

References

1. Ganesan, P., Rajini, V.: A method to segment color images based on modified fuzzy-possibilistic-c-means clustering algorithm. In: Recent Advances in Space Technology Services and Climate Change (RSTSCC), 2010. IEEE (2010)
2. Gonzalez, R.C.: Digital Image Processing, 2nd edn. Prentice Hall of India (2006)
3. Awad, M., Chehdi, K., Nasri, A.: Multi-component image segmentation using a hybrid dynamic genetic algorithm and fuzzy C-means. IET Image Process. 3(2), 52–62 (2009)
4. Awad, M., Chehdi, K., Nasri, A.: Multi-component image segmentation using genetic algorithm and artificial neural network. IEEE Geosci. Remote Sens. Lett. 4(4), 571–575 (2007)
5. Goldberg, D.E.: Genetic Algorithms in Search. Optimization and Machine Learning. Addison-Wesley, New York (1989)

6. Bhanu, S. Lee, S., Ming, J.: Adaptive image segmentation using a genetic algorithm. IEEE Trans. Syst. Man Cybern. 1543–1567 (1995)
7. Zhang, H.Z., Xiang, C.B., Song, J.Z.: Application of Improved Adaptive Genetic Algorithm to Image Segmentation in Real-time. Optics and Precision Engineering, pp. 333–336 (2008)
8. Farmer, M.E., Shugars, D.: Application of genetic algorithms for wrapper-based image segmentation and classification. In: IEEE Congress on Evolutionary Computation, pp. 1300–1307 (2006)
9. Feitosa, R.Q., Costa, G.A.O.P., Cazes, T.B.: A genetic approach for the automatic adaptation of segmentation parameters. In: OBIA06 (2006)
10. Aria, E., Saradjian, M., Amini, J., Lucas, C.: Generalized occurrence matrix to classify IRS-1D images using neural network. In: Proceedings of XXth ISPRS Congress, Turkey, pp. 117–123 (2004)

Comparative Analysis of Cuckoo Search Optimization-Based Multilevel Image Thresholding

Sourya Roy, Utkarsh Kumar, Debayan Chakraborty, Sayak Nag, Arijit Mallick and Souradeep Dutta

Abstract The entropy image thresholding technique is much in demand today for image segmentation. Furthermore, population algorithm aided thresholding techniques have been proven previously to be extremely effective in producing better results. In this work, we have concentrated on the minimum cross-entropy criterion for image segmentation. The objective of this work is to demonstrate the capability of Cuckoo Search Optimization-based Minimum Cross-Entropy Technique. The algorithm has been compared against old algorithms GA and PSO. Results have been assimilated in this work. The results have clearly demonstrated the competence of Cuckoo Search Optimization algorithm in assisting Cross Entropy-based thresholding procedure.

Keywords Cuckoo search optimization · Minimum cross-entropy thresholding · PSNR · Levy flight · Metaheursitic algorithm · Threshold · Image segmentation

S. Roy (✉) · U. Kumar · D. Chakraborty · S. Nag · A. Mallick · S. Dutta
Department of Instrumentation and Electronics Engineering, Jadavpur University, Salt Lake Campus, LB-8, Sector 3, Kolkata, West Bengal, India
e-mail: souroy099@gmail.com

U. Kumar
e-mail: utkarsh.iee.ju@gmail.com

D. Chakraborty
e-mail: debayan.jaduniv@gmail.com

S. Nag
e-mail: sayak.nag9@gmail.com

A. Mallick
e-mail: aribryan@gmail.com

S. Dutta
e-mail: duttasouradeep39@gmail.com

© Springer India 2015
L.C. Jain et al. (eds.), *Intelligent Computing, Communication and Devices*,
Advances in Intelligent Systems and Computing 309,
DOI 10.1007/978-81-322-2009-1_38

1 Introduction

In image processing, image is treated as a two-dimensional signal, $f(x, y)$ [1]. The pixels are smallest indivisible part of the image. Every pair of x and y, the function points a particular pixel, and returns the intensity value of image at that point. The aim of image processing techniques is to draw information from the pixel and/or its surrounding pixels, and then generate a new value for the given pixel. Image segmentation is one of the basic image processing techniques. The process divides pixels into groups as per threshold values assigned for segmentation and then assigns every group a fixed intensity value irrespective of the previous intensity held in the pixels. There are various methods of determining the level/quality of segmented image. One of the widely accepted methods is minimum cross-entropy technique (MCET) [2–4]. In this method, the entropy between original image and the segmented image (cross entropy) is calculated. Lower the entropy value, better segmentation has been performed.

Metaheuristic computing is used to solve nonlinear problems containing multiple constraints, where a linear approach would be either extremely difficult or impossible to implement. Metaheuristic algorithm aided image segmentation has found to be a very efficient. Algorithms like PSO have been used previously with segmentation process to make it more proficient [5]. In this work, we have demonstrated the ability of one such algorithm, the CSO in optimizing the segmentation process. Its performance standards have been compared against old algorithms GA and PSO. The capability of CSO is clearly presented by the obtained results.

2 Minimum Cross-Entropy Technique

Cross Entropy was proposed by Kullback [2] and is defined by

$$D(F, G) = \sum_{i=1}^{N} f_i \log \frac{f_i}{g_i} \tag{1}$$

where, $F = \{f_1, f_2, \ldots, f_N\}$ and $G = \{g_1, g_2, \ldots, g_N\}$ are two probability distributions on the same set. What the MCET algorithm does is that it selects those thresholds for which the cross entropy between the original image and the resulting image is minimum. Let I be the original image and $h(i), i = 1, 2, \ldots, L$ be the corresponding histogram with L being the number of gray levels. Thus

$$I_t(x, y) = \begin{cases} \mu(1, t) & I(x, y) < t \\ \mu(t, L + 1) & I(x, y) \geq t \end{cases} \tag{2}$$

where, I_t is the image obtained after thresholding the original image using t as the threshold value.

$$\mu(a, b) = \sum_{i=0}^{b-1} ih(i) / \sum_{i=a}^{b-1} h(i) \tag{3}$$

The calculation of the cross entropy is then done by

$$D(t) = \sum_{i=1}^{L} ih(I)\log(i) - \sum_{i=1}^{t-1} ih(i) \log[\mu(1, t)] - \sum_{i=t}^{L} ih(i)\log[\mu(t, L+1)]^* \tag{4}$$

The selection of optimal threshold t^* is done, by the MCET, in such a way that the cross entropy by Eq. (4) is minimum.

$$t^* = \operatorname{argmin}_t\{D(t)\} \tag{5}$$

Given the first term is constant, the objective function thus becomes,

$$\begin{aligned}
\eta(t) &= -\sum_{i=1}^{t-1} ih(i) \log[\mu(1, t)] - \sum_{i=1}^{L} ih(i) \log[\mu(t, L+1)] \\
&= -\left[\sum_{i=1}^{t-1} ih(i) \log\left(\frac{\sum_{i=1}^{t-1} ih(i)}{\sum_{i=1}^{t-1} h(i)}\right)\right] - \left[\sum_{i=t}^{L} ih(i) \log\left(\frac{\sum_{i=t}^{L} ih(i)}{\sum_{i=t}^{L} h(i)}\right)\right] \\
&= -m^1(1, t) \log\left[\frac{m^1(1, t)}{0 m^0(1, t)}\right] - m^1(t, L+1) \log\left[\frac{m^1(t, L+1)}{m^0(t, L+1)}\right]
\end{aligned} \tag{6}$$

where, $m^0(a, b) = \sum_{i=a}^{b-1} h(i)$ and $m^1(a, b) = \sum_{i=a}^{b-1} ih(i)$ are the zero-moment and first-moment on partial range of the image histogram, respectively.

To ensure that the computational time is as less as possible the two dummy thresholds $t_0 = 1, t_{c+1} = L + 1$ are used, and an assumption of n thresholds, denoted by $t_1, t_2, t_3, \ldots, t_c$, is made. As a result, the objective function then becomes

$$\eta(t_1, t_2, \ldots, t_n) = -\sum_{i=}^{c+1} m^1(t_1, t_2, \ldots, t_i) \log\left[\frac{m^1(t_{i-1}, t_i)}{m^0(t_{i-1}, t_i)}\right] \tag{7}$$

Using our proposed algorithm, we have tried to obtain this optimum c-dimensional vector $[t_1, t_2, t_3, \ldots, t_c]$ for which Eq. (7) will be minimum [2–4].

3 Applied Algorithms

3.1 Genetic Algorithm

Simple Genetic Algorithm procedure:

- An initial population of strings is chosen to begin with.
- An evaluation for quality level of every single element is initiated.
- A certain quality-level achievement or a limiting time period is imposed as a termination condition for this evaluation.
- New elements are formed by crossover or mutation operations as offspring.
- Again an evaluation for quality check for the newly bred elements is initiated. Elements with least quality level are then replaced by better newly bred elements.

3.2 Particle Swarm Optimization

The particle swarm optimization procedure

- A population of particles is created whose position, velocity, and quality levels are constantly monitored.
- Each particles' position are determined according to the objective function.
- A particles position is updated if its current position is better than its previously determined position.
- Determine the best particle according to the particle's previous best positions.
- Particles' velocity is updated according to:
- $velocity_{ex} = velocity^0 + \varphi \cdot \left(p_{-}^{best}pos^0\right) + \emptyset \cdot \left(g^{best} - pos^0\right)$
- where, $velocity^0$ = current velocity, p^{best} = current best position, pos^0 = current position, g^{best} = global best position, $\varphi, \emptyset$ = random values in range: [0, 1]. Move particles to their new positions according to: $pos = pos^0 + velocity_{ex}$
- Loop to step 2 still stopping criteria are reached.

3.3 Cuckoo Search Optimization

The Cuckoo Search Algorithm draws its inspirations from the unique breeding pattern of cuckoos. In the algorithm, the eggs laid by cuckoos represent the solutions at particular iteration. With every advancing loop, the number of solutions (eggs) keeps on decreasing [6–8]. The new number of solutions with every passing iteration is determined by the Levy flight distribution, which is based and inspired from random flight pattern of birds. When the required threshold is reached, the algorithms return the values of the then available solutions which represent the optimized value of function parameters. The pseudo code of CSO:

- Objective function $f(x) = (x1,...,)T$;
- Initial a population of n host nests xi $(i = 1, 2, ... n)$;
- While $(t < \text{MaxGeneration})$ or (stop criterion);
- Get a cuckoo (say i) randomly by Levy flights; evaluate its quality/fitness Fi;
- Choose a nest among n (say j) randomly;
- If $(Fi > Fj)$, replace j by the new solution; end;
- Abandon a fraction (pa) of worse nests (And build new ones at new locations via Levy flights);
- Keep the best solutions (or nests with quality solutions)
- Rank the solutions and find the current best;
- End while [6–8]

4 Methodology and Results

All the three algorithms GA, PSO, and CSO start the segmentation process by selecting a threshold value at random, based on which first segmentation is performed. The entropy is then calculated between original and segmented image and the value is fed back to the algorithm. Depending upon the entropy value, the algorithm either chooses a new threshold value or stops the iterations to generate the best possible segmented image. The generated images for comparative analysis have been presented as follows. Benchmark parameters have been tabulated after the images.

4.1 Original Images

See Figs. 1 and 2.

4.2 Bi-Level Thresholding

See Figs. 3, 4, 5, 6, 7 and 8.

4.3 Tri-Level Thresholding

See Figs. 9, 10, 11, 12, 13 and 14.

Fig. 1 Cameraman

Fig. 2 Kids

4.4 Four Level Thresholding

See Figs. 15, 16, 17, 18, 19 and 20.

4.5 Five Level Thresholding

See Figs. 21, 22, 23, 24, 25, 26 and Table 1.

4.6 Tabulation of the Parametric Values of Generated Image Results

Fig. 3 GA aided MCET

Fig. 4 PSO aided MCET

Fig. 5 CSO aided MCET

Fig. 6 GA aided MCET

Fig. 7 PSO aided MCET

Fig. 8 CSO aided MCET

Fig. 9 GA aided MCET

Fig. 10 PSO aided MCET

Fig. 11 CSO aided MCET

Fig. 12 GA aided MCET

Fig. 13 PSO aided MCET

Fig. 14 CSO aided MCET

Fig. 15 GA aided MCET

Fig. 16 PSO aided MCET

Fig. 17 CSO aided MCET

Fig. 18 GA aided MCET

Fig. 19 PSO aided MCET

Fig. 20 CSO aided MCET

Fig. 21 GA aided MCET

Fig. 22 PSO aided MCET

Fig. 23 CSO aided MCET

Fig. 24 GA aided MCET

Fig. 25 PSO aided MCET

Fig. 26 CSO aided MCET

Table 1 Parametric Values of generated image results

Level of threshold	Image ALGORITHM	Threshold values Cameraman	Kids	PSNR Cameraman	Kids	Time (s) Cameraman	Kids
Bi-level	GA	69	57	20.1344	14.8606	.535	.709
	PSO	66	60	20.0438	15.0129	.621	.721
	CSO	66	63	20.0438	15.0685	.043	.028
Tri-level	GA	57; 133	41; 154	23.9681	19.9358	.603	.753
	PSO	51; 137	51; 137	23.9134	20.3472	.655	.722
	CSO	53; 139	46; 160	24.0168	20.4041	.071	0.093
Four level	GA	29; 75; 127	30; 83; 165	24.4019	22.1728	1.121	.995
	PSO	30; 83; 144	15; 72; 166	25.6654	21.8650	.950	1.101
	CSO	34; 95; 150	27; 91; 167	25.8035	22.2426	.170	.176
Five level	GA	29; 56; 83; 153	16; 47; 88; 178	25.8050	22.6489	1.521	1.323
	PSO	29; 76; 125; 158	10; 51; 106; 171	27.4734	23.4628	1.556	1.667
	CSO	35; 86; 130; 166	10; 51; 106; 171	27.5158	23.6722	385	.265

5 Conclusion

CSO aided segmentation produces better segmented images than GA or PSO aided segmentation. The quality of the image can be judged from their PSNR values. The CSO algorithm is also very fast compared to GA and PSO thus saving on computational time while also producing more efficient images. This relatively more efficient method can be applied in various spheres of medical imaging [9], machine vision, fault detection, feature detection, etc. With growing industrial standards an algorithm which is both faster and efficient in producing results is in huge demand. CSO aided segmentation can be applied in the aforementioned disciplines and even more now that it has been demonstrated that it is better than any of the previous applied algorithm for optimized image segmentation.

The work from this point may progress in two ways. The capability of CSO can be tested with other spheres of image processing [5, 10, 11] or with the development/creation of a better optimization technique, the new algorithm can be applied with MCET, and it may generate better results than CSO.

References

1. Gonzalez, R.C., Woods, R.E., Eddins, S.L.: Digital image processing using MATLAB, 2nd edn. Prentice Hall, New Jersey (2002)
2. Kullback, S., Leibler, R.A.: On information and sufficiency. Ann. Math. Stat. **22**(1), 79–86 (1951)
3. Pal, N.R.: On minimum cross-entropy thresholding. Pattern Recognit. **29**(4), 575–580 (1996)
4. Brajevic, I., Tuba, M., Bacanin, N.: Multilevel image thresholding selection based on the cuckoo search algorithm. In: Proceedings of the 5th International Conference on Visualization, Imaging and Simulation (VIS'12), Sliema, Malta (2012)
5. Yin, P.-Y.: Multilevel minimum cross entropy threshold selection based on particle swarm optimization. Appl. Math. Comput. **184**(2), 503–513 (2007)
6. Yang, X.-S., Deb, S.: Cuckoo search via Levy flights. In: Proceedings of World Congress on Nature & Biologically Inspired Computing (NaBIC 2009), December 2009, pp. 210–214, India. IEEE Publications, USA (2009)
7. Yang, X.-S., Deb, S.: Engineering Optimisation by Cuckoo Search. Int. J. Math. Model. Numerical Optim. **1**(4), 330–343 (2010)
8. Brajevic, I., Tuba, M.: Cuckoo search and firefly algorithm applied to multilevel image thresholding. In: Cuckoo Search and Firefly Algorithm. Studies in Computational Intelligence, vol. 516, pp. 115–139. Springer, Heidelberg (2014)
9. Manjunath, A.P., Rachana, C.S., Ranjini, S.: Retinal vessel segmentation using local entropy thresholding. In: Emerging Research in Electronics, Computer Science and Technology. Lecture Notes in Electrical Engineering, vol. 248, pp. 1–8 (2014)
10. Agrawala, S., Pandaa, R., Bhuyana, S., Panigrahib, B.K.: Tsallis entropy based optimal multilevel thresholding using cuckoo search algorithm. Swarm Evol. Comput. **11**, 16–30 (2013)
11. Bhandari, A.K., Singh, V.K., Kumar, A., Singh, G.K.: Cuckoo search algorithm and wind driven optimization based study of satellite image segmentation for multilevel thresholding using Kapur's entropy. Expert Syst. Appl. **41**(7), 3538–3560 (2014)

An Intelligent Method for Moving Object Detection

Mihir Narayan Mohanty and Subhashree Rout

Abstract Detection of activities of moving objects is a challenging problem for its promising applications. Emerging research topic on computer vision includes with detection on many applications to reduce the computation cost, simple and faster the object. In this paper, we present a motion control method for mobile robots in indoor environments based on color object detection. Probing over a digitized image of robots taken at top view to uniquely identify them is not quite an easy task. The recognition process involves scanning a digitized image and characterizing it, which is made difficult by varying illumination, position, and rotation. Furthermore, the vision system is plagued with inherent difficulties that cannot be completely controlled. Effects such as lighting and shadows, lens focus, and even quantum electrical effects in the sensor chip combine to make it essentially impossible to guarantee that the color being tracked down would remain constant as the robot traverses the exploration field. Among the different recognition cues, like shape, size, position, and motion, this paper focuses on color as the primary discriminating feature. After identification, the robots are operated wirelessly by interfacing with wireless module and motor driver.

Keywords Object detection · Color detection · Intelligent method · Fuzzy logic

M.N. Mohanty (✉)
ITER, Sikshya O Anusandhan University, Bhubaneswar, Odisha, India
e-mail: mihirmohanty@soauniversity.ac.in

S. Rout
Department of Computer Science and Application, Vani Vihar, Utkal University,
Bhubaneswar, Odisha, India
e-mail: subhashree22662@gmail.com

© Springer India 2015
L.C. Jain et al. (eds.), *Intelligent Computing, Communication and Devices*,
Advances in Intelligent Systems and Computing 309,
DOI 10.1007/978-81-322-2009-1_39

343

1 Introduction

In the vision-based system, the moving object has the functions like velocity, control, position control, obstacle avoidance, and so on. Here, moving objects are considered as the movable robots in the indoor environment. Along with the detection, the control is taken care, keeping the vision toward the vehicles in traffic. This control operation may be implemented to those vehicles for particular applications. The host computer processes vision data and calculates next behaviors of robots according to strategies commands to the robots using a RF modem. The robots make their moves according to these commands keeping away from obstacles autonomously. The robots have sensors for position control (encoders) and obstacle avoidance (IR sensors). The calculation load of the host computer is much less compared with the former system.

The next step in the video analysis is tracking, which can be simply defined as the creation of temporal correspondence among detected objects from frame to frame. This procedure can provide temporal identification of the segmented regions and generates cohesive information about the objects in the monitored area such as trajectory, speed, and direction.

The algorithms are implemented and tested on different robot platforms (detecting motion of external objects from a moving robot is the subject of active research). There are two independent motions involved: (i) the motion of the robot and (ii) the motions of moving objects in the environment.

To extract the maximum benefit from this recorded digital data, detection of any moving object from the scene is needed without engaging any human eye to monitor things all the time. Real-time segmentation of moving regions in image sequences is a fundamental step in many vision systems.

The paper is organized as follows. Section 2 describes the related literature, and Sect. 3 explains the method of detection. Section 4 exhibits the result; simultaneously, Sect. 5 concludes the piece of work.

2 Related Literature

Object recognition, detection, and tracking in video has found a wide area of application in the field of transport, retail, physical security, environmental monitoring, smart space, and ambient intelligence. Wide range of algorithms using various types of data sets are being used by many researchers. A number of methods have been used for detection and recognition by authors [1–8].

In [1], authors used an approach for video detection and tracking. Their algorithm was based on indexing to search a tracked vehicle. A comprehensive quantitative analysis was performed using real surveillance data. Wu et al. in [2] used an adaptive method for detecting and tracking vehicles in environments like sunny, shadowing, and in dark during night. The method was based on dynamic segmentation. Shan et al. [3] has taken the support of probabilistic method.

Moving target's fundamental frequency (MTFF) method has been developed for the application of motion parameters estimation and is useful for the pitch extraction of moving targets. Zhu et al. in [5] used visual tracking method for measuring the speed of a moving vehicle within a structured environment using stationary stereo cameras. It was integrated within the framework of particle filtering for tracking processes. Adaptive method was proposed to track congested vehicles. On most of the work, the algorithm is based on TMD method [6].

Hidetomo Sakaino et al. [7] used an extended Markov Chain Monte Carlo (MCMC) method for tracking and an extended hidden Markov model (HMM) method for learning/recognizing multiple moving objects in videos.

3 Methodology

It proposes a way to devise an algorithm deciding whether coordinated motions exist for the mobile robots along their paths so that each robot can reach its own goal without colliding with the other ones. The model uses an n-dimensional coordination space exploiting the cylindrical structure of the diagram obstacles. It can help determine the robot groups that can cause a deadlock (avoiding any movement in that path).

To detect the surrounding objects, it includes range or position of the objects, and sometimes, it considers the movement of the object. This can be said to be passive detection as it is not required to fit the equipment to the objects being detected. The major advantage of this method is all objects are visible and can be detected potentially.

In the overall architecture, the overhead camera sends the image to the computer via USB connection. The image signal is traced by the program and processed, and the information is extracted on the basis of which the control signals are given to the transceiver circuit via serial port connection .The transceiver circuit then sends the signals (byte) serially to the receiver present on the robot wirelessly to the microcontroller. The microcontroller present on the robot has been programmed to interpret the signal and activate the particular pin's triggering movement which may be of forward/backward, left/right, etc. After successful execution of a particular instruction, the position of the robot has now changed on the arena, so the entire coordinates have to be revisited; hence, the above process continues till the robot reaches its destination.

3.1 Image Thresholding

Due to many factors like effect of lightning, the values of the components on the whole surface of a particular colored object are not same; therefore, ranges are defined to identify one color in the image from other color [10, 11].

A video can be imagined as a continuous flow of images, i.e., many frames of images played one after another with a high speed. In preprocessing stage, image filtering to remove the noise from the video for a better analysis and image enhancement for resizing and many others to serve to respective problems have been performed.

Thresholding result of the image follows the procedure of image enhancement by the help of zero padding technique. In this case, the surface of a particular color is given a constant value of three components, i.e., after finding range for a color, the range is replaced by a single color with constant values of RGB. In zero padding, it leaves the colored pixels of our concern, i.e., obstacle, destination, robot circles, and all other pixels that are padded with zero. So all other pixels are converted to perfect black and so other small noisy light effects and other colors of background are eliminated.

3.2 Detection of Geometrical Center

For tracking of the machine (robot) and other objects like destination and obstacles, it is required to find the geometrical center of colored figures that identify them. In dealing with moving robots, it is necessary to find the center dynamically; that is, the center is found out continuously at run time. The flowchart for transmission and reception is shown in Fig. 1.

For navigation purpose, it is connected through the UART communication. The robots in this experiment have predefined destination and obstacles but do not have any predefined or fixed path for reaching to them or avoiding the obstacles. For navigation, basic coordinate geometry concepts are used. In this case, the two robots, its front and back obstacles, and arena consist of different colors.

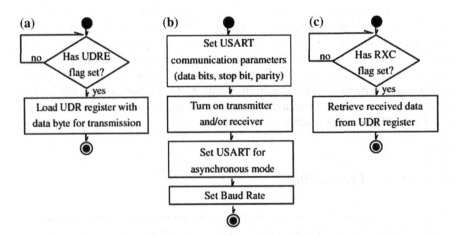

Fig. 1 Flowchart for transmission and reception

The movement of robot is described as follows.

i. Move straight: both wheels moving forward (in front direction)
ii. Move left: right wheel moving forward, whereas left wheel is either stopped or moving backward.
iii. Move right: left wheel moving forward, whereas right wheel is either stopped or moving backward.
iv. Move back: both wheels moving backward (in back direction).

The navigation algorithm consists of two parts

i. Destination following algorithm.
ii. Obstacle avoidance (repelling) algorithm.

Both these algorithms are combined and are run dynamically, i.e., continuously to determine the final path of the robots. It follows the rule-based system.

Algorithm:

For one-robot system

Step 1: First centers of the front and back are found out.
Step 2: line of orientation of robot was found out by joining the centers.
Step 3: center of destination and obstacle is found out.
Step 4: position of destination is checked with respect to line of orientation of robot.
Step 5: robot is made to move toward the destination.
Step 6: continuously, centers of robot, destination, and obstacle are determined along with the line of orientation of robot.
Step 7: distances of robot to destination and obstacle are continuously measured.
Step 8: if robot gets near to the obstacle (distance from obstacle decreases below a certain distance), robot is moved backward and moved away from obstacle.
Step 9: after robot reaches the destination, the robot is stopped.

3.3 For Two-Robot System

The algorithm for two-robot system is same as that of the one-robot system except here there are two different destinations one for each robot. And here, one robot is an obstacle for the other robot. The algorithm continues till both the robots reach their respective destinations. The method is fuzzy rule-based system and is explained as follows [12]. The elements in membership function have a varying degree of membership in a set. The membership values are obtained by mapping the values obtained for a particular parameter onto a membership function. The membership functions are listed in Table 1. Here, triangular membership function is used. It is defined by

Table 1 Membership values
of membership function

Linguistic variable	Fuzzy set
Direction	{Left, Right, Front}
Distance	{low, medium, high}
Handoff decision	{no change, Detected}

Fig. 2 Fuzzy inference
system

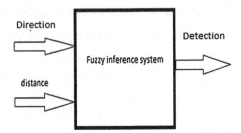

$$\text{trimf}(x; a, b, c) = \max\left(\min\left(\frac{x-a}{b-a}, \frac{c-x}{c-b}\right), 0\right) \quad (1)$$

The fuzzified data will be passed to the inference engine and is matched against
a set of fuzzy rules using fuzzy techniques to produce output fuzzy sets. By using
the fuzzy sets of the inputs, the IF-THEN rules are implemented. In the next step,
the output fuzzy sets are passed to the defuzzifier for computation of output.
Centroid method of defuzzification returns the center of area under the curve. If
you think of the area as a plate of equal density, the centroid is the point along the
x-axis about which this shape would balance.

The fuzzy rules provide knowledge base to the system and results for accurate
detection (Fig. 2).

4 Result and Discussion

The decision of robot movement is determined according to the following conditions. For destination, the following algorithm is used (Fig. 3).

If dist1<30: stop
Else if dist3+dist1-dist2<0.70: move straight
Else

Yc1 < Yc2	d1 > 0	d > 0	Turn Right		Yc1 > Yc2	d1 > 0	d > 0	Turn Left
		d < 0	Turn Left				d < 0	Turn Right
	d1 < 0	d > 0	Turn Left			d1 < 0	d > 0	Turn Right
		d < 0	Turn Right				d < 0	Turn Left
Yc1 == Yc2	Xc1 > Xc2		d > 0	Turn Left				
			d < 0	Turn Right				
	Xc1 < Xc2		d > 0	Turn Right				
			d < 0	Turn Left				

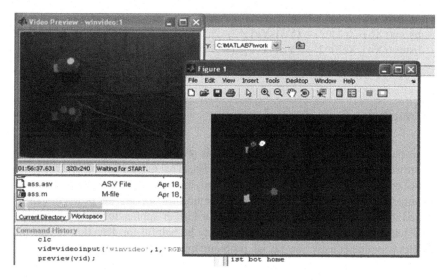

Fig. 3 Detected output based on color

To avoid the obstacle, the following algorithm is developed.

Step 1: If distb − dista < 45 and dista < 45 then move backward, pause for .5 sec and move left, pause for .5 sec
Step 2: check for destination
Center of robot:

Xc1= x-coordinate of center of 1st circular patch, i.e., head of robot
Xc2= x-coordinate of center of 2nd circular patch, i.e., tail of robot
Yc1= y-coordinate of center of 1st circular patch, i.e., head of robot
Yc2= y-coordinate of center of 2nd circular patch, i.e., tail of robot

Center of destination:

Xg1= x-coordinate of center of destination
Yg1= y-coordinate of center of destination
Center of obstacle
Xo1= x-coordinate of center of obstacle
Yo1= y-coordinate of center of obstacle

Slopes and distances

m1= ((Yc1-Yc2)/(Xc1-Xc2)): slope of orientation line of robot
m2= ((Yc1-Yg1)/(Xc1-Xg1)): slope of line from head of robot to destination

d=Yg1-m1*Xg1-Yc1 + m1*Xc1.
dist2=((Xg1-Xc1)^2 + (Yg1-Yc1)^2)^.5;
dist1=((Xg1-Xc2)^2 + (Yg1-Yc2)^2)^.5;

dist3=((Xc2-Xc1)^2 + (Yc2-Yc1)^2)^.5;
dista=((Xo1-Xc1)^2 + (Yo1-Yc1)^2)^.5;
distb=((Xo1-Xc2)^2 + (Yo1-Yc2)^2)^.5;

5 Conclusion

In this paper, a multiple detection approach is presented. Initial phase of the work is hardware design for the robots as vehicles. The movement of the robots is controlled and detected using the proposed method. It has been successfully implemented and verified. This proposed method for solving the problem can be applied for vehicle detection in traffic jam for further applications. As the number of vehicles grow day by day, the method may be helpful for detection of specific vehicle; simultaneously, the social problem may be solved by this alternative method. Still, there is the future work for equal color condition, and the object must be detected.

References

1. Feris, R.S., Siddiquie, B., Petterson, J., Zhai, Y., Datta, A., Brown, L.M., Pankanti, S.: Large-scale vehicle detection, indexing, and search in urban surveillance videos. IEEE Trans. Multimedia **14**(1), 28–42 (2012)
2. Wu, B.-F., Juang, J.-H.: Adaptive vehicle detector approach for complex environments. IEEE Trans. Intell. Transp. Syst. **13**(2), 817–827 (2012)
3. Shan, M., Worrall, S., Nebot, E.: Probabilistic long-term vehicle motion prediction and tracking in large environments. IEEE Trans. Intell. Transp. Syst. **14**(2), 539–552 (2013)
4. Gopalan, R., Hong, T., Shneier, M., Chellappa, R.: A learning approach towards detection and tracking of lane markings. IEEE Trans. Intell. Transp. Syst. **13**(3), 1088–1098 (2012)
5. Zhu, J., Yuan, L., Zheng, Y.F., Ewing, R.L.: Stereo visual tracking within structured environments for measuring vehicle speed. IEEE Trans. Circuits Syst. Video Technol **22**(10), 1471–1484 (2012)
6. Dan, T., Lei, J., Yang, Y.: A robust approach for congested vehicles tracking based on tracking-model-detection framework. IEEE Trans. Circuits Syst. Video Technol. **23**(10), 820–824 (2013)
7. Sakaino, H.: Video-based tracking, learning, and recognition method for multiple moving objects. IEEE Trans. Circuits Syst. Video Technol. **23**(10), 1661–1674 (2013)
8. O'Malley, R., Jones, E., Glavin, M.: Rear-lamp vehicle detection and tracking in low-exposure color video for night conditions. IEEE Trans. Intell. Transp. Syst. **11**(2) (2010)
9. Cucchiara, R., Piccardi, M., Mello, P.: Image analysis and rule-based reasoning for a traffic monitoring system. IEEE Trans. Intell. Transp. Syst. **1**(2), 119–130 (2000)
10. Kar, S.K., Mohanty, M.N.: Statistical approach for color image detection. In: IEEE International Conference on Computer Communication and Informatics (ICCCI), 2013, pp. 1–4, 4–6

11. Mohanty, M.N., Kar, S.K., Mohanty, B.: Color detection and resistor evaluation in real time environment. In: IEEE International Conference on Control, Instrumentation, Energy and Communication, University of Calcutta, Calcutta, WB, India (2014)
12. Pattnaik, L., Mohanty, M.N., Mohanty, B.: An intelligent method for handoff decision in next generation wireless network, SEMCCO, In: Swarm, Evolutionary, and Memetic Computing, LNCS 7677, pp. 465–475, © Springer, Berlin (2013)

Prediction of Monthly Rainfall in Tamilnadu Using MSARIMA Models

S. Meenakshi Sundaram and M. Lakshmi

Abstract One of the most important problems in the hydrological cycle is the prediction of rainfall. Many researchers are working hardly into it, but are still unable to get a perfect model because of its unsure and unexpected variation. Understanding the variability is very much needed because of its vast applications in real-life scenario. The prediction of seasonality is very much essential with respect to the nature of the data. In this paper, we have framed a Seasonal Autoregressive Integrated Moving-Average model with the help of a data set of sea surface temperature for a period of 59 years with a total of 708 readings.

Keywords Seasonality · Sea surface temperature · Prediction

1 Introduction

Time series analysis is an important tool in modeling and forecasting. Even though there are a lot of models, Box and Jenkins ARIMA models are very popular and very effective in the environmental analysis. The association between the southwest and northeast monsoon rainfall (NEMR) over Tamilnadu has been examined for a 100-year period from 1877 to 1976 through a correlation analysis by Dhar and Rakhecha [1]. The average rainfall series of Tamilnadu for the northeast monsoon months of October to December and the season as a whole were analyzed for trends, periodicities, and variability using standard statistical methods by Dhar et al. [2]. Balachandran et al. [3] examined the local and teleconnective association between NEMR over Tamilnadu and global surface temperature anomalies (STA) using the monthly gridded STA data for the period 1901–2004. The trends, periodicities, and variability in the seasonal and annual rainfall series of Tamilnadu were analyzed by Dhar et al. [4].

S. Meenakshi Sundaram (✉) · M. Lakshmi
Faculty of Computing, Sathyabama University, Chennai, India
e-mail: sundarambhu@rediffmail.com

© Springer India 2015
L.C. Jain et al. (eds.), *Intelligent Computing, Communication and Devices*,
Advances in Intelligent Systems and Computing 309,
DOI 10.1007/978-81-322-2009-1_40

Fig. 1 Geographical
location of area of study,
Tamilnadu

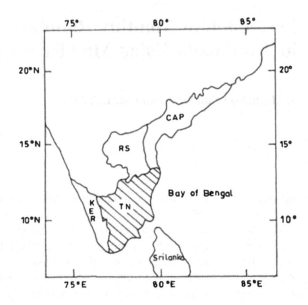

2 Study Area and Materials

Tamilnadu stretches between 8°5′ and 13°35′ N by latitude and between 78° 18′ and
80°20′ E by longitude (Fig. 1). Tamilnadu receives rainfall in both the southwest and
northeast monsoon. Agriculture is more dependent on the northeast monsoon. The
rainfall during October to December plays an important role in deciding the fate of
the agricultural economy of the state. Another important agroclimatic zone is the
Cauvery river delta zone, which depends on the southwest monsoon. Tamilnadu
should normally receive 979 mm of rainfall every year. Approximately 33 % is from
the southwest monsoon, and 48 % is from the northeast monsoon (Fig. 1).

A dataset containing a total of 59-year (1950–2008) monthly rainfall totals of
Tamilnadu was obtained from Indian Institute of Tropical Meteorology (IITM),
Pune, India. The monthly sea surface temperature of Nino 3.4 indices was obtained
from National Oceanic and Atmospheric Administration, United States, for a
period of 59 years (1950–2008) with 708 observations.

3 Methodology

3.1 Box–Jenkins Multivariate Seasonal ARIMA Model

ARIMA models are powerful tools in predicting a time series and are widely used
in hydrology cycles, mainly surface water processes and stream flow events. It is a
useful technique in rainfall prediction due to ease of development and

implementation. The ARIMA models are a combination of autoregressive models and moving-average models [5]. The autoregressive models AR(p) base their predictions of the values of a variable x_t on a number p of past values of the same variable with number of autoregressive delays $x_{t-1}, x_{t-2}, ..., x_{t-p}$ and include a random disturbance etc. The moving-average models MA(q) generate predictions of a variable x_t based on a number q of past disturbances of the same variable prediction errors of past values $e_{t-1}, e_{t-2}, ..., e_{t-q}$. The combination of the autoregressive and moving-average models AR(p) and MA(q) generates more flexible models called ARMA (p, q) models. The stationarity of the time series is required for the implementation of all these models. In 1976, Box and Jenkins proposed the mathematical transformation of the non-stationary time series into stationary time series by a difference process defined by an order of integration parameter d. This transforms ARMA (p, q) models for non-stationary-transformed time series into the ARIMA (p, d, q) models. The ARIMA models are built with strategies estimation, diagnosis, and prediction. The model is valid through the graphs ACF and PACF. The estimated parameters are always tested using statistics.

Next in the diagnosis process, the autocorrelation of the residuals from the estimated model should be sufficiently small and should resemble white noise. Once the model is identified, it is used to predict the monthly rainfall series. To understand the meteorological information and integrate it into planning and decision-making process, it is important to study the temporal characteristic and predict lead times of the rainfall of a region. The seasonal ARIMA (p, d, q)(P, D, Q)$_s$ model is defined as

$$\phi_p(B)\Phi_P(B^s)\nabla^d\nabla_s^D y_t = \Theta_Q(B^s)\theta_q(B)\varepsilon_t \tag{1}$$

where

$$\phi_p(B) = 1 - \phi_1 B - \cdots - \phi_p B^p, \theta_q(B) = 1 - \theta_1 B - \cdots - \theta_q B^q$$
$$\Phi_P(B^s) = 1 - \Phi_1 B^s - \cdots - \Phi_P B^{sP}, \Theta_Q(B^s) = 1 - \Theta_1 B^s - \cdots - \Theta_Q B^{sQ} \tag{2}$$

ε_t denotes the error term, ϕ's and Φ's are the non-seasonal and seasonal autoregressive parameters, and θ's and Θ's are the non-seasonal and seasonal moving-average parameters.

4 Results and Discussions

A time series is said to be stationary if its underlying generating process is based on constant mean and constant variance with its autocorrelation function (ACF) essentially constant through time. The ACF is a measure of the correlation between two variables composing the stochastic process, which are k temporal lags

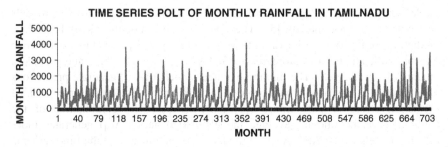

Fig. 2 Time series plot of monthly rainfall in Tamilnadu (1950–2008)

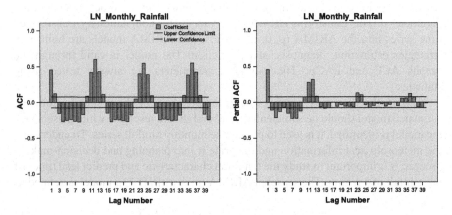

Fig. 3 ACF and PACF correlograms

far away, and the partial autocorrelation function (PACF) measures the net correlation between two variables, which are k temporal lags far away.

The simple time series plot of monthly rainfall in Tamilnadu is given in Fig. 2.

The stationarity of the rainfall data is visualized in the above Fig. 2. If the ACF is zero within two lags, we say that the time series is stationary. From Fig. 3, it is clear that the monthly rainfall is stationary with seasonality.

Seasonality is defined as a pattern that repeats itself over fixed intervals of times. For a stationary data, seasonality can be found by identifying those autocorrelation coefficients of more than two or three time lags that are significantly different from zero. Since the monthly rainfall series consists seasonality of order $s = 12$, the model considered here is a seasonal multivariate ARIMA model (Table 1).

The ACF and PACF correlograms (Fig. 3) and the coefficients are analyzed carefully, and the tentative multivariate ARIMA model chosen is ARIMA $(1, 0, 1)(1, 1, 1)_{12}$. The parameter's estimates are tabulated in Table 2.

After fitting the appropriate ARIMA model, the goodness of fit can be examined by plotting the ACF of residuals of the fitted model. If most of the autocorrelation

Table 1 ACF and PACF coefficients for the first five lags

Lag	ACF coefficient	PACF coefficient
1	0.457	0.457
2	0.123	−0.108
3	−0.155	−0.215
4	−0.268	−0.127
5	−0.250	−0.069

Table 2 Parameter's estimates of ARIMA $(1, 0, 1)(1, 1, 1)_{12}$ model

Model	Parameters	Parameter's estimates
ARIMA $(1, 0, 1)$ $(1, 1, 1)_{12}$	AR	0.475
	MA	0.392
	AR, seasonal	0.072
	MA, seasonal	0.965

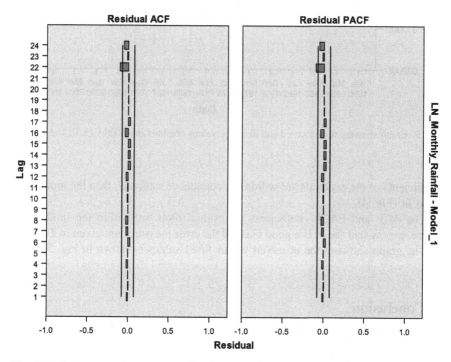

Fig. 4 Residual plots of ACF and PACF of ARIMA $(1, 0, 1)(1, 1, 1)_{12}$ model

Table 3 Error measures obtained for the model ARIMA $(1, 0, 1)(1, 1, 1)_{12}$

Error measures	Value
RMSE	1.082
MAPE	16.081
R squared	0.547

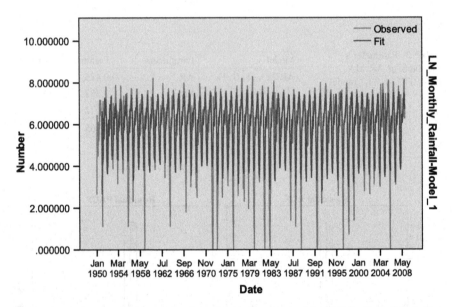

Fig. 5 Graph showing the observed and the fitted values obtained in ARIMA $(1, 0, 1)(1, 1, 1)_{12}$ model

coefficients of the residuals are within the confidence intervals, then the model is a good fit (Fig. 4).

The ACF and PACF coefficients in residual plots are within the limits, and hence, we say that the fit is a good one and the error measures are given in Table 3.

The graph showing the observed versus fitted values is shown in Fig. 5.

5 Conclusion

In recent days, due to the ever-increasing demand for rainfall, the related predictions are vital. Effective prediction models are useful to obtain initial and earlier ideas for the management of rainfall. The seasonal autoregressive model is a very effective tool with a predictor sea surface temperature as evidenced from the least error measures.

References

1. Dhar, O.N., Rakhecha, P.R.: Foreshadowing northeast monsoon rainfall over Tamilnadu. Mon. Weather Rev. **3**(1), 109–112 (1983)
2. Dhar, O.N., Rakhecha, P.R., Kulkarni, A.K.: Fluctuations in northeast monsoon rainfall of Tamilnadu. Int. J. Climatol. **2**(4), 339–345 (1982)
3. Balachandran, S., Asokan, R., Sridharan, S.: Global surface temperature in relation to northeast monsoon rainfall over Tamilnadu. J. Earth Syst. Sci. **115**(3), 349–362 (2006)
4. Dhar, O.N., Rakhecha, P.R., Kulkarni, A.K.: Trends and fluctuations of seasonal and annual rainfall of Tamilnadu. Proc. Indian Acad. Sci. (Earth Planet. Sci.) **91**(2), 97–104 (1982)
5. Box, G.E.P., Jenkins, G.M., Reinsel, G.C.: Time Series Analysis—Forecasting and Control, 3rd edn. Pearson Education, Inc (1994)

An Ontology-Based Approach for Test Case Reuse

Satish Dalal, Sandeep Kumar and Niyati Baliyan

Abstract Building software test cases is a costly and time-consuming process; hence, good-quality test cases are stored in libraries, for possible reuse. Existing methods for search and retrieval of reusable test cases are not flexible enough, since they do not consider semantics. We propose a test case reuse approach based on the semantic Web technique—ontology, which improves flexibility and reusability during test case generation. Moreover, the addition of semantics to test cases handles users' queries better.

Keywords Ontology · Software testing · Test cases · Reuse

1 Introduction

Test cases are stored as reusable components in libraries, after software testing is performed successfully. We can reduce testing cost and time and also improve the quality of software through component reuse [1].

Currently used methods of retrieval of prebuilt components are based on attribute–value pair, keyword, and predefined rules among others. None of these considers semantics of components while retrieving them. Although some techniques such as mapping and dictionary search provide some semantics, yet they do not provide enough flexibility to handle users' complex queries and reasoning.

S. Dalal (✉) · S. Kumar · N. Baliyan
Department of Computer Science and Engineering, Indian Institute of Technology, Roorkee, India
e-mail: satishdalal2007@gmail.com

S. Kumar
e-mail: sandeepkumargarg@gmail.com

N. Baliyan
e-mail: niyati.baliyan@gmail.com

© Springer India 2015
L.C. Jain et al. (eds.), *Intelligent Computing, Communication and Devices*,
Advances in Intelligent Systems and Computing 309,
DOI 10.1007/978-81-322-2009-1_41

The notion of ontology is successfully leveraged in the testing phase of software development life cycle (SDLC). Ontology can be thought of as a vocabulary which is used to describe a specific domain and relationships that exist among the concepts of that domain [2]. We propose a test case reuse approach based on ontology, which is more flexible and handles user queries better.

2 The Proposed Approach

In our approach, ontology is used to search and retrieve the required concepts with the help of semantics existing between these concepts. Here, we will search for concepts, subconcepts, and data type properties. In addition, we will check for possible reuse of test cases. Our approach consists of four steps—build ontology, find concepts, find data properties, and check range of data properties. The output obtained from each step serves as an input to the next step. Each of these steps is described briefly, below.

2.1 Build Ontology

Our problem definition starts with identification of the software artifact for which test cases are to be built and reused. Next, we search for a possible conceptual matching in existing ontologies; in case of failure, we build our own ontology for the same. Building ontology is a simple process if we know about the software components. This step can also be automated by using software requirements [3].

In order to describe further steps, we will use ontology as shown in Fig. 1, i.e., 'test' ontology and 'sample' ontology. 'Email,' 'username,' and 'password' are the data type properties for 'login' concept. We will apply our approach to check for reusable test cases for 'authentication' concept.

2.2 Find Concepts

In this step, we will search for the concept for which we have to build test cases in other ontologies. A concept may contain subconcepts; therefore, we will search in the entire inheritance relationship. In the above example, we have the following concepts in the 'test' ontology set.

$$T_{con} = \{\text{Authentication, Login}\}$$

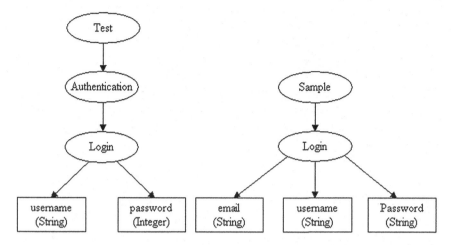

Fig. 1 Ontology for authentication concept

We will search for these concepts in 'sample' ontology set by finding the concepts of 'sample' ontology and taking intersection with concepts in the 'test' ontology set.

$$S_{con} = \{Login\}$$
$$T_{con} \cap S_{con} = \{Login\}$$

If no ontology concept or subconcept is found to be matching, then we will search some other ontology for a possible match. On the other hand, if a match is found, then we will move to the third phase of finding properties.

In Fig. 1, 'login' is common in both sets. Therefore, in the next phase, we will find the data type properties for 'login' concept in both ontologies.

2.3 Find Properties

After successfully finding a concept match in the previous step, we will search for data type properties match in the matching concepts. Since the matching concepts may further be divided into subconcepts, we will look for data type properties of matching concepts as well as of their subconcepts.

As shown in Fig. 1, after this step, the 'login' concept is common in both ontologies. We will search for the data type properties and extract common properties.

$$T_{\text{prop}} = \{\text{username, password}\}$$
$$S_{\text{prop}} = \{\text{email, username, password}\}$$
$$T_{\text{prop}} \cap S_{\text{prop}} = \{\text{username, password}\}$$

If the properties match, one of the three cases may occur: full match, partial match, or no match. In case of 'full match' and 'partial match,' we will continue to the next phase, whereas in case of 'no match,' we will stop our algorithm and move to the previous step.

2.4 Check Range

The range of a property decides which set of values that property can take. For two identical data type properties, one having integer range while the other having character range, one may get wrong results. In this step, we will check the range of data type properties which are common to both matching concepts. If range of properties matches, then we will go to the next step; otherwise, we will search again.

It is possible that some of the properties have the same range, while others do not. Hence, in this step, we will find those properties which have the same range. In our ontology of Fig. 1, the properties having same range are 'username' and 'password'.

3 Implementation

We have implemented our approach in Java using Jena API [4] that handles ontology. Firstly, ontology is created in Protégé ontology editor [5]. Next, Jena library is used to handle RDF/XML data in Java.

The inputs to our system are the ontology created using Protégé, and the concept for which we have to search for test case reuse. Ontology may be any file with .rdf or .owl extension. Here, we design one 'test' ontology and five different 'example' ontologies. The aim is to report all possibilities of test case reuse in the 'authentication' concept present in 'test' ontology against the five 'example' ontologies.

The output produced by our system is the data type properties for which we can reuse test cases, as shown in Fig. 2.

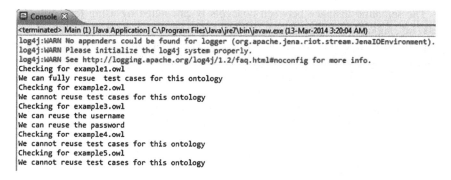

Fig. 2 Output screen containing test case reuse information

4 Evaluation

A summary of few techniques that use ontology in testing is given in Table 1.

An agent is assigned for each activity in the testing process. Software testing ontology is developed for use in communication and is also used by test case generation agent to remove redundant test cases and select consistent test cases [6].

Operations of Web service are transformed into Petri-Net ontology, using which a Petri-Net model is created to generate test cases and to test the Web service [7].

An ontology-based test case framework is presented. Firstly, test objectives are generated by reasoning performed on software artifact ontology. Next, redundant test objectives are removed and abstract test case is created for non-redundant test objectives. Finally, executable test cases are generated using software ontology [8].

Table 1 Comparison of ontology-based approaches used in testing

Approach	Goal	Methodology	Features
Multi-agent system [6]	Automation of testing process	Semantic search	Use of agents
			Reusability of test cases not leveraged
Petri-Net model [7]	Test cases for web service	Semantic search	Reuse of test cases for Web service could have been explored
Ontology-based framework [8]	Test case generation	Semantic search	Reuse of test cases could have been explored
Test case reuse based on ontology [9]	Reuse of test cases	Semantic distance	Searches for exact concept match
			Range of data type properties is not considered
Our approach	Reuse of test cases	Semantic search	Searches for exact concept match as well as subconcept match
			Range of data type properties is considered

The semantic distance between concepts is calculated by using four different sets, i.e., superclass difference set, subclass difference set, intension different set, and extension difference set. If the value is greater than some threshold value, one can say that reuse of test cases for one concept is possible for other concepts [9].

The aforementioned techniques mainly focus on assisting the testing process, except one that focuses on reuse of test cases [9]. Our approach also promotes software test case reuse; moreover, it handles user queries better and is more flexible.

5 Conclusion

Software testing involves building and execution of test cases and consumes almost half of the SDLC time and cost. Test cases are reused in an effort to improve testing efficiency over traditional methods which are inflexible because they do not consider the semantics. Ontology is widely used for information sharing and reuse. This paper discusses an ontology-based approach for search and retrieval of test cases, semantically, which provides benefit of reuse, among others. In future, automatic archiving of reusable test cases searched by our approach could be implemented.

References

1. Frakes W.: Systematic software reuse: a paradigm shift. In: Proceedings of Third International Conference on software Reuse. Advances in Software Reuse. Los Alamitos California. IEEE Computer Society Press (1994)
2. Antoniou, G., Van Harmelen, F.: A Semantic Web Primer. MIT press, Cambridge (2004)
3. Moshirpour, M., Mireslami, S., Alhajj, R., Far, B.H.: Automated ontology construction from scenario based software requirements using clustering techniques. In: IEEE 13th International Conference on Information Reuse and Integration (IRI), IEEE, pp. 541–547 (2012)
4. Jena Ontology API, http://jena.apache.org/documentation/ontology/
5. Protégé, http://protege.stanford.edu/products.php#web-protege
6. Maamri, R., Sahnoun, Z.: MAEST: multi-agent environment for software testing. J. Comput. Sci. 3(4), 249 (2007)
7. Wang, Y., Bai, X., Li, J., Huang, R.: Ontology-based test case generation for testing web services. In: Eighth International Symposium on Autonomous Decentralized Systems, ISADS'07. IEEE. pp. 43–50 (2007)
8. Nasser, V.H., Du, W., MacIsaac, D.: An ontology-based software test generation framework. SEKE, pp. 192–197 (2010)
9. Cai, L., Tong, W., Liu, Z., Zhang, J.: Test case reuse based on ontology. In: 15th IEEE Pacific Rim International Symposium on Dependable Computing, PRDC'09. IEEE, pp. 103–108 (2009)

A Decision-Driven Computer Forensic Classification Using ID3 Algorithm

Suneeta Satpathy, Sateesh K. Pradhan and B.N.B. Ray

Abstract Rapid evolution of information technology has caused devices to be used in criminal activities. Criminals have been using the Internet to distribute a wide range of illegal materials globally, making tracing difficult for the purpose of initiating digital investigation process. Forensic digital analysis is unique and inherently mathematical and generally comprises more data from an investigation than is present in other types of forensic investigations. To provide appropriate and sufficient security measures has become a difficult job due to large volume of data and complexity of the devices making the investigation of digital crimes even harder. Data mining and data fusion techniques have been used as useful tools for detecting digital crimes. In this study, we have introduced a forensic classification problem and applied ID3 decision tree learning algorithm for supervised exploration of the forensic data which will also enable visualization and will reduce the complexity involved in digital investigation process.

Keywords Digital crime · Digital investigation · Computer forensics · Data fusion · Data mining · ID3 · Visualization

1 Introduction

High-performance computing power with fast and ultrafast broadband connectively is becoming more and more accessible and available for use. Such computing powers and networks have become an indispensable tool for smooth operation of businesses, government, and even our personal lives. Log data and

S. Satpathy (✉)
Department of Computer Application, CEB, BPUT, Bhubaneswar, India
e-mail: suneetasatpathy@rediffmail.com

S.K. Pradhan · B.N.B. Ray
Department of Computer Application, Utkal University, Bhubaneswar, India

© Springer India 2015
L.C. Jain et al. (eds.), *Intelligent Computing, Communication and Devices*,
Advances in Intelligent Systems and Computing 309,
DOI 10.1007/978-81-322-2009-1_42

other data files extracted from personal, network, and cloud computing platforms can provide a lot of information about an individual's interests, patterns of behavior, and even their whereabouts at a specific space and time. As computers, laptops, tablets, and other smart devices become more widely used and prevalent, the chances of such devices and networks being involved in criminal activity will also naturally increase [1]. These activities have spawned the concept of cyber crime, which refers to illegal computer-mediated activities that can be conducted through global electronic networks, [2, 3]. A variety of cybercrime-related activities can include but not necessarily limited to is the distribution of pirate software and child pornography, etc. It is due to this increase in crimes and incidents, the field of computer forensics [2–4] has rapidly emerged, and research is being conducted into ways of improving the quality and efficiency of digital forensic investigations. However, such investigations can vary drastically in degrees of complexity. Increasingly, disclosure and discovery are two legal procedures and involve data generated by computers, stored on computers, and can only be deciphered by computers [1, 2]. Electronic disclosure process is the review and production of evidentiary materials retrieved in different digital formats. Many of the documents created today exist only in digital form. This has made paper disclosure process almost archaic. Nevertheless, the disclosure and discovery of computer evidence in civil proceedings does present some unique challenges that paper evidence would not. Among the most common challenges are but not limited to:

- Locating, identifying, and classifying volume of data
- Preserving data subject to discovery
- Retrieving documents that have been deleted from the computers
- Retrieving embedded files
- Conducting on-site inspections
- Contracting expert assistance

Addressing these challenges require a variety of tools and application of techniques to detect and defend against crimes. Data Fusion along with data mining techniques can help address these challenges. A fusion-based digital investigation tool [5] was developed using JDL data fusion model [6–10] to assist digital forensic investigators [11]. We have also demonstrated the use of the investigation tool in misuse of internet investigation justified by statistical hypothesis testing [12]. This paper is an extension of our previous work to accurately classify the files in seized hard drives with the help of data mining decision tree ID3 algorithm.

2 Challenges in Forensic Digital Investigation

Forensic digital investigation is a process to examine digital evidence that develops and tests theories, which can be entered into a court of law, to answer questions about events that occur [2, 3]. The purpose of such a type of investigation is to find evidence related to the events and present them to the investigator. Some of the challenges of such type of investigations are [13]:

- Storage media are steadily growing in size, so forensic investigations [2, 3, 14] are becoming more time consuming and complex as the volumes of data requiring analysis continue to grow.
- Forensic investigators are finding it increasingly difficult to use current tools to locate vital evidence within the massive volumes of data [14, 15]. The data offered by computer forensic tools can often be misleading due to the dimensionality, complexity, and amount of the data presented. These tools focus on digital evidence recovery, i.e., on recovering residual data from a piece of media. So they are not ideal for

  - Reduction of duplicate data
  - Identification of correlations among data
  - Discovery and sorting of data into groups based on similarities of what?
  - and visual presentation of data as groups of facts previously unknown or left unnoticed
  - Discovery or identification of patterns in data that may lead to reasonable predictions.

- Log files are often large in size and multidimensional, which makes the forensic investigation and search for supporting evidence more complex.

3 Exploring the Forensic Data with Fusion-Based Investigation Tool and ID3 Decision Tree Algorithm

Forensic analysis of a single machine is becoming increasingly cumbersome because of increase in size and sophistication of information storage. It has posed major challenges to the investigating agencies to detect criminal activities and identify the criminal [1]. A digital investigation tool based on data fusion has been developed by grouping and merging the digital investigation activities that provide the same output into an appropriate phase [5]. The grouping processes balance the investigation process and the output of the investigation process is also mapped into the domain of data fusion [5]. The data fusion application [6, 7] along with decision mining rules formed to deal with misuse of Internet in the workplace case study has also enabled graphical representation in computer forensics [11]. The present study demonstrates the application of decision tree ID3 algorithm [9] to classify same forensic data in seized hard drives to provide more accurate result. Fusion-based digital investigation tool phases [5] can be summarized into three steps as shown in Fig. 1.

1. **File Selection Process** (Data collection and preprocessing). Data preprocessing involves data cleaning and reduction. The objective of data reduction is to

Step-1

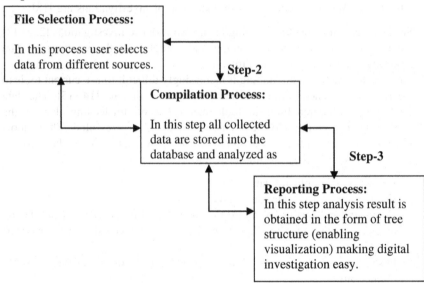

Fig. 1 Steps of fusion-based investigation tool

reduce the volume of the representation of the data set. The hard drives of computers from an organization are studied for forensic classification. As the first principle of digital investigation is never to work on the original, Forensic toolkit (FTK) [16] was used to create an image of the seized hard drives. Once the image was created, files were extracted from the hard disk and analyzed. Since the case was to deal with accessing, storing, and distributing illegal files (e.g., photos and videos), our main focus was to extract all the image files and video files. FTK toolkit was used to collect all the image files and audio and video files even if the file extension was found to be altered. The following attributes in Table 1 were considered for forensic classification.

Table 1 Symbolic attributes used for case investigation

Attribute	Possible values
File type	Image (bmp, jpeg, gif, tiff, png),MP3 files, Other Files
File creation date	Older files (with earlier creation date), New files
File creation time	Early hours of morning (12 a.m.–6 a.m.)
	Day time (6 a.m.–7 p.m.)
	Night (7 p.m.–6 a.m.)
File creation day	Beginning of the week (Monday, Tuesday)
	Middle of the week (Wednesday, Thursday)
	End of week (Friday, Saturday, Sunday)

Source	Type	USER	Creation Date	Creation Day	File Name	Creation Time
Server	jpg	USER 2	12/8/11	Thursday	2-colorized-ocean-0...	2:23 PM
Server	jpg	USER 2	12/8/11	Thursday	050204-F-2907C-3...	5:58 PM
Server	jpg	USER 2	12/8/11	Thursday	100121-F-5927B-0...	1:52 PM
Server	jpg	USER 2	12/8/11	Thursday	110322-F-PM825-0...	1:55 PM
Server	jpg	USER 2	12/8/11	Thursday	Blue_ocean.jpg	2:22 PM
Server	jpg	USER 2	12/8/11	Thursday	Cargo-Ship-004.jpg	1:55 PM
Server	jpg	USER 2	12/8/11	Thursday	Cargo-Ship-015.jpg	1:55 PM
Server	jpg	USER 2	12/8/11	Thursday	FEMA_-_42208_-_...	1:53 PM
Server	jpg	USER 2	12/8/11	Thursday	images.jpg	1:55 PM

Fig. 2 Data file structure

The structure of the data file in Fig. 2 (screenshot of data file was taken after it was loaded in the investigation tool) was analyzed by the fusion-based investigation tool [5].

The first line in the data file should contain the column headings and each data record in the data file must appear on a new line. A data record can be referred to as an input pattern.

2. **Compilation Process** (Data transformation). The following compilation on file creation day and file creation time was completed in low-level fusion step [5] of the investigation tool.

- A 24 h cycle was classified into three sections:

  - Early Morning (12–6 a.m.)
  - Late night (7 p.m.–12 a.m.)
  - Working hour (6 a.m.–7 p.m.)

- One week cycle was classified into three sections:

  - Monday and Tuesday—early week
  - Wednesday and Thursday—middle week.
  - Friday, Saturday, and Sunday—late week.

3. **Reporting Process**—Result analysis and evidence report generation was completed in high level and decision level processing [5] step of the investigation tool. Generally, this step involves training and testing of the tool as well as validation and analysis. It draws upon methods and algorithms developed from various fields (such as data mining, machine learning, and pattern recognition) [9] to detect interesting patterns. It develops a background description of relations between entities. It effectively extends and enhances the completeness, consistency, and level of abstraction of the situation description produced by refinement.

In the present case study, we used ID3 decision tree algorithm to classify forensic data in digital investigation and in evidence reporting since decision tree

Table 2 ID3 decision tree algorithm

The ID3 algorithm [2], [11] used for classification problem is as follows.
ID3 (Learning Sets S, Attributes Sets A, Attribute Values V)
Return Decision Tree.
 	
Begin
Load learning set first, create decision tree root node 'r_Node', add learning set S into root node as its subset.
 	
For r_Node, we compute
Entropy (r_ Node .subset) first
 	
If Entropy (r_ Node. subset) == 0, then
r_ Node subset consists of records all with the same value for the_categorical attribute, return a leaf node with decision attribute : attribute value;
 	
If Entropy (r_ Node .subset)! =0, then
compute information gain for each attribute left(have not been used in_splitting), find attribute A with Maximum(Gain(S,A)). Create child nodes of this r_ Node and add to r_Node in the decision tree.
 	
For each child of the r_Node, apply ID3(S,A,V) recursively until we reach node that has entropy = 0 or reach leaf node.
End ID3.

learning algorithm [9] has been successfully used in expert systems in capturing knowledge. We performed specified tasks by using inductive methods to the given values of attributes of an unknown object to determine appropriate classification according to decision tree rules. A cost sensitive decision tree learning algorithm [17–19] was also used for forensic classification problem. This is commonly used for gaining information for the purpose of decision making.

4 Decision Tree ID3 Algorithm

ID3 is a simple decision tree learning algorithm developed by Ross Quinlan in the year 1983 [9, 20]. With the help of ID3 algorithm (Table 2), one can construct the decision tree by employing a top-down, greedy search through the given data sets to test each attribute at every tree node. To find an optimal way to classify a learning set, the information gain metric function Gain (S, A) is used to find the most balanced splitting.

Entropy is a measure in the information theory, which characterizes the impurities of an arbitrary collection of examples. If the target attribute takes on c different values, then the entropy S relative to this c-wise classification is defined as in Eq. 1.

$$\text{Entropy}(s) = \sum_{i=1}^{i=n} -P_i \log_2 P_i \qquad (1)$$

where Pi is the proportion/probability of S belonging to class i. Logarithm is base 2 because entropy is a measure of the expected encoding length measured in bits. Information gain measures the expected reduction in entropy by partitioning the examples according to this attribute. The information gain Gain (S, A) of an attribute A, relative to the collection of examples S, is defined as in Eq. 2.

$$\text{Gain}(S, A) = \text{Entropy}(S) - \sum_{v \in Values(A)} \frac{|S_v|}{|S|} Entropy(S_v) \tag{2}$$

where $Values(A)$ is the set of all possible values for attribute A, and Sv is the subset of S for which the attribute A has value v. This can be used to measure rank attributes and build the decision tree where each node is located, and the attribute with the highest information gain among the attributes not yet considered in the path from the root.

The learning set for Forensic classification problem was prepared according to the data file structure shown in Fig. 2 from which columns, file type, file creation time, file creation day, and decision columns were taken for learning. In the learning sample, the attribute column "decision" had two values as "positive" and "negative."

The learning sample was divided into two sets

1. Set S1 corresponded to "positive" as decision values and
2. Set S2 corresponded to "negative" as decision values.

There were 25 samples in S1 and 16 samples in S2. To compute the information gain of each attribute, the expected information of the sample class was computed in Eq. 3 by using Eq. 1.

$$\begin{aligned} \text{I}(S1, S2) = \text{I}(25, 16) &= -(25/41) * \log(25/41) - (16/41) * \log(16/41) \\ &= 0.15367 \end{aligned} \tag{3}$$

Then information gain for each attribute was computed in Eqs. 4, 5 and 6 as.

1. **File_Type**—All image files were categorized as image files category and rest of the mp3, audio, and video files as other files category. The expected information gain in connection with each distribution was computed in Eq. 4 by using Eq. 2.

$$\text{File\_Type} = \text{``imagefiles''}, \ S11 = 20, \ S21 = 9, \ \text{and}$$
$$\text{File\_Type} = \text{``other files''}, \ S12 = 0, \ S22 = 12 \ \text{then}$$

$$\text{Gain (File\_Type) is calculated as} = 0.039454 \tag{4}$$

2. **File_Creation_Day**- Weekdays are divided into three sections: early week, middle week, and late week and information gain value was calculated in Eq. 5 by using Eq. 2.

File_Creation_Day = "earlyweek i.e. mon, tue", $S11 = 2$, $S21 = 3$

File_Creation_Day = "middle week i.e. wed, thurs", $S12 = 4$, $S22 = 5$

File_Creation_Day = "late week i.e. fri, sat, sun", $S13 = 20$, $S23 = 7$

Then Gain (File_Creation_Day) was calculated as $= 0.0091195$ (5)

3. **File_Creation_Time**—File creation time is divided into three parts: early morning, working hour, late night, for which information gain is calculated as in Eq. 6 using Eq. 2.

File_Creation_Time = Early morning = "12 a.m.–6 a.m.", $S11 = 10$, $S21 = 5$

File_Creation_Time = Working hour = "6 a.m.–7 p.m.", $S12 = 4$, $S22 = 0$

File_Creation_Time = Latenight = "7 p.m.–12 a.m.", $S13 = 18$, $S23 = 4$

Then Gain (File_Creation_Time) was calculated as $= -0.00458$ (6)

From the above calculation (Eqs. 4, 5 and 6), Gain (File_Type) had the largest value which could help the investigator to conclude the followings:

1. The attribute File_Type could perform an important function in composing data to subclasses as shown in tree structure in Fig. 3.
2. From (Fig. 3), one rule could be created in each route from root to leaf. The unique behavior of one user hard drive is shown by positive and negative files classification to detect the illegal use.

Finally, conclusion could be drawn as when the File_Type was other than image files, and then chances of criminal level of that hard drive could be negative. But when the File_Type were image files and files were created or accessed during weekends (Friday, Saturday, and Sunday during 6 a.m.–7 p.m. and during Monday, Tuesday Wednesday, and Thursday during late in the night (7 p.m.–12 a.m.) or during (12 a.m.–6 a.m.)) then chances of criminal level of that hard drive is positive. The seven decision mining rules (Table 3) already assumed and hypothesized in our paper [11, 12] for classifying the files into positive and negative were justified from the tree structure (Fig. 3).

The conclusion drawn from above digital investigation decision tree structure justified that ID3 decision tree algorithm could work fairly well on classifying the complex and large voluminous forensic data into a simpler analysis report. It will also help the investigators to easily draw the conclusion from simple and easy to visualize decision tree structure and prepare the evidence report to present it in the court of law to justify the evidence against crime.

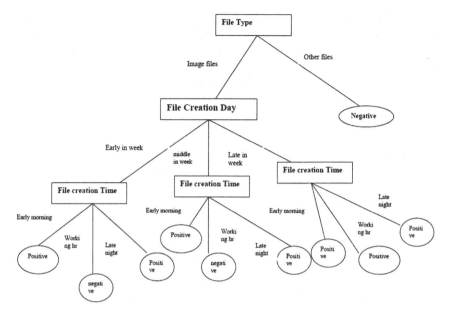

Fig. 3 Decision tree

Table 3 Decision mining rules

1. If an image file was created/modified/accessed early in the week (mon, tue) during 12–6 a.m. and 7 p.m.–12 a.m. (early morning, late night) then it was classified as suspicious
2. If it an image file was created/modified/accessed early in the week (mon, tue) during 6 a.m.– 7 p.m. (working hr) then it was classified as not suspicious
3. If an image file was created/modified/accessed middle in the week (wed, thurs) during 12–6 a.m. and 7 pm–6 a.m. (early morning, late night) then it was classified as suspicious
4. If an image file was created/modified/accessed middle in the week (wed, thurs) during 6 a.m.– 7 p.m. (working hr) then it was not classified as suspicious
5. If an image file was created/modified/accessed late in the week (fri, sat, sun) during 12–6 a.m. and 7 p.m.–12 a.m. (early morning, late night) then it was classified as suspicious
6. If an image file was created/modified/accessed late in the week (fri, sat, sun) during 6 a.m.– 7 pm (day time working hour) then it was classified as suspicious
7. If the logical file size was detected to be large and if it was downloaded during working hours on any day of the week, it would need further investigation. Same rule was applicable for mp3 files downloaded at any time of the day of the week

5 Conclusion

With the rapid proliferation of Internet technologies and applications, cyber crime has become a major concern for the law enforcement community. Cyber criminals have been distributing messages, illegal materials on the internet and engage in illegal activities. The extensive growth of internet and the lack of truly secure systems make it an important field of research in computer forensics. Data fusion and data mining has proven itself in the area of information security. So in the context of digital information sea in the current digital information technology, decision tree ID3 application in computer forensic could help to pick out the right behavioral information that reflected the intent of the criminal, the crime type, and security requirement in a timely, comprehensive, and accurate manner.

References

1. Lipson, H.: Tracking and tracing cyber attacks: technical challenges and global policy issues. CERT Coordination Center, Nov (2002)
2. Casey, E. (ed.): Handbook of Computer Crime Investigation. Academic Press, Waltham (2001)
3. Casey, E.: Digital Evidence and Computer Crime, 2nd ed. Elsevier Academic Press (2004)
4. Brezinski, D., Killalea, T.: Guidelines for evidence collection and archiving. RFC3227 (2002)
5. Satpathy, S., Pradhan, S. K., Ray, B.N.B.: A digital investigation tool based on data fusion in management of cyber security systems. Int. J. Inf. Technol. ad Knowledge management, vol 2(2) (2010)
6. David L. Hall, Sonya A.H.: Mathematical Techniques in Multisensor Data Fusion, 2nd ed., Artech House (2004)
7. Hall, D.L., Linas, J.: An introduction to multisensor data fusion. In Proceedings of The IEEE, vol 85, Jan (1997)
8. Waltz, E.J.: Linas Multisensor Data Fusion, Artech House. Boston (1990)
9. Han, J., Kamber, M.: Data mining: concepts and techniques, 2nd ed. (2005)
10. Introduction to data fusion, or information fusion, http://www.data-fusion.org
11. Satpathy, S., Pradhan, S.K., Ray, B.N.B.: Rule based decision mining with JDL data fusion model for computer forensics: a hypothetical case study. Int. J. Comput. Sci. Inf. Sec., 9(12) (2011)
12. Satpathy, S., Pradhan, S. K., Ray, B.N.B.: Application of data fusion methodology for computer forensics dataset analysis to resolve data quality issues in predictive digital evidence, Int. J. Forensic Comput. Sci., 7(1) (2012)
13. Meyers, M., Rogers, M.: Computer forensics: the need for standardization and certification, Int. J. Digital Evi. 3 (2004)
14. Beebe, N, Clark, J.: Dealing with terabyte data sets in digital investigations. Advances in Digital Forensics, pp. 3–16, Springer, (2005)
15. Danielsson, J.: Project Description A system for collection and analysis of forensic evidence. Appl. NFR, Apr (2002)
16. Access Data Corporation, http://www.accessdata.com
17. Qin, I.U.: Data mining method based on computer forensics-based ID3 algorithm. In: Proceedings of IEEE Conference on Information Management and Engineering, pp. 340–343 (2010)
18. Davis, J.V., Rossbach, C.J., Ramadan, H.E., Witchel, E.: Cost-sensitive decision tree learning for forensic classification. In Proceedings of the 17th European Conference on Machine Learning, pp. 622–629, Berlin, Germany (2006)
19. Mendoza1, M., Zamora, J.: Building decision trees to identify the intent of a user query. In Proceedings of the 13th International Conference on Knowledge-Based and Intelligent Information and Engineering Systems, Santiago, pp. 285–292 (2009)
20. Adriaans, P., Zantige, D.: Data Mining. Addison Wesley, Harlow England (1997)

Non-ideal Iris Segmentation Using Wavelet-Based Anisotropic Diffusion

Chinmay Singh, Madhusmita Sahoo and Bibhu Prasad Mohanty

Abstract In current scenario, biometric identification is becoming more popular. The performance of iris recognition depends on iris segmentation. Iris segmentation plays a very vital role for iris recognition. In this work, different scheme for non-ideal iris image segmentation is investigated. In non-ideal iris image eyelash and eyelid occlusion, reflection exists which is absent in the ideal iris image. So non-ideal iris segmentation is challenging. The use of anisotropic diffusion to the iris image also removes the eyelash and eyelid occlusions. This piece of work has been subdivided into two different areas: first to finding iris center and inner boundary and secondly to find exterior boundary.

Keywords Anisotropic diffusion · Image segmentation · Thresholding · Closing · Binarization

1 Introduction

In image segmentation, a digital image can be subdivided into multiple segments. Based on two basic properties, that is, discontinuities and similarity, segmentation can be carried out. The main objective is to extract meaningful regions which will be useful for analysis. For public security or personal identification, iris recognition is often preferred than any other biometric techniques. The texture patterns of human eye are stable and different from one person to another. So iris provides better security. Segmentation of non-ideal iris is more challenging, because during image acquisition, some undesirable disturbances are created. The main cause behind disturbances are body motion, reflection of spectacles, improper illumination, and

C. Singh · M. Sahoo (✉) · B.P. Mohanty
Department of ECE, ITER (Faculty of Engineering),
Siksha O Anusandhan University, Bhubaneswar, India
e-mail: madhusmitasahoo@soauniversity.ac.in

© Springer India 2015
L.C. Jain et al. (eds.), *Intelligent Computing, Communication and Devices*,
Advances in Intelligent Systems and Computing 309,
DOI 10.1007/978-81-322-2009-1_43

eyelash and eyelid occlusion, off-angled iris, pupil center deviation due to eye movement, etc. So it becomes difficult to segment iris region properly in a non-ideal iris image [1–3].

Iris image can be segmented using many methodologies such as using Hough transform and level set method. In this piece of work, we have proposed to detect the centroid of the iris image by using thresholding and morphological operation. At the same time, we apply polar transformation for interior boundary detection. Exterior boundary is detected using following steps, such as region selection, applying polar transform on selected region, and finally region-based segmentation.

The paper is organized as follows. In Sect. 2, level set method and Hough transform for image segmentation are discussed. Section 3 includes proposed methodology. Section 4 includes results and analysis. We have concluded the paper in Sect. 5 with a brief summary of work in future direction.

2 Level Set and Hough Transform Method

The level set technique [4, 5] is based on active contours model and evolution. The contours model works on split and merge concept and useful for simultaneous detection of several objects and both interior and exterior boundaries. The boundary representation is done by level-set method [4], and depending upon the threshold value selected adaptively, the boundary either merges with the previous one or splits further at each evolution. During every evolution, the level set function is periodically reinitialized to a signed distance function using reinitialization technique.

The Hough transform is a standard algorithm [6] which is useful for shape detection. Here, using edge detection algorithms, edge map is generated and using Hough transform, disjoint edge points are connected. For every pixel, P is calculated and Hough space is generated. Then, local maxima in the parameter space are found and maximum point in the Hough space interprets the radius and center coordinates of circle. It determines center coordinates and the radius r.

If pixel is an edge pixel, then for all values of r, (x_c, y_c) and P are calculated using following equation.

$$x_c = x - r\cos\theta, \quad y_c = y - r\cos\theta$$
$$P(x_c, y_c, r) = P(x_c, y_c, r) + 1 \tag{1}$$

For every edge point, the corresponding accumulator element is incremented in the accumulator array. At the end of this process, accumulator (P) is checked and the local maxima in the accumulator (parameter space) give corresponding (x_c, y_c) and r which are the center coordinate and radius of the detected circle [6].

The above schemes for iris segmentation are excellent and accurate in their respective domain but fail for non-ideal iris images.

3 Proposed Method

3.1 Wavelet Decomposition

Using DWT [7], an image can be decomposed into different levels. In this work, the detailed and approximate coefficients are determined taking "db" wavelet and dwt function. The 1st step that we use in our approach is wavelet decomposition, which is a simple approach to reduce the iris data due to its coarser level. Here, we use wavelet decomposition to reduce computational complexity of iris data.

3.2 Anisotropic Diffusion

Perona and Malik proposed a numerical method for selectively smoothing digital images. In anisotropic diffusion, flow is not only proportional to the gradient, but is also controlled by a function $g(|\nabla u|)$. Regions with low $|\nabla u|$ are plains. By choosing a high diffusion coefficient, the noise can be reduced. Regions with high $|\nabla u|$ can be found near edges. In order for those edges to be preserved, a low diffusion coefficient is chosen accordingly. This leads to the function $[0, \infty] \rightarrow [0,1]$, $g(0) = 1$, $\lim s \rightarrow \infty\ g(s) = 0$, which is monotonically decreasing.

$$u(0, x) = u_0(x) \tag{2}$$

where t is the time parameter, $u(x, y, 0)$ is the original image, and $\nabla u(x, y, t)$ is the gradient version of image at time "t".

$$g(|\nabla u|) = \frac{1}{1 + |\nabla u|^2 / \lambda^2} \tag{3}$$

λ is always greater than 0.

In the above equation, g is smooth non-increasing function with $g(0) = 1$, $g(x) \geq 0$, and $g(x)$ tending to zero at infinity. The idea is that the smoothing process obtained by the equation is "conditional"; that is, if $\nabla u(x)$ is large, then diffusion will be low, and therefore, the exact localization of the "edges" will be kept. If $\nabla u(x)$ is small, then the diffusion will tend to smooth still more around x.

Thus, the choice of g corresponds to a sort of thresholding which has to be compared to the thresholding of $|\nabla u|$ used in the final step of classical theory.

Perona and Malik discretized their anisotropic diffusion equation as

$$u_{t+1}(s) = u_t(s) + \frac{\lambda}{|\eta_s|} \sum_{p \in \eta_s} g_k(|\nabla u_{s,p}|) \nabla u_{s,p} \tag{4}$$

S denotes the pixel position in the discrete 2D grid, t denotes the iteration step, g is the conduction function, and λ is the gradient threshold parameter that determines the rate of diffusion. λ is a scalar quantity which determines the stability, and it is usually less than 0.25. η denotes the spatial neighborhood of pixel (x, y). $\eta_s =$ [N S E W], where N, S, E, and W are the north, south, east, and west neighbors of pixel S. η is equal to 4 (except for the image borders). The symbol ∇u is now representing a scalar defined as the difference between neighboring pixels to each direction.

Gradient ∇u in four different directions can be calculated as follows

$$
\begin{aligned}
\nabla u_N(x, y) &= u(x, y - 1, t) - u(x, y, t) \\
\nabla u_s(x, y) &= u(x, y + 1, t) - u(x, y, t) \\
\nabla u_E(x, y) &= u(x + 1, y, t) - u(x, y, t) \\
\nabla u_W(x, y) &= u(x - 1, y, t) - u(x, y, t)
\end{aligned}
\tag{5}
$$

If any eyelash occlusion, off-axis gaze, and reflection are present in an iris image, then anisotropic diffusion is employed which will smoothen those occlusions and simultaneously preserve the edges. So here, we applied this approach and select the proper iteration time for removal of occlusion and so as to preserve pupil boundary. The diffusion is controlled by evolution parameter, which is inversely proportional to iteration time.

3.3 Thresholding, Morphological Operas, and Mapping

After diffusion, except the pupil part, the contrast of iris data is very distinctive. Dark part of pupil has very low values of intensity, so if histogram is plotted, a valley will be formed and from the histogram, a threshold value can be chosen which differentiates pupil from the iris image. Thus, applying appropriate threshold value, pupil is detected. Threshold value is different for different images. So by selecting the threshold value adaptively, a particular algorithm is capable of detecting various iris images. After detecting the pupil region, the next step is to remove the reflection. For this purpose, morphological operation is used. The opening and hole-filling by darkness operation is employed to remove the reflection. The pupil image is mapped to its original level, and this can be achieved using nearest interpolation method, and then, that image is added to wavelet decomposed image. Idwt is applied to get original image.

3.4 Centroid Calculation and Interior Boundary Detection

In Fig. 1, we have presented the block schematic of our proposed method of wavelet-based interior boundary detection. We have followed the steps as suggested by H.L. Wan et al., [8] by modifying the procedure suitable for wavelet domain.

The image resulted from idwt is taken, and $x1 = 0$, $x2 = 0$, $y1 = 0$, are $y2 = 0$ is initialized. Then, the summation of black pixels in each row is determined. The first row that has the maximum summation is stored in $x1$, and the last row that has the maximum summation is stored in $x2$. Then, the summation of black pixels in each column is obtained. The first column that has the maximum summation is put in $y1$, and the last column that has the maximum summation is put in $y2$. Then, x_c and y_c are evaluated using Eq. (6)

$$x_c = (x1 + x2/2), \ y_c = (y1 + y2/2) \tag{6}$$

First image is converted into its polar transform form. Then, Sobel operator is used which determines maximum of each column in polar-transformed image and those points are treated as candidate points. Then, all distances from each candidate point to the pupil center (x_c, y_c) are calculated, and then, histogram is obtained. The distance that has maximum probability is the radius. In this manner, the internal boundary is detected.

3.5 Exterior Boundary Detection

For exterior boundary detection, the iris image undergoes wavelet decomposition, anisotropic diffusion and morphological operation as described in Sects. 3.1, 3.2 and 3.3 respectively. After that, a region is selected within an angular range,

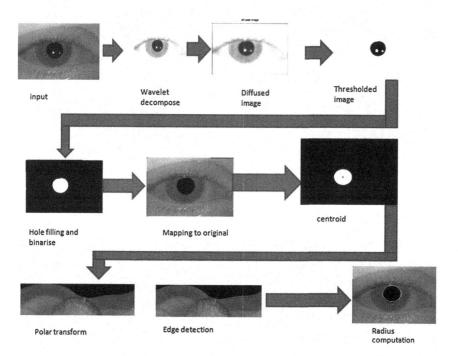

Fig. 1 System overview for interior boundary detection

located in the top or bottom of iris region. It should be ensured that eyelid occlusions is absent in that region. Then, polar transform of that region is taken. After that, region-based segmentation is carried out. A line is initialized as exterior boundary near to interior boundary of iris. The line is working on active contour mechanism. The line separates the image into two regions; then, mean values of the two regions are computed. Then, the line is pushed far away from interior boundary and both new mean values are obtained. If the difference of the previous mean values is smaller than that of the new ones, the line continues to move. Else, the previous line is specified as exterior boundary [8].

4 Results and Conclusion

It is implemented using MATLAB7 software. CASIA database is used for iris images.

As shown in Fig. 2 using Hough transform, it is difficult to segment exterior boundary of non-ideal iris, but it is efficiently detected using proposed scheme.

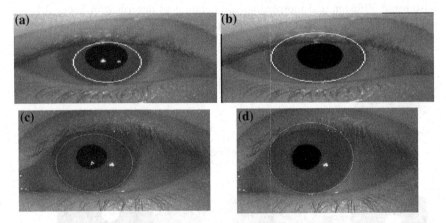

Fig. 2 Exterior boundary detection (**a**, **c**) using Hough transform and (**b**, **d**) proposed scheme

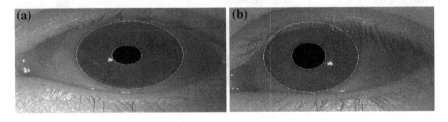

Fig. 3 Exterior boundary detection using proposed scheme

Figure 2 shows comparisons of exterior boundary detection of iris image using Hough Transform and the proposed scheme. From the result it is visually clear that our approach efficiently detect the exterior boundary satisfactorily. The exterior boundary is detected using simple active contour mechanism.

In Fig. 3, we have presented the interior and exterior boundary segmentation simultaneously using the proposed wavelet-based scheme. The coordinates for boundaries for Fig. 3b are $x\_in = 246.3143$, $y\_in = 239.0476$, $r\_in = 54.6576$, $x\_out = 251$, $y\_out = 240$, $r\_out = 158$. This research proposes a new effective and fast algorithm to segment the non-ideal iris images captured under unconstrained imaging conditions. Experimental results on CASIA iris database indicate high accuracy compared with previous existing algorithms.

References

1. Kang, B., Park, K.: A robust eyelash detection based on iris focus assessment. Pattern Recogn. Lett. **28**, 1630–1639 (2007)
2. Kong, W., Zhang, D.: Detecting the eyelash and reflection for accurate iris segmentation. Int. J. Pattern Recogn. Artif. Intell. 1025–1034 (2003)
3. Daugman, J.: New methods in iris recognition. IEEE Tran. Syst. Man Cybern.-Part B Cybern. **37**(5), 1167–1175 (2007)
4. Roy, K., Bhattacharya, P., Suen, C.Y.: Iris segmentation using variational level set method. Opt. Laser Eng. **49**(1), 578–588 (2011)
5. Ross, A., Shah, S.: Segmenting non-ideal irises using geodesic active contours. In: Proceedings of Biometrics Symposium, 2006
6. Jan, F., Usman, I., Khan, S.A., Malik, S.A.: Iris localization based on the Hough transform, a radial-gradient operator, and the gray-level intensity. Int. J. Optik (Elesvier) **124**, 5976–5985 (2013)
7. Mallat, S.: A compact multiresolution representation: the wavelet model. In: Proceedings IEEE Computer Society Workshop on Computer Vision. IEEE Computer Society Press, Washington, D.C., pp. 2–7 (1987)
8. Wan, H.-L., Li, Z.-C., Qiao, J.-P., Li, B.-S.: Non-ideal iris segmentation using anisotropic diffusion. Published in IET Image Processing (2012). doi:10.1049/iet-ipr

Detection of Video Objects in Dynamic Scene Using Local Binary Pattern Subtraction Method

Prashant Kumar, Deepak K. Rout, Abhishek Kumar, Mohit Verma and Deepak Kumar

Abstract In this paper, the problem of video object detection in dynamic scene has been addressed. The dynamism is referred to the changes in the scene of interest, due to swaying of tree branches, leaves, fluctuation of surface in case of water bodies, variation of scene illumination, etc. The problem is formulated in a fixed camera scenario and with unavailability of reference frame (background model). The local binary pattern (LBP) is a very strong element used in object detection algorithms. In the literature, many methods exist, where the LBP histograms of current frame and previous frames are combined and used for background subtraction, to get the foreground detected. This histogram computation and construction of a final histogram for the background subtraction method is a very time-consuming and complex process. The complexity can be reduced to a large extent by using our proposed window-based LBP subtraction (WBLBPS) method. Moreover, the efficacy of the proposed method in terms of correct classification is quite satisfactory as compared to the other LBP-based methods.

Keywords Local binary pattern · Object detection · Background subtraction · WBLBPS method · Dynamic scene

1 Introduction

Moving object detection is a fundamental task in computer vision and vision-based system design. Starting from the field of automation and robotics to virtual reality, computer vision has been a tangible area of focus for many researchers. Video moving target tracking is an important part of artificial intelligence techniques and

P. Kumar · D.K. Rout (✉) · A. Kumar · M. Verma · D. Kumar
Department of Electronics and Telecommunication Engineering,
C. V. Raman College of Engineering,
Bidya Nagar, Mahura, Janla, Bhubaneswar 752054, India
e-mail: deepak.y.rout@gmail.com

© Springer India 2015
L.C. Jain et al. (eds.), *Intelligent Computing, Communication and Devices*,
Advances in Intelligent Systems and Computing 309,
DOI 10.1007/978-81-322-2009-1_44

is the key to solve many computational vision problems. It is very difficult to detect the object when scene is dynamic. This is because there are many uncertainties in the actual scene, such as branches shaking, surface fluctuations, or the illumination changes in the scene. Throughout this paper, the camera is assumed to be fixed and thus, the dynamism in the background is not due to the movement of the camera. The availability of reference frame is assumed to be not present. Background is supposed to have less change as compared to the object. Thus, the problem addressed is to detect the moving object in such a dynamic scene, in the absence of reference frame and absence of any prior knowledge of the object of interest. Many researchers [1–11] have developed many useful algorithms and techniques for the detection of object under such circumstances. In Xue et al. [1], experiments on challenging sequences indicated that their method can produce comparably better results while using less computation time, compared to the existing LBP-based method. Hence, in this paper, we have taken this method as reference method for the comparative analysis.

In this paper, an algorithm has been devised to detect moving objects in a video sequence under dynamic scene. The problem is addressed when the reference frame is not available. A window-based local binary pattern (LBP) subtraction method has been proposed, which takes care of all type of cases irrespective of the availability of reference frame and irrespective of the a priori knowledge of the moving object. This method takes the spatial LBP (SLBP) plane of the frame under consideration and computes the spatio-temporal LBP by taking the previous frame into account. It could be observed that the dynamism present due to illumination variation, swaying of the leaves, branches, etc. can be handled efficiently using our proposed method.

The rest of the paper is organized as follows:

Section 2 describes the proposed WBLBPS method. Section 3 illustrates the experimental results. The paper ends with a conclusion in Sect. 4.

2 Proposed Window-Based LBP Subtraction Method

The efficacy of the algorithm largely depends on the accurate measurement of the LBPs. The region taken for computation of the LBP plays a vital role in the efficiency of the method. The LBP can be computed in the spatial domain as well as in temporal domain. A spatio-temporal framework is developed to detect the LBP, which is shown in Fig. 1. The WBLBPS algorithm is as follows:

1. Convert the colour frame to greyscale.
2. Calculate the SLBP of tth frame by comparing the pixel of consideration with all the other pixels present in the same region.
3. Calculate the STLBP of tth frame by comparing the pixel of consideration with all the other pixels present in the same region, as well as the pixels present in the corresponding region of the previous frame, i.e., $(t - 1)$th frame.

Fig. 1 The spatio-temporal framework for computation of LBP

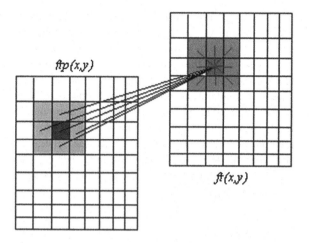

4. Subtract the result of step 2 from step 1 using the window-based method.
5. If the difference is less than a threshold, then consider the pixel to be a background pixel, whereas if the difference is higher than the threshold, then the pixel is classified as foreground.

The conversion of RGB to greyscale image is carried out by the following formulation.

$$f(x, y) = 0.299f(x, y, z)_R + 0.587f(x, y, z)_G + 0.144f(x, y, z)_B \qquad (1)$$

where $f(x, y, z)_R, f(x, y, z)_G,$ and $f(x, y, z)_B$ are the red, green, and blue planes of the colour video sequence, respectively. This conversion is carried out for the sake of reducing the computational complexity on the system. The net processing time also reduced drastically.

LBP is a greyscale invariant texture description. The operator labels the pixels of an image by thresholding the eight neighbourhood of each pixel with the centre value and considering the result as an eight-bit binary number (LBP code), which is obtained as

$$\text{SLBP}(x_C, y_C) = \sum_{p=0}^{l} s(f_{t,p} - f_{t,c})2^p \qquad (2)$$

where f_C corresponds to the grey value of the central pixel (x_c, y_c) and f_p to the grey values of the eight neighbouring pixels. The function $s(x)$ is defined as follows:

$$s(x) = \begin{cases} 1 & x \geq 0 \\ 0 & x < 0 \end{cases} \qquad (3)$$

LBP features are robust to monotonic greyscale changes and very fast to compute which are very important for background subtraction in real-time application.

The grey cell in $f_t(x, y)$ shows the pixel of consideration, whose LBP has to be evaluated, and yellow cells represent the region within which the LBP has to be computed in spatial framework. The grey cells in $f_{t-1}(x, y)$ show the cells with respect to which the temporal LBPs are computed for the pixel in $f_t(x, y)$. The black colour lines from $f_t(x, y)$ to $f_{t-1}(x, y)$ show the temporal link between the pixel of consideration in the current frame, and the red lines show the spatial link between the pixel whose LBP has to be computed and the pixels present in the region of consideration. The joint computation of the LBP in spatial as well as temporal direction is considered as the spatio-temporal framework for LBP computation. The spatio-temporal framework considered can be seen as below. In Fig. 1, $ftp(x, y)$ represents $f_{t-1}(x, y)$ and $ft(x, y)$ represents $f_t(x, y)$.

The STLBP is obtained as

$$STLBP(x_C, y_C) = \frac{1}{2} \left(\sum_{p=0}^{l} s(f_{t,p} - f_{t,c})2^p + \sum_{p=0}^{l} s(f_{t-1,p} - f_{t,c})2^p \right) \quad (4)$$

where f_C corresponds to the grey value of the central pixel (x_c, y_c) and f_p to the grey values of the eight neighbouring pixels of the previous frame. The function $s(x)$ is defined in Eq. (3).

The SLBP, at time t, and the spatio-temporal LBP (STLBP) are calculated. Then, a window of mxm is taken for both LBPs. The difference between the two LBPs is calculated on window basis. This is carried out by taking a window of desired dimension and then computing the average LBP value for the specific window. The average LBP value of the STLBP is then subtracted from the average LBP value of SLBP. The window is then moved to the next pixel. This process has to be continued unless until all the pixels get classified.

$$DLBP(x_c, y_c) = SLBP(x_c, y_c) - STLBP(x_c, y_c) \quad (5)$$

$$f_S(x, y) = \begin{cases} \text{Object} & \text{if} \quad DLBP(x, y) \geq Th \\ \text{Background} & \text{otherwise} \end{cases} \quad (6)$$

The difference of these two average values is then compared with a threshold (manually chosen) to classify the object and background pixels. If the difference is greater than a threshold, then it is detected as object otherwise background.

3 Experimental Results

The efficacy of the proposed algorithm has been tested in different situations. Although the window size depends upon the object shape and extent of dynamism, in our experiments and simulations, we have taken the window size to be 7×7, as it gives very good result, in accordance with the time complexity and accuracy. The thresholds used in all the video sequences are chosen manually by hit and trial basis.

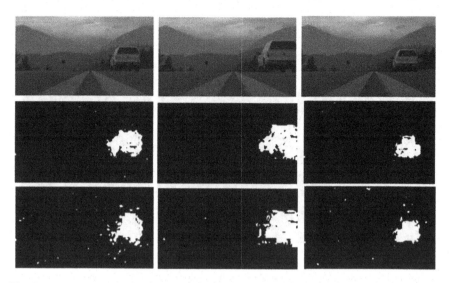

Fig. 2 Comparison results of car sequence. The *top row* is the original 5753rd, 5766th and 5772nd frames. The *second row* is the result obtained by our proposed method, and the *third row* shows the results obtained using Xue's method

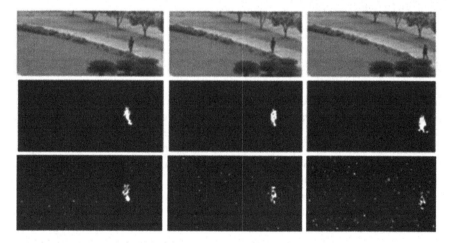

Fig. 3 Comparison results on campus sequence. The *top row* is the original 1534th, 1600th, and 1618th frames. The *second row* is the result obtained using WBLBPS method. The *third row* is the result obtained using Xue et al. [1] method

The dynamism in Fig. 2 is due to the movement of clouds in the sky region, which is handled very well by our method. The dynamism considered in the campus video, which is shown in Fig. 3, is the swaying of leaves and branches of the tree. Our proposed algorithm yields quite satisfactory results as compared to

Fig. 4 Comparison of results on helicopter sequence. The *top row* is the original 51269th, 51276th and 51316th frames. The *second row* is the results obtained by our proposed method, and the results obtained by Xue et al. [1] method is shown in the *third row*

the Xue's method. Figure 4 shows the helicopter video sequence. Our results completely outperform the Xue's method.

4 Conclusion

In this paper, the problem of moving object detection in dynamic scene without the availability of reference frame and prior knowledge of object has been addressed. To deal with this problem, first, a spatial LBP feature map is constructed from the current frame and then, a spatio-temporal framework-based LBP feature map is constructed. Then, the two LBPs are compared. In case both the LBPs are similar, then the pixel at that position is considered as the background pixel, else it is classified as foreground pixel. It is very fast to compute. The proposed dynamic background modelling and subtraction method based on the above formulation is very robust to dynamic movement in natural scenes such as swaying vegetation, waving trees and rippling water. It achieves detection of moving objects with high accuracy and suppresses most of the false detections by traditional methods. The proposed method can be used in real-time visual surveillance applications.

References

1. Xue, G., Song, L., Sun, J., Wu, M.: Hybrid center-symmetric local pattern for dynamic background subtraction. In: IEEE International Conference on Multimedia and Expo (ICME), pp. 11–15. July 2011
2. Stauffer, C., Grimsom, W.E.L.: Adaptive background mixture models for real-time tracking. In: IEEE CS Conference on Computer Vision and Pattern Recognition, vol. 2, pp. 246–252 (1999)
3. Elgammal, A., Duraiswami, R., Harwood, D., Davis, L.S.: Background and foreground modelling using nonparametric kernel density estimation for visual surveillance. Proc. IEEE **90**(7), 1151–1163 (2002)
4. Rout, D.K., Puhan, S.: A spatio-temporal framework for moving object detection in outdoor scene. In: Computer in Communication and Information Securities CCIS, vol. 270, pp. 494–502. Springer, Berlin, Heidelberg (2012)
5. Deepak K., Rout, Sharmistha Puhan.: "Video Object Detection using Inter-frame Correlation Based Background Subtraction", IEEE Recent Advances in Intelligent Computational Systems (RAICS), pp. 167–171 (2013)
6. Heikkila, M., Pietikainen, M.: A texture based method for modelling the background and detecting moving objects. IEEE Trans. Pattern Anal. Mach. Intell. **28**(4), 657–662 (2006)
7. Heikkila, M., Pietikainen, M., Schmid, C.: Description of interest regions with local binary patterns. Pattern Recogn. **42**(3), 425–436 (2009)
8. Zhang, S.: Dynamic background modelling and subtraction using spatio-temporal local binary patterns. In: 15th IEEE International Conference on Image Processing (ICIP), pp. 1556–1559 (2008)
9. Zhao, G., Pietikainen, M.: Dynamic texture recognition using volume local binary patterns. In: Workshop on Dynamical Vision, pp. 165–177 (2007)
10. Zhong, B.: Texture and motion pattern fusion for background subtraction. In: 11th Joint Conference on Information Sciences, pp. 1–7 (2008)
11. Zhao, Y., ; Wang, B., Xu, X., Liu, Y.: A moving object detection method based on level set in dynamic scenes. In: International Symposium on Intelligent Signal Processing and Communications Systems (ISPACS), pp. 4–7. Nov 2012

Automatic Lane Detection in NH5 of Odisha

P. Kanungo, S.K. Mishra, S. Mahapatra, U.R. Sahoo, U.S.Kr. Sah and V. Taunk

Abstract The efficacy of any intelligent transportation systems depends on efficiency of the lane detection system. This paper addressed the lane detection problem during the daytime in the NH5 of Odisha, India. In NH5, the contrast between road and lane is very low because of dust and mud layers from the side of the road and at many places, the lane markings are not visible due to natural or unnatural processes of erosion. Therefore, most of the proposed lane detection algorithms that are for the foreign roads failed to correctly detect the lanes. In this paper, we proposed a lane detection model and a new thresholding approach for correct detection of lanes in the NH5 between Khandagiri and Khurdha, Odisha, India.

Keywords Road scene · Lane detection · Driver assistance system · Inverse perspective mapping · Gray conversion · Segmentation · Thresholding · Peak–valley detection

P. Kanungo (✉) · S.K. Mishra · S. Mahapatra · U.R. Sahoo · U.S.Kr. Sah · V. Taunk
Image Analysis and Computer Vision Lab, Department of Electronics and
Telecommunication Engineering, C. V. Raman College of Engineering, Bhubaneswar, India
e-mail: pkanungo@gmail.com

S.K. Mishra
e-mail: sandeep.mishra2911@gmail.com

S. Mahapatra
e-mail: sjm0005@yahoo.in

U.R. Sahoo
e-mail: uddesh.8460@gmail.com

U.S.Kr. Sah
e-mail: devsah87@yahoo.com

V. Taunk
e-mail: varsha.taunk@gmail.com

© Springer India 2015
L.C. Jain et al. (eds.), *Intelligent Computing, Communication and Devices*,
Advances in Intelligent Systems and Computing 309,
DOI 10.1007/978-81-322-2009-1_45

1 Introduction

Vision-based automated guided vehicle (AGV) needs to solve two important task: lane detection and lane tracking. Lane detection is the first and most important task than the lane tracking [1]. The main problems that must be faced in the detection of road boundaries or lane markings are: (1) the presence of shadows, producing gray level or color variation on the road surface, and thus altering its texture (2) the presence of other vehicles on the path, partly occluding the visibility of the road.

Alberto Broggi [2] used the edge curvature and direction features to improve the performance of a lane detection algorithm. Assidiq et al. [3] used edge as a feature and used a pair of hyperbolas that are fitting to the edges are extracted by Hough transform. Most of the literature used the inverse perspective mapping to create a bird's-eye view of the road scene [4, 5]. Borkar et al. [6] proposed a solution to address the problem of scattered shadows, illumination changes, pressure of neighboring vehicles, etc., during a lane detection. Borkar et al. [7] developed another method based on filter to detect lane marking features and then using the polar randomized Hough transform to detect the orientation of each line to detect the lanes in a road scene. Recently, Tran et al. [8] and Wang et al. [9] performed different methods for lane detection. Whereas, Tran et al. proposed the method for night scene by considering the region of interest in a scene.

Almost all these methods applied on the foreign road scenes, where the contrast between the road and lanes is very high and the roads are quite neat and clean. Because of that, most of the literature considered the above two problems during the lane detection phase. Whereas in most of the road scenes in the Odisha in India, particularly the road scenes considered in this project between Khandagiri, Bhubaneswar and Khurdha bypass, it has been observed that the contrast between the road and lane is very poor due to the dust and mud surface covered on the road and most of the lanes are eroded due to natural or unnatural process. Therefore, for lane detection in these conditions, we added two new problems including the existing two problems that are: (1) the contrast between lane marking and road is very low due to dust surface on the road and (2) missing of lane marks due to the natural or unnatural tampering.

In this paper, we developed a model where we considered the segmentation processes first before the inverse perspective map (IPM) to decrease the misclassification process in the subsequent processes and to avoid other post processing approaches. Therefore, we proposed a global thresholding approach to increase the efficiency of the binarization processes by minimizing the misclassification error.

2 Inverse Perspective Mapping

The road can be analyzed from the car view image or from a bird-eye view after a perspective transformation assuming that the road is flat. But the problem associated with analyzing from the car view is that the width of the lane markings changes according to the distance from the camera due to the perspective effect introduced by the acquisition conditions. But after the removal of perspective effect or inverse perspective mapping [4, 5, 10], each pixel represents the same portion of road, allowing homogeneous distribution of information among all the pixels. To remove the perspective effect, it is necessary to know the specific acquisition conditions and the scene represented in the image. This constitutes the a priori knowledge. The inverse perspective mapping [11] is done as follows:

$$x = h * \cot\left\{\theta - \beta + u\left(\frac{2\beta}{m-1}\right)\right\} * \cos\left\{\gamma - \alpha + v\left(\frac{2\alpha}{n-1}\right)\right\} + l$$

$$y = h * \cot\left\{\theta - \beta + u\left(\frac{2\beta}{m-1}\right)\right\} * \sin\left\{\gamma - \alpha + v\left(\frac{2\alpha}{n-1}\right)\right\} + d \qquad (1)$$

$$z = 0$$

where, u and v represent the image coordinate system, and x, y, and z be the world coordinate system where (x, y, and 0) indicates the road surface. l, d, and h are the coordinates of the camera in the world coordinate system, while θ and γ are the camera's tilt and pan angles, respectively. α and β are the horizontal and vertical aperture angles, and m and n indicate the height and width of an image.

3 Proposed Automatic Lane Detection Method

The process flow of the proposed automatic lane detection method is illustrated in Fig. 1. First, the video is captured using a Canon HDCAM HFM31 (25fps) camera, which was located inside a vehicle.

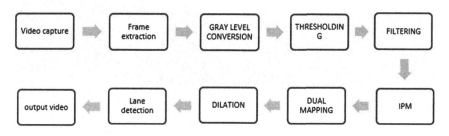

Fig. 1 Process flow of the proposed system

The extracted frames are resized and converted into a grayscale based on the proposed model, which is not the same as the conventional color to gray conversion process, as of (2)

$$I = w_1 * R + w_2 * G + w_3 * B \tag{2}$$

where, I is intensity value of grayscale image, and w_1, w_2, w_3 are the weights given to the R, G, B components of the image, respectively. In this work, $w_1 = w_2 = 0.1$ and $w_3 = 0.8$ because in test road scene videos, the variation in the red and green components to the left and right of the road markings in the image is quite high whereas the variation in the blue component is very low.

This gray image (frame) is subjected to a segmentation and filtering process. The gray-level image is segmented using the proposed weighted peak–valley thresholding method to give the appropriate result for the lane detection purpose. The proposed thresholding method is discussed in Sect. 4. The brightness value of a generic pixel (x, y) of the segmented image is compared to its horizontal left and right neighbors at a distance C. The application of this filtering further reduces the white portions of the image, which may be due to sky or barricades or other noise. The filtering process is carried out using following expression:

$$g(x,y) = \begin{cases} f(x,y), & \text{if}(D_{+c}(x,y) > 0 \text{ and } D_{-c}(x,y) > 0) \\ 0, & \text{otherwise} \end{cases} \tag{3}$$

where, $D_{+c}(x,y) = f(x,y) - f(x,y+c)$, $D(x,y) = f(x,y) - f(x,y-c)$, and $g(x,y)$ is the filtered binary image and $f(x,y)$ is the segmented image. The original image, the processed thresholded image, and the corresponding proposed filtered image are shown in Fig. 2a, b, and c, respectively.

The IPM operation is performed according to the process described in Sect. 2. The result is shown in Fig. 3a.

The dual mapping or reverse IPM is performed on the IPM output image. The result of dual mapping is shown in Fig. 3b. The dual mapping [10] can be performed using the equations:

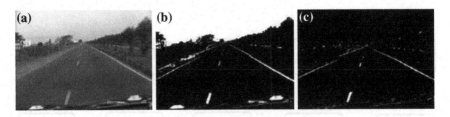

Fig. 2 **a** *Gray* image, **b** binarized image, **c** filtered image

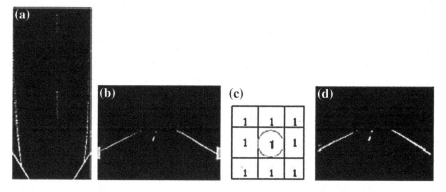

Fig. 3 **a** IPM image of Fig. 2c, **b** Dual-mapping result of (**a**), **c** kernel used for dilation, and **d** dilation result of (**b**)

$$u(x, y, 0) = \frac{\arctan\left[h\sin\frac{\gamma(x,y,0)}{y-d}\right] - (\theta - \alpha)}{\left[\frac{2\alpha}{n-1}\right]} ; \ v(x, y, 0)$$
$$= \frac{\arctan[(y - d)/(x - l)] - (\gamma - \alpha)}{\left[\frac{2\alpha}{n-1}\right]} \tag{4}$$

The image obtained after dual mapping is dilated using a 3×3 kernel to give improved lane detection result. The result of dilation is shown in Fig. 3d.

The portions of dilated image having white pixel values correspond to the lane marking. Thus, if a pixel in dilated image is white, we make the corresponding pixel on the original frame red (or any suitable color). After processing all pixels, lane markings on the original frame become red.

4 Weighted Peak–Valley Thresholding (WPVT) Method

In this proposed method, the maxima points (peaks) and minima points (valley) of the corresponding histograms are evaluated and the threshold is evaluated. The algorithm is developed as follows:

1. Evaluate the featured histogram $h(x)$ by smoothing the original histogram.
2. Consider a gray-level value x. If $h(x + 1) < h(x) > h(x - 1)$ and frequency of the gray value x is more than a threshold (in this work, the threshold is $0.001 \times M \times N$. Where, M and N are height and width of the image, respectively), then x is a valid peak.
3. There must be some minimum distance between peaks. We have taken this distance to be 20 gray-value distance. If more than one maxima is present in the distance of 20 intensity values, largest of them will be taken as valid peak and rest of the peaks are discarded.

Table 1 Thresholding values for different thresholding methods

Test images	Otsu's method	Mean-sigma global thresholding	Weighted Peak–valley global thresholding	Manual
Image 1	162	203	212	201
Image 2	104	133	145	138
Image 3	133	141	149	154
Image 4	167	182	167	176
Image 5	154	177	182	194
Image 6	175	193	157	177

4. Consider a gray-level value x in between last two peaks. If $h(x+1) > h(x) < h(x-1)$, then at gray value x, there is a minima. Due to the above restricting conditions, number of minima between two peaks may be more. Then, the threshold value is evaluated by

$$T = k1 * \min + k2 * \text{avg} \qquad (5)$$

where "min" is the minima having lowest gray-level value and "avg" is the average intensity of all minimas. $k1$ and $k2$ are the weights applied to them, respectively $k1 + k2 \le 1$. We have taken $k1 = 0.55$ and $k2 = 0.38$ for calculating the threshold.

Table 1 shows the comparison of various thresholding methods on various road scenes and corresponding threshold values. From the table, it has been observed that the threshold values obtained using proposed method are much closer to that of manual thresholding method in comparison with OTSU's and mean-sigma thresholding method [9].

5 Simulation and Results

In this section, different Indian lane images are taken into consideration to perform lane detection using the proposed method and Bertozzi et al.'s method [10]. The results of both these methods are compared in the end. All the experiments were performed on a COMPAQ 620 notebook: ×64, Intel(R) Core (TM) 2Duo CPU, 3GB RAM, 2MB L2 Cache. The algorithm is implemented using Open CV libraries. The acquisition parameters for camera are as follows: $\alpha = 24.09°$, $\beta = 53.72$, $\theta = 13.26$, $\gamma = 0$, $D = 70$ cm.

The proposed lane detection approach is tested with three different scenes as shown in Fig. 4: row (a). The figures in the first and second column of row (a) represent a straight road with different left, right, and horizon. The road scene in

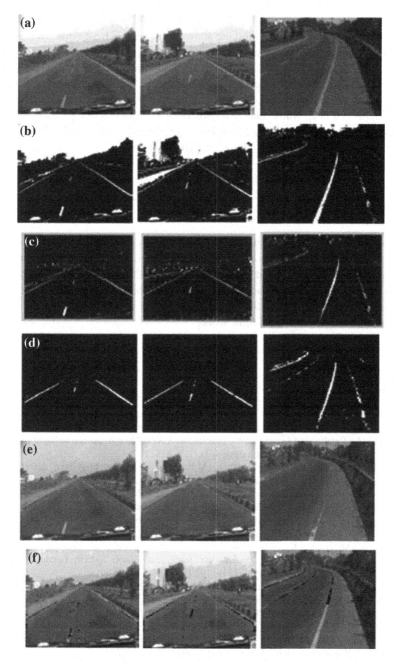

Fig. 4 **a** Original *gray*-level images, **b** corresponding segmented images using WPVT method, **c** corresponding filtered binary images, **d** dilated DIPM outputs, **e** lane detection output using proposed method, and **f** lane detection output using GOLD method

column 3 of Fig: 4 row (a) shows a curved lane. The corresponding binary images using our proposed WPVT method are shown in row (b). These binary images are passed through a filtering process. After the filtering, IPM and DIPM are evaluated, followed by dilation. The filtered image and dilated DIPM results are shown in Fig. 4 row (c) and (d), respectively. The proposed lane detection, shown in red color, is shown in Fig. 4 row (e). We compared our approach with the Bertozzi et al.'s approach [10], whose lane detection results are shown in Fig. 4 row (f). It is clearly observed that our proposed method is better than Bertozzi et al.'s approach for lane detection in NH5 road scene of Odisha, India.

6 Conclusion

The problem of NH5 in Odisha has been studied. A new method is proposed to address this problem. To improve the efficiency, a new thresholding algorithm and filtering method are proposed. It has been found that the proposed method is simple and reduced most of the post processing methods, basically used by many literatures. This algorithm's efficiency is less in case of curve road scenes, which we intend to improve in our future work.

References

1. Kluge, K., Lakshmanan, S.: A deformable template approach to lane detection. In: Proceedings of IEEE Intelligent Vehicles, pp. 54–59. Detroit, MI (1995)
2. Broggi, A.: Parallel and Local Feature Extraction: A Real Time Approach to Road Boundary Detection. IEEE Trans. Image Process. **4**(2), 217–223 (1995)
3. Assidiq, A.A.M., Khalifa, O.O., Islam, M.R., Khan, S.: Real time lane detection for autonomous vehicles. In: Proceedings of the International Conference on Computer and Communication Engineering, pp. 82–88. Kuala Lumpur, Malaysia (2008)
4. Borkar, A., Hayes, M., Smith, M.T.: Robust Lane Detection And Tracking With ransac and kalman filter. In: Proceedings of the International Conference on Image Processing, pp. 3261–3264 (2009)
5. Collado, J.M., Hilario, C., de la Escalera, A., Armingo, J.M.: Adaptative road lanes detection and classification. Advanced Concepts for Intelligent Vision System Lecture Notes in Computer Science, vol. 4179, pp. 1151–1162 (2006)
6. Borkar, A., Hayes, M., Smith, M.T.: An efficient method to generate ground truth for evaluating lane detection systems. In: Proceedings of the International Conference on Acoustics Speech and Signal Processing, pp. 1090–1093 (2010)
7. Borkar, A., Hayes, M., Smith, M.T.: Polar randomized Hough transform for lane detection using loose constraints of parallel lines. In: Proceedings of the International Conference on Acoustics Speech and Signal Processing, pp. 1037–1040 (2011)
8. Tran, T.T., Son, J.H., Uk, B.J., Lee, J.H., Cho H.M.: An Adaptive Method for Detecting Lane Boundary in Night scene. ICIC 2010, LNAI 6216, pp. 301–308 (2010)
9. Wang, H., Ren, M., Shao, S.: Lane Markers Detection based on Consecutive Threshold Segmentation. J. Info. Comput. Sci. **6**(3), 207–212 (2011)

10. Bertozzi, M., Broggi, A.: GOLD: a parallel real-time stereo vision system for generic obstacle and lane detection. IEEE Trans. Image Process. **7**, 62–81 (1998)
11. Lin, C.-T., Shen, T.-K., Shou, Y.-W.: Construction of fisheye lens inverse perspective mapping model and its applications of obstacle detection. EURASIP J. Adv. Signal Process. **2010**, 1–23 (2010)

Mammogram Image Segmentation Using Hybridization of Fuzzy Clustering and Optimization Algorithms

Guru Kalyan Kanungo, Nalini Singh, Judhisthir Dash and Annapurna Mishra

Abstract Mammogram images have the ability to assist physicians in detecting breast cancer caused by cells abnormal growth. But due to visual interpretation, false results can be obtained. In this paper, to reduce false results, image segmentation is carried out to find breast cancer mass. Image segmentation using Fuzzy clustering: K means, FCM, and FPCM shows result better than other existing methods but initialization problem and sensitivity to noise do not make them to achieve better accuracy. Various extension of the FCM for segmentation is developed. But most of them modify the objective function which changes the basic FCM algorithm. Hence efforts have been made to develop FCM algorithm without modifying objective function for better segmentation. We have proposed a technique GA-ACO-FCM, which is the hybridization of optimization tools: genetic algorithm and ant colony optimization with fuzzy C means .GA-ACO-FCM is suitable to overcome initialization problem of FCM and shows better results with achieving high accuracy.

Keywords Image segmentation · Mammography · K-means · Fuzzy C-means · FPCM and GA-ACO-FCM

G.K. Kanungo (✉) · J. Dash · A. Mishra
Department of Electronics and Telecommunication Engineering, Silicon Institute
of Technology, Bhubaneswar, India
e-mail: gurukalyan@rocketmail.com

N. Singh
Department of Applied Electronics and Instrumentation Engineering, Silicon Institute
of Technology, Bhubaneswar, India

© Springer India 2015
L.C. Jain et al. (eds.), *Intelligent Computing, Communication and Devices*,
Advances in Intelligent Systems and Computing 309,
DOI 10.1007/978-81-322-2009-1_46

1 Introduction

1.1 Imaging System Used for Breast Cancer Detection

Breast image analysis can be performed using X-rays, magnetic resonance, nuclear medicine, or ultrasound [1].

(a) *X-Ray Mammography*

X-Ray Mammography is commonly used in clinical practice for diagnostic and screening purposes [2]. Mammography provides high sensitivity on fatty breast and excellent demonstration of micro-calcifications; it is highly indicative of an early malignancy [3, 4].

(b) *MRI of the Breast*

Magnetic Resonance Imaging is the most attractive alternative to Mammography for detecting some cancers which could be missed by mammography. In addition, MRI can help radiologists and other specialists determine how to treat breast cancer patients by identifying the stage of the disease [5].

1.2 Research Goal

Since screening mammography is currently the main test for early detection of breast cancer, a huge number of mammograms need to be examined by a limited number of radiologists, resulting misdiagnoses due to human errors by visual fatigue. In the previous work, we were not able find out the total cancer affected area. It was only able to find out the masses of the tumor. Currently, there are several image processing methods proposed for the detection of tumors in mammograms. In this paper, we have proposed a new technique for cancer mass detection of the mammogram image.

2 Literature Review

1. Nalini Singh, Ambarish G Mohapatra, Gurukalyan Kanungo from Silicon Institute of Technology, Bhubaneswar, India in their paper "Breast Cancer Mass Detection in Mammograms using K-means and Fuzzy C-means Clustering" described a computational method that modeled a type of breast cancer. In this method mainly two algorithms are used which are K-means Clustering and Fuzzy C-means Clustering to find the cancer infected cells in the processed image [6].
2. S. Saheb Basha, Dr. K. Satya Prasad in their paper "Automatic Detection of Breast Cancer Mass in Mammograms Using Morphological Operators And

Fuzzy C–Means Clustering" described a process of automatic determination of the breast cancer mass in mammograms using morphological operators and fuzzy c–means clustering [7].

3. Poulami Das, Debnath Bhattacharyya, Samir K. Bandyopadhay, and Tai-hoon Kim in their paper "Analysis and Diagnosis of Breast Cancer" described a graphical process to determine the cell is infected or not and suggest for further pathological test or no need of the test. In this process, they took 18 invasive breast cancer tissues from different 18 patients and 8 noncancerous falsely detected breast tissues from 8 different normal females. Each of the 24-bit BMP image was converted to 256-color gray scale image which were again converted to bi-color (using Pixel Clustering on Threshold Value, T) and then the cells were represented in a spatial domain. From it, a graph was drawn which was compared with other graphs to suggest whether further pathological test was required or not [8].

3 Methods for Breast Cancer Detection

3.1 K-Mean Clustering Algorithm

In statistics and machine learning, k-means clustering is a method of cluster analysis which aims to partition "n" observations into "k" clusters in which each observation belongs to the cluster with the nearest mean [9]. For a given set of observation $(x_1, x_2, \ldots \ldots x_n)$, where each observation is a d-dimensional real vector, then k-means clustering aims to partition the "n" observations in to "k" sets $(k < n)$. k is positive integer number. The grouping is done by minimizing the sum of squares of distances between data and the corresponding cluster centroid. Thus, the purpose of K-mean clustering is to classify the data.

(a) k initial "means" (in this case $k = 3$) are randomly selected from the dataset.
(b) k clusters are created by associating every observation with the nearest mean. The partitions here represent the Voronoi diagram generated by the means.
(c) The centroid becomes the new means.
(d) Steps (b) and (c) are repeated until convergence has been reached.

Problems of K means:
Data points that is almost equally distant from two or more clusters. Such special data points can represent hybrid-type or mixture objects, which are (more or less) equally similar to two or more types. A crisp partition arbitrarily forces the full assignment of such data points to one of the clusters, although they should (almost) equally belong to all of them.

3.2 Fuzzy C-Mean Algorithm

The Fuzzy C-means algorithm, also known as fuzzy ISODATA, is one of the most frequently used methods in pattern recognition. Fuzzy C-means (FCM) is a method of clustering which allows one piece of data to belong to two or more clusters [9]. It is based on the minimization of objective function to achieve a good classification. "J" is a squared error clustering criterion, and solutions of minimization are least squared error stationary point of "J" in Eq. (1).

$$j_m = \sum_{i=1}^{k} \sum_{j=1}^{c} u_{ij} \left\| x_i - c_j \right\|^2 \tag{1}$$

where, $1 \leq m \leq \infty$

Where "m" is any real number greater than 1, and is the degree of membership of in the cluster "j", and is the dimensional measured data, and is the dimension center of the cluster, and is any norm expressing the similarity between any measured data and the center. Fuzzy partitioning is carried out through an iterative optimization of the objective function shown above, with the update of membership u_{ij} in Eq. (2) and the cluster centers c_j by Eq. (3)

$$u_{ij} = \frac{1}{\sum_{k=1}^{c} \left[\frac{\|x_i - c_j\|}{\|x_i - c_k\|} \right]^{\frac{2}{m-1}}} \tag{2}$$

$$c_j = \frac{\sum_{i=1}^{k} u_{ij} . x_i}{\sum_{i=1}^{k} u_{ij}} \tag{3}$$

The iteration will stop when

$$\max_{ij} = \left\{ \left| u_{ij}^{k+1} - u_{ij}^{k} \right| \right\} < \in \tag{4}$$

where $\in$ is the termination criterion between 0 and 1, whereas k is the iteration step. This procedure converges to a local minimum or a saddle point of J_m.

The fuzzy c means algorithm composed of following steps.

1. Initialize $U = [U_{ij}]$ matrix, $U^{(0)}$
2. At k-step calculate the center vectors $C^{(k)} = [C_j]$ with $U^{(k)}$.
3. Update $U^{(k)}$, $U^{(k+1)}$

$$u_{ij} = \frac{1}{\sum_{k=1}^{c} \left[\frac{\|x_i - c_j\|}{\|x_i - c_k\|} \right]^{\frac{2}{m-1}}}$$

4. If $\left\| u_{ij}^{k+1} - u_{ij}^{k} \right\| < \in$ then STOP, otherwise return to step 2.

Working of FCM: FCM, starts with an initial guess for the cluster centers, which are intended to mark the mean location of each cluster. The initial guess for these cluster centers is most likely incorrect. Next, FCM assigns every data point a membership grade for each cluster. By iteratively updating the cluster centers and the membership grades for each data point, FCM iteratively moves the cluster centers to the right location within an image.

Disadvantage of FCM: Although often desirable, the "relative" character of the probabilistic membership degrees can be misleading. Fairly high values for the membership of datum in more than one cluster can lead to the impression that the data point is typical for the clusters, but this is not always the case. Consider, for example, the simple case of two clusters shown in Figs. 1 and 2. Datum x_1 has the same distance to both clusters and thus it is assigned a membership degree of about 0.5. This is plausible. However, the same degrees of membership are assigned to datum x_2 even though this datum is further away from both clusters and should be considered less typical. Because of the normalization, however, the sum of the memberships has to be 1. Consequently, x_2 receives fairly high membership degrees to both clusters. For a correct interpretation of these memberships, one has to keep in mind that they are rather degrees of sharing than of typicality, since the constant weight of 1 given to a datum must be distributed over the clusters. A better reading of the memberships, avoiding misinterpretations, would be "If the datum x_i has to be assigned to a cluster, then with the probability u_{ij} to the cluster i." The

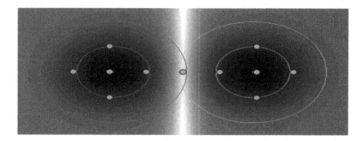

Fig. 1 A symmetric dataset with two clusters

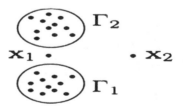

Fig. 2 Situation in which the probabilistic assignment of membership degrees is counterintuitive for datum x_2

normalization of memberships can further lead to undesired effects in the presence of noise and outliers. The fixed data point weight may result in high membership of these points to clusters, even though they are a large distance from the bulk of data. Their membership values consequently affect the clustering results, since data point weight attracts cluster prototypes. By dropping the normalization constraint in the following definition, one tries to achieve a more intuitive assignment of degrees of membership and to avoid undesirable normalization effects.

3.3 Fuzzy Possibility C Means Algorithm

To overcome difficulties of the PCM i.e., very sensitive to good initialization and coincident clusters may result, because the columns and rows of the typicality matrix are independent of each others. So due to coincident of clusters lead to undesirable detection of cancer. Pal defines a clustering technique that integrates the features of both Fuzzy a possibilistic c-means called Fuzzy possibilistic c-Means (FPCM). Membership and typicality's are very significant for the accurate characteristic of data substructure in clustering difficulty. FPCM generates memberships and possibilities at the same time, together with the usual point prototypes or cluster center for each cluster. An objective function in the FPCM depending on both membership and typicality's are represented as:

$$J_{FPCM}^m(U,T,A,X) = \sum_{i=1}^{c} \sum_{i=1}^{n} \left(\mu_{ij}^m + t_{ij}^n \right) \|x_j - c_i\|^2 \tag{5}$$

Constraints:

$$\text{membership} : \sum_{i=1}^{c} \mu_{ij} = 1, \forall j = 1, 2, \ldots, n$$

$$\text{typicality} \sum_{j=1}^{n} t_{ij} = 1, \forall i = 1, 2, \ldots, c$$

3.4 Proposed Segmentation Method

GA-ACO-FCM

In many occasions, the mammogram images are affected with noise. Fuzzy c means clustering-based segmentation does not give good segmentation result under such condition and initialization is another problem of FCM. Various extension of the FCM for segmentation is developed in recent years. But most of them modify the objective function hence changing the basic FCM algorithm present in MATLAB toolboxes. Hence efforts have been made to develop FCM algorithm without modifying its objective function for better segmentation.

To denoise mammogram image, in preprocessing phase median filter is used. The median filter is a nonlinear digital filtering technique used to remove noise. Such noise reduction is a typical preprocessing step to improve the results of later processing. It preserves edges while removing noise.

GA-ACO: ACO is used to find the shortest path with help of distance value, the output of the ACO is given as a input to the GA. The set of paths are obtained during the ACO process are inputs to the GA. The genetic algorithm undergoes the selection, crossover, and mutation process and it gives the result. The result contains only one path which is optimal among the shortest path.

The pseudo code for the hybrid is start with the current node, it whether the current node is the destination node, if it is a destination then save the path, otherwise check for other neighboring node with the distance value and find fitness. Here note one thing that every node have a certain value, this values are refer as a distance in ACO at a same time it is used as a cost in GA. For finding a path, we used a ACO and to find the fitness, optimal path among these can found by GA. The ACO is used not only to find the path but also it is used to maintain the routing table as very simple and it is easy to understand [10].

The reasons for using genetic algorithms are:

They are parallel in nature. They explore solution space in multiple directions at once. GA is well suited for solving problems, where the solution space is huge and time taken to search exhaustively is very high. They perform well in problems with complex fitness. If the function is discontinuous, noisy, changes over time or has many local optima, then GA gives better results. Genetic algorithm has ability to solve problems with no previous knowledge (blind). For this reason, we hybrid the ACO with GA to find the best initial value [9].

Steps:

Initialization

(a)   Set initial parameters: variable states, function, input
(b)   Set initial pheromone trails value
(c)   Each ant is individually placed on initial state with empty memory.

While (not termination)

(a)   Construct Ant Solution:
(b)   Calculate attractiveness of next move
(c)   Apply Local Search GA (Avoid local minima)
(d)   If there is an improvement-. Update Trails
(e)   Calculate evaporation through genetic operators
(f)   Select the population with a probability based on fitness.

End While

Then the optimized initial value is used for initialization of FCM. Then repeat all the steps of standard FCM as above in Sect. 3.2.

4 Results and Discussions

The various experiment carried out on the above said imagery in MATLAB v7.12. In this work, there are two classes: one is benign (noncancerous) and the other is malignant (cancerous) images were taken. In mammogram images, the background and the gray level of the normal breast tissue are very low. Thus having one cluster center of value "zero" i.e., minimum pixel values of image would cluster those regions into one cluster. Considering the abnormal or cancerous image a cluster center value "255" i.e., maximum pixel values of image would cluster those into one region. The cancerous region is separated from the fatty breast regions using clustering techniques. For the normal tissue, there is no evidence of the separation in three clusters i.e., distance of pixel value of normal mammogram image to any cluster is same. The complete process and the standard results are summarized in subsequent Fig. 3, 4, and 5.

Segmentation Evaluation:

The accuracy of a measurement system is the degree of closeness of measurements of a quantity to that quantity's actual (true) value.

$$\text{Accuracy} = \frac{(TP + TN)}{(TP + TN + FP + FN)} \tag{6}$$

Fig. 3 Flow chart of segmentation algorithm

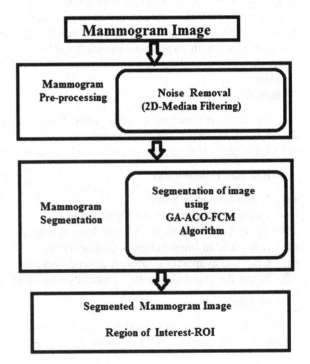

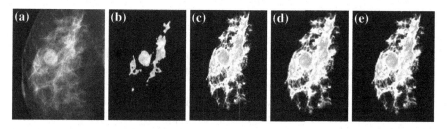

Fig. 4 **a** The original cancer image results from three clusters with **b** k-means algorithm. **c** FCM **d** FPCM **e** GA-ACO-FCM

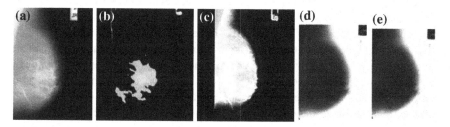

Fig. 5 **a** The original cancer image results from three clusters with **b** k-means algorithm. **c** FCM **d** FPCM **e** Proposed GA-ACO-FCM

Table 1 Comparison of accuracy of different methods

Method	Accuracy in percentage
K Means	63.00
FCM	73.00
FPCM	87.67
GA-ACO-FCM	92.52

True positive (TP) = correctly identified False positive (FP) = incorrectly identified

True negative (TN) = correctly rejected False negative (FN) = incorrectly rejected

A false positive is when the outcome is incorrectly classified as "yes" (or "positive"), when it is in fact "no" (or "negative"). A false negative is when the outcome is incorrectly classified as negative when it is in fact positive. True positives and true negatives are obviously correct classifications. Mammogram images are taken from DDSM and mini MIAS, UK database. From Fig. 5b K means shows presence of cancer in normal image which is false report. From a mixture 322 images from MIAS database and 65 images from DDSM database, the accuracy of K means only 63 %, FCM has 73 % where as proposed GA-ACO-FCM shows good accuracy i.e., 92.52 %. The percentage of accuracy of different fuzzy clustering methods performed on 387 images are given in the (Table 1).

5 Conclusion

Breast cancer is one of the major causes of death among women. So early diagnosis through regular screening and timely treatment has been shown to prevent cancer. As per report raw mammogram images, which are visually interpreted by radiologist having accuracy only 63 %, which in result mislead the treatment of patient. In this paper, we have presented a novel approach to identify the presence of breast cancer mass and calcification in mammograms using Fuzzy Clusterings: K-means, Fuzzy C-Means, FPCM, and proposed GA-ACO-FCM clustering for clear identification of mass. Combining these we have successfully detected the breast cancer area in raw mammograms images with accuracy 92.52 %. The results indicate that this system can facilitate the radiologist to detect the breast cancer in the early stage of diagnosis as well as classify the total cancer affected area. This will help doctor to take or analyze in which stage of cancer the patient have and according to which he/she can take necessary and appropriate treatment steps. This proposed method is low cost as it can be implemented in general computer. A real-time system can be implemented using suitable data acquisition software and hardware interface with digital mammography systems.

Acknowledgment The authors thank to Prof. A.K. Tripathy, Dr. B.B. Mishra, and Dept. of Electronics and Telecommunication for their valuable tips and suggestion on this topic and programming.

References

1. Dong, A., Wang, B.: Feature selection and analysis on mammogram classification. In: Communications, Computers and Signal Processing, pp. 731–735 (2009)
2. Thangavel, K., Mohideen, A.K.: Semi-supervised k-means clustering for outlier detection in mammogram classification. In: IEEE International Conference on Trendz in Information Sciences & Computing (TISC), pp. 68–72 (2010)
3. Cahoon, T.C., Sutton, M.A., Bezdek, J.C.: Breast cancer detection using image processing techniques. In: IEEE 9th International Conference on Fuzzy Systems, vol. 2, pp. 973–976 (2000)
4. Barrea, A.: Local fuzzy C-means clustering for medical spectroscopy images. Appl. Math. Sci. **5**(30), 1449–1458 (2011)
5. Maitra, I.K., Nag, S., Bandyopadhyay, S.K.: Identification of abnormal masses in digital mammography images. Int. J. Comput. Graph. **2**(1), 17–30 (2011)
6. Singh, N., Mohapatra, A.G., Kanungo, G.K.: Breast cancer mass detection in mammograms using K-means and fuzzy C-means clustering. Int. J. Comput. Appl. **22**(2), 15–21 (2011)
7. Prasad, K.S., Basha, S.S.: Automatic detection of breast cancer mass in mammograms using morphological operators and fuzzy C-means clustering. J. Theor. Appl. Info. Technol. **5**, 704–709 (2009)
8. Das, P., Bhattacharyya, D., Bandyopadhyay, S.K., Kim, T.H.: Analysis and diagnosis of breast cancer. Int. J. u- and e-Serv. Sci. Technol. **2**(2), 1–12 (2009)

9. Peng, Y., Hou, X., Liu, S.: The k-means clustering algorithm based on density and ant colony. In: IEEE International Conference in Neural Networks and Signal Processing, vol. 1, pp. 457–460 (2003)
10. Azadeh, A., Keramati, A., Panahi, H.: A hybrid GA-ant colony approach for exploring the relationship between IT and firm performance. Int. J. Bus. Info. Syst. Arch. **4**(5), 542–563 (2009)

A New ICA-Based Algorithm
for Diagnosis of Coronary Artery Disease

Zahra Mahmoodabadi and Saeed Shaerbaf Tabrizi

Abstract Large amount of data in recent years have pushed experts to use data mining techniques in all fields. Data mining is a process in which useful information from raw data is obtained; it can also be used in classification problems. Lately, in some systems, especially in medical systems, experts have tried to combine data mining techniques and evolutionary algorithms to get accurate results. One of the most critical diseases, which has a considerable mortality rate in the world, is coronary artery disease. To improve the diagnosis of this dangerous disease in the early stages, we proposed a system which uses data mining techniques and an evolutionary algorithm called Imperialist Competitive algorithm (ICA). The proposed system used an algorithm based on the decision tree to reduce the data dimension and to produce valid rules. Then, a fuzzy system is created. Tuning fuzzy membership functions were done using ICA and Improved ICA to optimize the results. Since the convergence speed is one of the important factors in an evolutionary algorithm, a change was made in this algorithm so that the convergence occurs more quickly. The results show that ICA and Improved ICA produce the same results in classification accuracy, but the convergence time is different. The proposed system gets an accuracy of 94.92 %, which is high in comparison with similar works.

Keywords Imperialist competitive algorithm · Fuzzy · Coronary artery disease · Data mining techniques

Z. Mahmoodabadi (✉) · S.S. Tabrizi
Imamreza University, Mashhad, Iran
e-mail: za.mahmoodabadi@gmail.com

S.S. Tabrizi
e-mail: shaerbaf@imamreza.ac.ir

© Springer India 2015
L.C. Jain et al. (eds.), *Intelligent Computing, Communication and Devices,*
Advances in Intelligent Systems and Computing 309,
DOI 10.1007/978-81-322-2009-1_47

415

1 Introduction

The modern world and lifestyle have effects on the health of many people. Today, the incidence of some diseases has increased due to lifestyle changes, dietary habit changes, and some behavior changes in general. Cardiovascular diseases are a group of diseases that occur due to the mentioned issues, and a great number of people have suffered from this disease. Coronary artery disease is the most common disease in the group which leads to considerable mortality rate every year. Each year, over 60,000 deaths from this disease occur in Europe alone. This disease is the largest killer disease in America so that one out of six deaths in the United States in the year 2006 has been due to CAD [1].

In the coronary artery disease, the coronary arteries become narrower and are deprived of adequate blood and oxygen from reaching the heart muscle. Factors such as high cholesterol, diabetes, blood pressure, and smoking over time lead to narrowing of the coronary arteries. Coronary artery disease is without any apparent signs and symptoms such as pain, shortness of breath, but it is not the reason for the health of the person [2].

In order to make diagnose heart disease, there are methods like echocardiography and myocardial scintigraphy [3], which are expensive and not always usable. Exercise testing is commonly used, and it is inexpensive, but it has low sensitivity and not very specific. It also may not be able to detect the disease in early stages. Currently, the most useful and reliable method for the detection of coronary occlusion is angiography [4]. This method, in addition to being time-consuming and costly, is an invasive procedure and may bear some unwanted risks for the patients. Therefore, a noninvasive or minimally invasive diagnostic method can be very important and beneficial.

In recent years, using a combination of intelligent algorithms and data mining techniques has been very common to improve detection of any disease. These methods, in addition to being noninvasive, help physicians to use large volumes of data much easier.

Data mining techniques involve gathering, preprocessing, and analysis of data. Intelligent algorithms often refer to algorithms which are inspired by natural phenomena to solve optimization problems. Included in such algorithms is the PSO algorithm which is inspired by the natural behavior of animals such as bees, birds, fish, etc. Algorithms like genetic algorithms and neural networks contribute to solving optimization problems by simulation of the behavior of genes and neurons, respectively. Unlike the above algorithms, the ICA[1] is the algorithm which is inspired by a social–political behavior called imperialist. This algorithm is invented based on the imperialists' behaviors and their competition to take over more countries. The number of countries of an empire represents its power.

We use a decision tree and a fuzzy expert system to create a rule-based system. Then, using ICA, we tune the membership functions of the fuzzy system and

---

[1] Imperialist Competitive Algorithm.

calculate the classification accuracy. Since convergence is one of the important factors of optimization problems, in this article, we propose an approach to improve ICA convergence speed. To estimate the performance of the proposed system, we use CAD dataset from UCI Repository.

This paper is organized as follows: Sect. 2 briefly reviews the related works in CAD diagnosis. In Sect. 3, the proposed method is explained. Section 4 presents the experimental results. Section 5 comes to a conclusion regarding the proposed system.

2 Related Works

In cardiovascular disease group, CAD is the most common disease, which leads to death. CAD is a chronic disease in which the coronary arteries progressively become hard and narrow. This disease is the most common heart disease in the United States and is the main cause of heart attacks [5]. Due to the dangerous nature of the disease, about one-third of all deaths in the world occur because of this complication [1].

Early detection and prevention of the disease is important, and it is the most significant area of medical research. For CAD, there are many approaches which have tried to improve the diagnosis of the disease. We just review some of them in this article.

CAD diagnosis using a decision support system was done in 2007. The system used fuzzy weighted rules and AIRS. The reported classification accuracy was 92.59 % [6]. Diagnosis of CAD had been done using ensemble neural networks by Das [7]. The system reported an accuracy of 89.01 %. In 2010, Babaoglu used exercise test data and designed a classification system using support vector machine. The accuracy reported was 79.17 % [8].

In 2011, Anooj used weighted fuzzy rules to predict the risk of heart disease [9]. The system was evaluated by classification accuracy, which was 57.85 % and sensitivity of 45.22 % and specificity of 68.75 %. Using an ensemble-based PSO and fuzzy expert system was the approach, proposed by Ghadiri and Saniee [4]. This approach leads to get the accuracy of 92.59 %, adjusting the membership functions using the PSO algorithm proposed by Muthukaruppan [10]. The unique property of the system was using PSO for tuning the membership functions which had an accuracy of 93.27 %.

3 Proposed System

3.1 Decision Tree Algorithm

The system rules are extracted from an algorithm based on decision tree. Decision trees are one of the most common tools in data mining techniques. The usage of this tool is the classification of data. The advantage of a decision tree is its interpretability which is determined by rules it produces. The structure of decision tree contains some nodes and branches. Each leaf node represents a class label. By scrolling down, a path to a leaf of a decision tree will have a classification rule.

In the proposed system to have a set of valid rules, the decision tree algorithm was performed on the dataset. Although decision tree is a useful tool, it may lead to complexity due to anomalies in data. To control this anomaly, we need to simplify the tree. Pruning decision trees can cause the tree to be simple. Decision tree pruning is a method that simplifies the tree by removing one or more parts of the tree (subtrees) and replacing them with a leaf. The leaf is labeled with the most frequent class among the replaced subtree [11].

3.2 Fuzzy Inference System

After creation of the decision tree and extraction of rules, the rules must be converted to fuzzy rules under a fuzzy system.

To create a fuzzy system, input variables and output variable must be determined. The variables selected by decision tree are considered as main variables, which have more effects on the result than others. For each variable, we define membership functions. Membership functions are defined based on the boundary values on the branches of the decision tree.

3.3 Membership Functions Optimizations

In the fuzzy system, triangular membership functions were used for the shape of membership functions. Each triangular membership function has three parameters: left (L), center (C), and right (R). Figure 1 shows the membership function's parameters. L', C', and R' refer to left, center and right of the membership functions, respectively.

The following equations are defined to adjust membership functions:

$$C' = (C + ki) \tag{1}$$

$$L' = (L + ki) - wi \tag{2}$$

$$R' = (R + ki) - wi \tag{3}$$

where ki and wi are adjustment coefficients; ki makes each membership function move to left or right with no distortion in the form. The membership function shrinks or expands through the parameter wi. These parameters take any integer either positive or negative value. Imperialist competitive algorithm (ICA) will be used to find the optimum values for ki and wi for the membership functions.

Figure 2 displays a flowchart of the proposed system.

Fig. 1 Membership function
parameters [10]

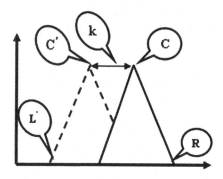

3.4 Improved ICA Versus ICA

ICA, like other evolutionary optimization algorithms, starts with an initial population. In this algorithm, each element of the population is called the country.
Countries are divided into two categories: colonies and colonizers. The colonizer,
depending on its power, dominates a number of colonies and controls them.
Assimilation and competing colonial policy constitutes the core of this algorithm.

According to the assimilation policy, imperialist countries using methods, such
as the construction of schools in their own language, attempt to colonize the
countries, by influencing the language of the country and its culture and customs.
In the algorithm, this policy is done by moving a colony of an empire, using a
special relationship. Figure 3 shows this movement.

According to this figure, the imperialist assimilates the colony in line with the
culture and language. As shown in this figure, the colony has moved to a new
position and is closer to the imperialist.

In this figure, the distance between the imperialist and the colony is shown by
d. And x is the random number with uniform distribution (or any other distribution). Thus, for x we have

$$x \sim U(0, \beta \times d) \tag{4}$$

where β is a number greater than one and close to 2. Considering $\beta = 2$ could be a
good choice. The existence of β coefficient makes the colony to move to the
imperialist, from different directions.

The problem with the normal ICA is that the assimilation process is done based
on a random relation in Eq. (4); that is, the movement and assimilation process do
not depend on the strength of the colony and imperialist. However, the angle and
the size of assimilation should be proportional to the strength of both sides.
Improved ICA proposes a relation which covers this problem.

To change the assimilation relation such that it depends on the strength of the
colony and imperialist, we define two variables of f^* and f^+ which stand for strength of
imperialist 1 and imperialist 2, respectively. Figure 4 displays the imperialists based

Fig. 2 Flowchart of the
proposed system

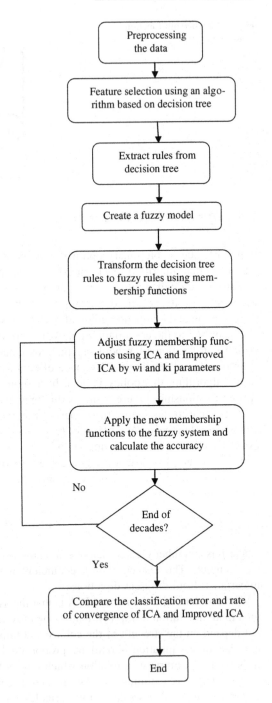

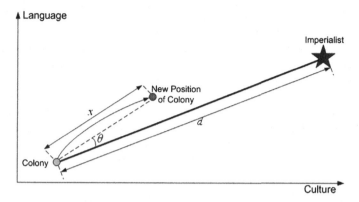

Fig. 3 Movement of the colony in its imperialist (assimilation policy) [12]

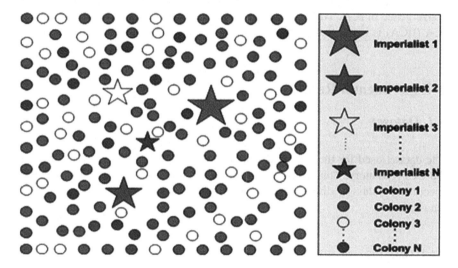

Fig. 4 Initial imperialists and their colonies [12]

on their strength. Imperialist 1 is the biggest one, and imperialist 2 is the second biggest one.

The proposed relation based on f^* and f^+ is as follows:

$$x_i = \frac{f^* - f_i}{f^* - f^+} \tag{5}$$

Where f_i specifies the strength of the colony ith.

In Eq. (5), for imperialist 2, x_i coefficient equals to 1; that is, for this colony, the movement does not change. However, for other colonies proportional to the distance they have to the imperialist, the strength of assimilation increased such that weaker colonies will be assimilated faster to the imperialist.

The idea is taken from the actual conditions in assimilation of the colonies where weaker colonies approximate their feature vector to the imperialist feature vector in a shorter time period.

With the same reasoning, it can be concluded that the weaker a colony is, the smaller will be the angle of movement to the imperialist. In other words, the possibility of keeping out from the imperialist features vector and withdrawal of assimilation is less for weak colonies. This concept is expressed by the following equation:

$$\theta_i = \frac{1}{x_i} \tag{6}$$

The proposed relations in the assimilation process lead to speeding up the convergence in ICA. The results show that the speed of convergence in improved ICA (IICA) is better than ICA.

4 Experimental Results

4.1 Dataset

The dataset used for the proposed system is CAD datasets which are taken from the UCI repository. This dataset has 13 input features and 1 output feature which specifies the class of disease. Table 1 shows the features and their types in CAD dataset.

4.1.1 Data Preprocessing

Data preprocessing in data mining is a significant step because it has a direct impact on classification accuracy. This step often involves dealing with missing values and outliers and normalizes data.

The real databases frequently include missing data, for many reasons like having the tests not to be performed entirely or having the unavailable data. Handling them is a very significant step because they could decline the accuracy of the classification. There are different methods for this purpose.

Some methods to deal with missing values are removing the attributes containing missing data and data imputation, which is defined as the process in which the missing data are estimated by appropriately computed values. We employed the second method, data imputation, to fill in missing values in this research. We replace the categorical values with the mode and numerical values with mean [13].

Table 1 Coronary artery disease dataset features

No.	Attribute name	Description
1	Age	Age in year
2	Sex	1 = male; 0 = female
3	CP	Chest pain type (1 = typical angina; 2 = atypical angina; 3 = non-angina pain; 4 = asymptomatic)
4	Trestbps	Resting blood pressure (in mm Hg on admission to the hospital)
5	Chol	Serum cholesterol in mg/dl
6	Fbs	(fasting blood sugar greater than 120 mg/dl) (1 = true; 0 = false)
7	Restecg	Resting electrocardiographic results (0 = normal; 1 = having ST-T wave abnormality; 2 = showing probable or de_ne left ventricular hypertrophy by Esres_criteria)
8	Thalach	Maximum heart rate achieved
9	Exang	Exercise induced angina (1 = yes; 0 = no)
10	Oldpeak	ST depression induced by exercise relative to rest
11	Slope	Slope of peak exercise ST segment (1 = upsloping; 2 = _at; 3 = downsloping)
12	Ca	Number of major vessels (0–3) colored by _ourosopy
13	Thal	(3 = normal; 6 = _xed defect; 7 = reversible defect)
14	Num	Diagnosis classes (0 = healthy; 1 = patient who is subject to possible heart disease)

In addition to filling in missing values, two other preprocessing steps, removing outliers and normalizing data, were performed. Distance-based outlier methods are used to detect the outlines using the K-nearest neighbor and Euclidean distance. With normalization step, the interval of all data was changed to between [0, 1]. Data were normalized with the Eq. (7).

$$\text{Normalize } (x) = \frac{x - X\_min}{X\_min - X\_max} \tag{7}$$

Reduction of data dimension is done using decision tree. For CAD dataset, decision tree selected 7 features from 13. The selected features are the following: trestbps, cp, thal, ca, chol, oldpeak, and thalach.

4.2 Evaluation of the Proposed System

The fuzzy inference system is designed using MATLAB software version 7.12. The average number of rules for the CAD dataset is about 11 rules with an average length of 4.

The 10-fold cross-validation approach was used to divide data into test and training sets. For evaluating the system, three measures were calculated using confusion matrix (CM), accuracy, sensitivity, and specificity. The general form of the matrix is as follows:

$$\text{C.M.} = \begin{pmatrix} \text{TP} & \text{FP} \\ \text{FN} & \text{TN} \end{pmatrix} \tag{8}$$

where True Positive (TP) is the number of instances in which classifier detects them positive and the detection is true. True Negative (TN) is the number of instances, which is detected negative and the instances are negative. False Negative (FN) is the number of instances in which classifier does not detect them as positive, but they are positive. False Positive (FP) stands for the number of instances that are detected positive, but they are not positive.

The following equations show the formula of accuracy, sensitivity (recall), and specificity.

$$\text{Accuracy} = \frac{\text{TP} + \text{TN}}{\text{TP} + \text{TN} + \text{FN} + \text{FP}} \tag{9}$$

$$\text{Sensetivity/Recall} = \frac{\text{TP}}{\text{TP} + \text{FN}} \tag{10}$$

$$\text{Specificity} = \frac{\text{TN}}{\text{TN} + \text{FP}} \tag{11}$$

The sample membership functions for cholesterol feature of CAD dataset is shown in Fig. 5. The shape of membership functions was determined before and after the optimizations.

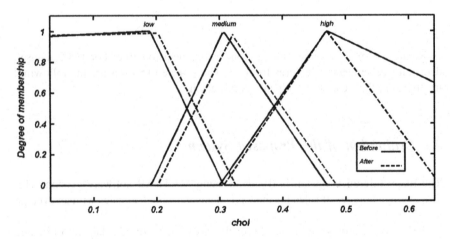

Fig. 5 The membership functions of Chol. feature before and after optimization

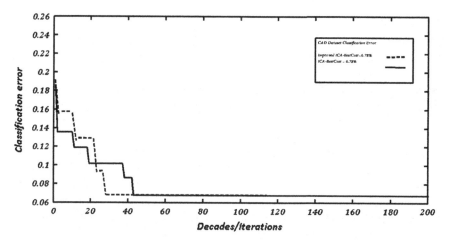

Fig. 6 Classification error by ICA and Improved ICA; classification accuracy = 94.92 %

Figure 6 shows the classification error using ICA and Improved ICA. Although the speed of convergence in Improved ICA is better than ICA, the classification error in both algorithms is the same.

The classification error obtained for CAD dataset was 94.92 %, which is an acceptable rate in comparison with similar methods. As it can be seen, the convergence in Improved ICA occurred in the 29th decade, while in ICA, the convergence occurred in the 43th decade. This proves that the assimilation process and the relation defined for it in the proposed system influence the convergence time, directly.

Table 2 compares the proposed method with similar methods.

Table 2 Comparison of the proposed method with similar works

Method	Accuracy (%)	Sensitivity (%)	Specificity (%)
Weighted fuzzy rules [9]	57.85	45.22	68.75
Decision tree [14]	78.9	72.01	84.48
Support vector machine [8]	79.17	–	–
K-NN [15]	81.5	–	–
Decision support system [16]	–	80	65
Ensemble Neural Network [7]	89.01	80.95	95.91
Fuzzy weighted rules and AIRS [6]	95.59	–	–
Ensemble-based PSO and fuzzy [4]	92.59	90.51	94.37
PSO and fuzzy [10]	93.27	93.2	93.3
ICA and fuzzy-Improved ICA and fuzzy (proposed)	94.92	94.11	92.30

5 Conclusion

In this study, a fuzzy expert system based on ICA and Improved ICA was developed in order to classify heart disease data. With this proposed approach, 94.92 % correct classification on the test set could be achieved. To determine the important attributes and obtain valid rules, decision tree tool was used. Fuzzy expert system was used to classify heart disease data. Membership function optimization was done using both ICA and Improved ICA.

ICA is an evolutionary algorithm proposed in 2007. This algorithm has acceptable results in optimization problems and is comparable with similar evolutionary algorithms like PSO. One of the most important factors in optimization algorithms is speed of convergence. The proposed approach suggested a new relation to improve this factor. The results show that the classification error in both algorithms is the same, but the speed of convergence in Improved ICA is better than in ICA.

References

1. American Heart Association, A., http://www.americanheart.org (2011)
2. Staff, M.C.: Coronary artery disease. Available from: http://www.mayoclinic.org/diseases-conditions/coronary-artery-disease/basics/symptoms/con-20032038 (2009)
3. Center, N.H.I.: How is coronary heart disease diagnosed. Available from: https://www.nhlbi.nih.gov/health/health-topics/topics/cad/diagnosis.html (2012)
4. Ghadiri Hedeshi, N., Saniee Abadeh, M.: An expert system working upon an ensemble PSO-based approach for diagnosis of coronary artery disease. In: The CSI International Symposiums on Artificial Intelligence and Signal Processing (AISP), vol. 1, pp. 77–82 (2011)
5. Das, R., Turkoglu, I., Sengur, A.: Effective diagnosis of heart disease through neural networks ensembles. Expert Syst. Appl. 36(4), 7675–7680 (2009)
6. Polat, K., Güneş, S.: A hybrid approach to medical decision support systems: combining feature selection, fuzzy weighted pre-processing and AIRS. Comput. Meth. Prog. Biomed. 88(2), 164–174 (2007)
7. Das, R., Turkoglu, I., Sengur, A.: Effective diagnosis of heart disease through neural networks ensembles. Expert Syst. Appl. 1, 7675–7680 (2009)
8. Babaoğlu, I., Fındık, O., Bayrak, M.: Effects of principle component analysis on assessment of coronary artery diseases using support vector machine. Expert Syst. Appl. 37(3), 2182–2185 (2010)
9. Anooj, P.K.: Clinical decision support system: risk level prediction of heart disease using weighted fuzzy rules. J. King Saud Univ. Comput. Inf. Sci. 24(1), 27–40 (2012)
10. Muthukaruppan, S., Er, M.J.: A hybrid particle swarm optimization based fuzzy expert system for the diagnosis of coronary artery disease. Expert Syst. Appl. 39(14), 11657–11665 (2012)
11. Jiawei, H., Kamber, M.: Data mining: concepts and techniques, 2nd edn. (2006)
12. Atashpas-Gargari, E., Lucas, C.: Imperialist competitive algorithm: an algorithm for optimization inspired by imperialistic competition. IEEE 1, 4661–4667 (2007)
13. Sajja, S.: Data mining of medical datasets with missing attributes from different sources, Youngstown State University (2010)

14. Tu, M.C., Shin, D., Shin, D.K.: Effective diagnosis of heart disease through bagging approach. In: 2nd International Conference on Biomedical Engineering and Informatics, China, vol. 2, pp. 1–4 (2009)
15. Setiawan, N.A., Venkatachalam, P.A., Ahmad Fadzil, M.H.: Rule selection for coronary artery disease diagnosis based on rough set. Int. J. Recent Trends Eng. **2**, 198–202 (2009)
16. Tsipouras, M.G., Exarchos, T.P. Fotiadis, D.I.: A decision support system for the diagnosis of coronary artery disease, vol. 2, pp. 279–284 (2006)

Static Hand Gesture Recognition Based on Fusion of Moments

Subhamoy Chatterjee, Dipak Kumar Ghosh and Samit Ari

Abstract A vision-based static hand gesture recognition algorithm which consists of three stages: pre-processing, feature extraction and classification are presented in this work. The pre-processing stage comprises of following three sub-stages: segmentation, which segments hand region from its background using YCbCr skin colour-based segmentation process; rotation, that rotates segmented gesture to make the algorithm, rotation invariant; Morphological filtering, that removes background and object noise. Non-orthogonal moments like geometric moments and orthogonal moments like Tchebichef and Krawtchouk moments are used here as features. To improve the performance of classification, two feature fusion strategies are proposed in this work: serial feature fusion and parallel feature fusion. A feed-forward multi-layer perceptron (MLP)-based artificial neural network classifier is proposed. A user-independent experiment is conducted on 1,500 gestures of 10 classes for 10 different users.

Keywords American sign language digits · Geometric moment · Tchebichef moment · Krawtchouk moment · Serial feature fusion · Parallel feature fusion · Artificial neural network

S. Chatterjee (✉) · D.K. Ghosh · S. Ari
Department of Electronics and Communication Engineering, National Institute of Technology, Rourkela 769008, Odisha, India
e-mail: subha.chat007@gmail.com

D.K. Ghosh
e-mail: dipakkumar05.ghosh@gmail.com

S. Ari
e-mail: samit.ari@gmail.com

© Springer India 2015
L.C. Jain et al. (eds.), *Intelligent Computing, Communication and Devices*,
Advances in Intelligent Systems and Computing 309,
DOI 10.1007/978-81-322-2009-1_48

1 Introduction

Static hand gesture recognition algorithms have been divided mostly into vision-based techniques [1, 2] and glove-based techniques [3, 4]. However, vision-based techniques are preferred to glove-based techniques because vision-based techniques are less complex compared to glove-based techniques. The vision-based gesture recognition systems are of two types: contour-based and shape-based. In a recent work [2], authors reported a robust static hand gesture recognition system using various non-orthogonal and orthogonal moment feature sets and minimum distance classifier. In real time system, hand gesture recognition algorithm should be user-independent. For user-independent case, none of the moments exhibits significant classification accuracy rate. Therefore, if different users are used for testing and training sets, the methodology adopted in [2] may not be suitable for static hand gesture recognition system. This work evaluates the non-orthogonal moment such as geometric moment and two popular orthogonal moments in user-independent gesture classification. The orthogonal moments considered are the: (1) Tchebichef and (2) Krawtchouk moments. To improve the performance of classification in user-independent situation, two feature fusion strategies [5]: parallel and serial feature fusion have been proposed. It shows an significant improvement in performance of classification for all the moments. The classification is done using the artificial neural network classifier. Rest of the paper is organized as follows: Sect. 2 presents the details of the proposed gesture recognition algorithm. Experimental results are discussed in Sects. 3 and 4 concludes the paper.

2 Methodology

The proposed static hand gesture recognition algorithm consists of following three stages: pre-processing, feature extraction and classification.

2.1 Pre-processing

Pre-processing of input gestures consists of 3 sub-stages: segmentation, rotation and morphological filtering. This phase detects and segments the skin colour region of hand from the captured image by YCbCr skin colour segmentation to detect hand regions [6]. The background is restricted such that the hand region is the largest object with respect to the skin colour. The threshold values of Cb, Cr and Y are proposed for skin colour segmentation as

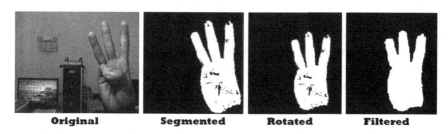

| Original | Segmented | Rotated | Filtered |

Fig. 1 Original, segmented, rotated and filtered hand gesture image of 'digit 6'

$$85 < \text{Cb} < 128, 129 < \text{Cr} < 185 \text{ and } \text{Th} < \text{Y} < 255 \tag{1}$$

where Th is the one third of the mean value of Y component.

The segmented hand gesture is rotated to make it rotation invariant [7]. A morphological filtering approach has been developed to obtain a smooth, closed and complete contour of a gesture by using a sequence of dilation and erosion operations [7] (Fig. 1).

2.2 Feature Extraction

Here we have used orthogonal and non-orthogonal moment features [2]. Moments have the ability to represent the global characteristics of the image shape. Here, moments based on discrete orthogonal polynomials like the Tchebichef [8] and the Krawtchouk [2] polynomials and a non-orthogonal moment namely geometric moment [9] have been used as features. These are directly defined in the image coordinate space and do not involve any numerical approximation like continuous moments.

Feature Fusion. To improve the classification performance in user-independent situation, two feature fusion strategies [5]: parallel and serial feature fusion, have been proposed. Suppose A and B are two feature spaces defined on pattern sample space. For an arbitrary sample, $\xi \in \phi$ the corresponding two feature vectors are $\alpha \in A$ and $\beta \in B$. The serial combined feature of ξ is denoted by $\lambda = (\alpha \ \beta)$. If feature vector α is n-dimensional and β is m-dimensional, then the serial combined feature of ξ is $(n + m)$-dimensional. In case of parallel fusion, two feature vectors are combined by a super-vector. Let α and β be two different feature vectors of the same sample ξ, then the super-vector $(\alpha + \beta)$ is used to represent the parallel combination of α and β. If the dimensions of α and β are not same, upsample the lower-dimensional one until its dimension equals to the other ones, before combination. For example, if $\alpha = (a_1, a_2, a_3)^T$ and $\beta = (b_1, b_2)^T$, β is first turned into $\beta = (b_1, b_1, b_2)^T$ and then the resulting combination is formed. The parallel combined feature space on ϕ is defined as $C = \{\alpha + \beta, \alpha \in A, \beta \in B\}$

and it is an n-dimensional vector space, where $n = \max(\dim\ (A),\ \dim\ (B))$. Sometimes dimension inequality in feature vectors leads to numerical unbalance between two feature vectors of same pattern. For example, if $\alpha = (10, 11, 10)^T$ and $\beta = (0.1, 0.2)^T$ are two feature vectors, α plays a more important role than β. To avoid it, a weighted combination form is adopted. The serial combination is formed by $\lambda = \{\alpha\ \theta\beta\}$, if $\dim(A) > \dim(B)$ or $\lambda = \{\theta\alpha\ \beta\}$ if $\dim(A) < \dim(B)$. Similarly, the parallel combination is formed by $\lambda = \{\alpha + \theta\beta\}$, if $\dim\ (A) > \dim\ (B)$ or $\lambda = \{\theta\alpha + \beta\}$ $\lambda = \{\theta\alpha\ \beta\}$ if $\dim\ (A) < \dim\ (B)$. Where the weight θ is called the combination coefficient and is given by $\theta = \theta = n^2/m^2$, where n and m are the dimension of A and B, respectively. To reduce unfavourable results from an unequal dimension, two feature vectors are normalized by dividing its maximum values as given below

$$\bar{\alpha} = \alpha/\max(\alpha) \tag{2}$$

$$\bar{\beta} = \beta/\max(\beta) \tag{3}$$

then the serial and parallel feature vectors are computed as given below

$$\bar{\lambda} = [\bar{\alpha}\ \theta\bar{\beta}] \tag{4}$$

$$\bar{\lambda} = \{\bar{\alpha} + \theta\bar{\beta} \tag{5}$$

2.3 Artificial Neural Network Classification

A feed-forward multi-layer perceptron (MLP) neural network classifier [10] with a single hidden layer has been designed, trained and tested using the feature sets described above. The dataset consists of 1,500 gestures of 10 classes, 15 samples each class of 10 users. The dataset is equally split into training and testing datasets of 750 gestures of 10 classes for 5 different users to make the system user-independent. The number of neurons in the hidden layer is empirically set to 200. A twofold operation is performed to evaluate the generalized performance of the system in user-independent condition. We have quantified our classifier performance using the most common matrices found in literature: Accuracy (A_c), Sensitivity (S_e), Specificity (S_p) and Positive Predictivity (P_p) [11].

3 Experimental Results and Discussion

The performances of the gesture recognition model are emulated on a static hand gesture database. It consists of 1,500 American Sign Language (ASL) digits colour images of 10 classes, 15 samples each class of 10 users. The dataset is equally split into training and testing dataset of 750 images for 5 different users to test the

Table 1 Comparison of gesture recognition performance for different features

Features	A_C	S_E	P_P	S_P
Geometric moment	95.24	76.20	76.06	97.36
Krawtchouk moment	98.31	91.53	91.87	99.06
Tchebichef moment	96.53	82.67	82.22	98.07
Geometric-Krawtchouk serial fusion	98.99	94.93	95.05	99.44
Geometric-Krawtchouk parallel fusion	98.84	94.2	94.32	99.36
Tchebichef-Krawtchouk serial fusion	98.71	93.53	93.85	99.28
Tchebichef-Krawtchouk parallel fusion	99.07	95.33	95.42	99.48

system independent of user. Training dataset is used to train the ANN network. However, testing set is used to test the performance. In this work, the classification performances are quantified using 4 metrics: accuracy, sensitivity, specificity and positive predictivity. To evaluate the performances of the system, the experiments are conducted for different moments. The experimental results are shown in Table 1. Experimental results show that Krawtchouk moment is the best in terms of all the performance metrics. In user-independent condition, neither of these moments has shown satisfactory classification performance. Geometric moment shows worst performance in terms of all these performance metrics. The recognition performances in term of sensitivity using Krawtchouk moment, Tchebichef moment and Geometric moment are 91.53, 82.67 and 76.20 %, respectively, as shown in Table 1. To improve recognition performance, we have implemented two feature fusion strategies as discussed in Sect. 2. When two feature vectors are combined, resultant feature consist attributes of both the features. For that reason misclassification rate has decreased in both serial and parallel fusion. Serial fusion of Krawtchouk-Tchebichef moments and Krawtchouk-Geometric moments have shown 93.53 and 94.93 % sensitivity as shown in Table 1. Parallel fusion of Krawtchouk-Tchebichef and Krawtchouk-Geometric moments have shown 95.33 and 94.20 % sensitivity, respectively, as shown in Table 1. It is clear that for both these two fusion strategies, recognition performance has increased significantly. From Table 1, it is clear that, for equal size features (Krawtchouk and Tchebichef moment), parallel feature fusion has given best result and for unequal size features (Krawtchouk and Geometric moment), serial feature fusion has given best result. Experimental study has been shown that parallel feature fusion of Krawtchouk-Tchebichef moment has given the best gesture recognition performance.

4 Conclusion

A novel feature extraction technique for static hand gesture recognition is proposed in this work, which overcomes the challenges of misclassification of geometrically closed gestures. In case of Geometric and Tchebichef moments, mismatches occur more than Krawtchouk moment. This is because geometric moment is a local feature and it only represents the statistical attributes of the shape. On the other hand, although Tchebichef moment is orthogonal, it does not show satisfactory result in user-independent condition. To overcome this mismatch problem, two feature fusion strategies (serial fusion and parallel fusion) have been proposed. Experimental results show that fusion strategies significantly improve the gesture recognition performance and parallel feature fusion of Krawtchouk-Tchebichef moment has given the best gesture recognition performance.

References

1. Gupta, S., Jaafar, J., Ahmad, W.F.W.: Static hand gesture recognition using local gabor filter. In: International Symposium on Robotics and Intelligent Sensors (IRIS 2012), vol. 41, pp. 827–832, Kuching, Sarawak, Malaysia (2012)
2. Priyal, S.P., Bora, P.K.: A robust static hand gesture recognition system using geometry based normalization and krawtchouk moments. Pattern Recogn. **46**(8), 2202–2219 (2013)
3. Kumar, P., Verma, J., Prasad, S.: Hand data glove: A wearable real-time device for human-computer interaction. Int. J. Adv. sci. Technol. **43** (2012)
4. Huang, Y., Monekosso, D., Wang, H., Augusto, J.C.: A concept grounding approach for glove-based gesture recognition. In: 7th International conference On Intelligent Environments (IE), pp. 358–361, Nottingham, UK (2011)
5. Yang, J., Yang, J.Y., Jhang, D., Lu, J.F.: Feature fusion: Parallel strategy versus serial strategy. Pattern Recogn. **36**(6), 1369–1381 (2003)
6. Aibinu, A.M., Shafie, A.A., Salami, M.J.E.: Performance analysis of ANN based YCbCr Skin Color Detection Algorithm. In: International Symposium on Robotics and Intelligent Sensors (IRIS), vol. 41, pp. 1183–1189, Sarawak, Malaysia (2012)
7. Ghosh, D.K., Ari, S.: A static hand gesture recognition algorithm using K-Mean based radial basis function neural network. In: 8th International Conference on Information, Communications and Signal Processing (ICICS), pp. 1–5, Singapore (2011)
8. Priyal, S.P., Bora, P.K.: A study on static hand gesture recognition using moments. In: International Conference on Signal Processing and Communications (SPCOM), pp. 1–5, IISC, Bangalore (2010)
9. Zou, Z., Premaratne, P., Monaragala, R., Bandara, N Premaratne, M.: Dynamic hand gesture recognition system using moment invariants. In: 5th International Conference on Information Automation for Sustainability (ICIAFs), pp. 108–113, Colombo, Sri Lanka (2010)
10. Haykin, S.: Neural networks. Prentice-Hall (1999, 2nd edn.)
11. Daamouche, A., Hamami, L., Alajlan, N., Melgani, F.: A wavelet optimization approach for ECG signal classification. Biomed. Signal Process. Control **7**(4), 342–349 (2012)

Real-Time Human Face Detection in Noisy Images Based on Skin Color Fusion Model and Eye Detection

Reza Azad, Eslam Ahmadzadeh and Babak Azad

Abstract Automatic human face detection is a challenging problem which has received much attention during recent years. In this paper, we propose a method that includes a denoising preprocessing step and a new face detection approach based on skin color fusion model and eye region detection. Preprocessing of the input images is concentrated on the removal of different types of noise while preserving the phase data. For the face detection process, firstly, skin pixels are modeled by using supervised training, and at the run time, the skin pixels are detected using optimal boundary conditions. Then by finding the eye location in the skin region, human face will be extracted. Experimental results obtained using images from the FEI, complex background, and noisy image database are promising in terms of detection rate and the false alarm rate in comparison with other competing methods. In addition, experimental results have demonstrated our method robust in successful detection of skin and face regions even with variant lighting conditions and poses.

Keywords Face detection · Noise removing · Skin detection · Color space transformation · Eye detection

R. Azad (✉)
Department of Electrical and Computer Engineering, The SRTTU,
Tehran, Iran
e-mail: rezazad68@gmail.com

E. Ahmadzadeh
Zand Higher Education Institute, Shiraz, Iran
e-mail: ahmadzadeh1358@yahoo.com

B. Azad
Technical and Engineering College, University of Mohaghegh Ardabili,
Ardebil, Iran
e-mail: babak.babi72@gmail.com

© Springer India 2015
L.C. Jain et al. (eds.), *Intelligent Computing, Communication and Devices*,
Advances in Intelligent Systems and Computing 309,
DOI 10.1007/978-81-322-2009-1_49

1 Introduction

With the ubiquity of new information technology and media, more effective and friendly methods for human–computer interaction (HCI) are being developed, which do not rely on traditional devices such as keyboards, mice, and displays. Furthermore, the ever-decreasing price/performance ratio of computing coupled with recent decreases in video image acquisition cost implies that computer vision systems can be deployed in desktop and embedded systems [1, 2]. The rapidly expanding research in face processing is based on the premise that information about a user's identity, state, and intent can be extracted from images and that computers can then react accordingly, e.g., by observing a person's facial expression [1]. In the last decade, face detection has been thoroughly studied due to its wide potential applications, including face recognition [3, 4], human–computer interaction [5], gender classification [6], and video surveillance [7]. Particularly, in the context of face recognition, the detection of faces along with the detection of some fiducial points, such as the eyes and mouth, is the first step of the face recognition system, and this step largely affects the performance of the overall system.

A first step of any face processing system is to detect the locations in images where faces are present. Actually, face detection is one of the visual tasks which humans can do effortlessly. However, in computer vision terms, this task is not easy. A general statement of the problem can be defined as follows: Given a still or video image, detect and localize an unknown number (if any) of faces. The solution to the problem involves segmentation, extraction, and verification of faces and possibly facial features from an uncontrolled background. As a visual front-end processor, a face detection system should also be able to achieve the task, regardless of illumination, orientation, and camera distance. Based on a survey [1] on face detection, existing face detection approaches in single image are grouped into four categories: knowledge-based, feature-based, template-based, and appearance-based methods. Some of the recent methods in these fields are [8–13].

In this paper, we propose an image preprocessing step different to variance normalization, to overcome the problem of noisy images and that of images contaminated with illumination artifacts. We use the image denoising method suggested by Kovesi in [14]. This method is able to preserve the important phase information of the images, based on the non-orthogonal and complex-valued log-Gabor wavelets. We apply this technique of denoising in conjunction with a fusion of skin detector approach which together leads to an outperforming result. Finally, we will use face ratio and eye region detection for face and non-face classifications. General block diagram of the proposed method is shown in Fig. 1.

The rest of the paper is organized as follows. Section 2 provides a discussion of the image denoising method based on a phase-preserving algorithm. The proposed method for face detection is detailed in Sect. 3. Section 4 deals with experimental results, and in Sect. 5, conclusion is given.

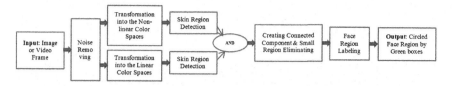

Fig. 1 Proposed method for face detection

2 Phase-Preserving Denoising of Images

A phase-preserving denoising method was proposed by Kovesi in [14]. It assumes that phase information of images is the most important feature and tries to preserve this information, of course by trying to keep the magnitude information, as well. Let M_ρ^e and M_ρ^o denote the even-symmetric and odd-symmetric wavelets, respectively, at a scale ρ, which are known as quadratic pairs. Considering the responses from each quadrature pair of the filters, a resultant response vector is defined as follows:

$$\left[e_\rho(x), o_\rho(x)\right] = \left[f(x) * M_\rho^e, f(x) * M_\rho^o\right] \tag{1}$$

where * denotes convolution and values $e_\rho(x)$ and $o_\rho(x)$ are the real and imaginary parts in the complex-valued frequency domain. The amplitude of the transform at a given wavelet scale is given by

$$A_\rho(x) = \sqrt{e_p(x)^2 + o_p(x)^2} \tag{2}$$

And the local phase is given by

$$\varphi_\rho(x) = \tan^{-1}\left[\frac{o_\rho(x)}{e_\rho(x)}\right] \tag{3}$$

Having one response vector for each filter scale, there will be an array of such vectors for each pixel x in a signal. The denoising process includes defining an appropriate noise threshold for each scale as well as reducing the magnitude of the response vectors, while maintaining the phase without any changes. The most important step of the denoising process is to determine the thresholds. For this end, Kovesi [14] used the expected response of the filters to a pure noise signal. If the signal is purely Gaussian white noise, then the position of the resulting response vectors from a wavelet quadratic pair of filters at some scale will form a 2D Gaussian distribution in the complex plane. Kovesi [14] showed that the distribution of the magnitude responses can be modeled by the Rayleigh distribution:

$$R(x) = \frac{x}{\sigma_g^2} \exp^{-x^2/2\sigma_g^2} \tag{4}$$

Also, the amplitude response from the smallest scale of the filter pair across the whole image will be the noise with Rayleigh distribution. Finally, by estimating the mean value μ_r and standard deviation σ_r of the Rayleigh distribution, the shrinkage threshold can be estimated. The thresholds are automatically determined and applied for each filter scale.

A number of parameters impact the quality of the denoised output image. The threshold of noise standard deviations to be rejected (k), the number of filter scales to be used (N_ρ), and the number of orientations (N_r) are the key parameters. We set the parameters $k = 3$, $N_\rho = 5$, and $N_r = 5$ in our experiments. These parameters result in an acceptable representation of small- and middle-sized faces. However, for large faces, it can lead to erroneous results. One approach is using a set of different parameters to obtain different images. Another approach is scaling the original images and then using the same parameters for conversions. We used the second approach for a better speedup. After a conversion to the denoised form, adaptive histogram equalization is used for test images. Figure 2 shows the discriminate advantage of using denoised images.

3 Proposed Method for Face Detection

General block diagram of the proposed method is shown in Fig. 1. In the proposed method, the input image is denoised at the first time, and then, denoised image is transformed into the linear and nonlinear color spaces. After color space transformation, skin region is specified by fusion of these color spaces. Then, face candidates are extracted by applying boundary conditions. Finally, by using face ratio and eye region detection, face and non-face regions are classified.

Fig. 2 Output of the denoised method **a** noisy image and **b** denoised image

3.1 Color Space Transformation for Skin Detection

Both linear and nonlinear color spaces are used in our paper for skin detection. To provide some direct information about these color spaces, we summarize below how typical color space transforms, including linear and nonlinear ones, are defined. From RGB to YUV, color spaces are linear transforms, in which the three components in both new spaces are defined simply by linear weighting of R, G, and B values and Y refers to illumination intensity defined as follows:

$$Y = \sum \lambda w_\lambda, \quad \lambda = R, G, B, \qquad \sum w_\lambda = 1, w_\lambda \geq 0$$
$$U = B - Y, \quad V = R - Y$$

$$(5)$$

Generally, these linear transforms can be defined as follows [15]:

$$\begin{bmatrix} Y \\ A \\ B \end{bmatrix} = \begin{bmatrix} w_r & w_g & w_b \\ a_{10} & a_{11} & a_{12} \\ a_{20} & a_{21} & a_{22} \end{bmatrix} \begin{bmatrix} R \\ G \\ B \end{bmatrix} \tag{6}$$

As hue is more effective in distinguishing different colors than illumination intensity, hue, saturation, value (HSV) and hue, intensity, saturation (HIS) transforms are taken as suitable color spaces that correspond to human visual perceptions and have been widely utilized in color clustering for image segmentation and coding [8]. The RGB to HSV transform can be defined as follows [15]:

$$V = \max(R, G, B)$$
$$S = \frac{V'}{V}$$
$$V' = V - M, \quad M = \min(R, G, B)$$

$$(7)$$

Let $r' = \frac{V-R}{V'}, g' = \frac{V-G}{V'}$ and $b' = (V - B)$ then H is given by [15]

$$H = \frac{1}{6} \begin{cases} 5 + b' & \text{if } R = V \text{ and } G == M \\ 1 - g' & \text{if } R = V \text{ and } G \neq M \\ 1 + r' & \text{if } G = V \text{ and } B == M \\ 3 - b' & \text{if } R = V \text{ and } B \neq M \\ 3 + g' & \text{if } R = V \text{ and } R == M \\ 5 + r' & O.W \end{cases} \tag{8}$$

3.2 Skin Region Detection

Since skin detection is a classification problem defined on color similarity, supervised clustering is applied to achieve the exact rules for effective skin color clustering and pixel classification. Through manually specifying representative skin and non-skin pixels, we can learn linear relationships between different components in the new color spaces. Finally, we obtain several main boundary conditions for skin pixel classification in different color spaces. Firstly, skin pixels are modeled by using the histogram-based approach, in which the probability or likelihood that each color represents skin is estimated by checking its occurrence ratio in the training data. In the equation given below, V_{skin} indicates volumes or total occurrences of all skin colors in manual ground truth of training data.

$$P(\text{Color}|\text{Skin}) = \text{sum}(\text{Color}|\text{Skin})V_{skin} \tag{9}$$

Then, boundary conditions in the skin model are extracted to allow more than 97.50 % of skin pixels covered. Using the boundary conditions, test images are segmented into skin and non-skin regions accordingly. For each color space, these boundary conditions are found as follows. For YUV space, the boundary conditions are found as follows:

$$\text{YUV Condition} : \begin{cases} 147 \leq V \leq 186 \\ 189 \leq U + 0.56V \leq 215 \end{cases} \tag{10}$$

Considering the illumination intensity variation, we have boundary conditions as follows:

$$\text{Illumination} : \begin{cases} Y > 84 \\ Y < 84, U > 108, \quad Y + U - V > 3 \end{cases} \tag{11}$$

In HSV space, we scale the H into [0, 255] and let $H = 255 - H$ if $H > 128$ [8]. We also find several boundary conditions for skin pixels in HSV space and are given below. Figure 3c shows the sample output of this step on entrance image.

$$\text{HSV Condition} : \begin{cases} S \leq 22, V \leq 2.4S \\ 159 \leq H + V \leq 390, \quad H + V > 13S \\ H > 0.2V, \quad H > 4.2S \end{cases} \tag{12}$$

3.3 Face Candidate Region Extraction

In this part, the candidate region has been identified by applying mathematical morphology, and then, among these areas, face locations are extracted. Mathematical morphology is one of the branches of image processing that argues about shape and appearance of object in images [16]. The erosion and dilation operators are basically operators of mathematical morphology that are used in this part to improve the skin detection image. For this respect, first, erosion action is applied in the skin detection image. The erosion action is defined as follows [16]:

$$A \ominus B = \left\{ X | (B)_x \subseteq A \right\} \tag{13}$$

A and B erosion is a collection of all points of X; if B is replaced in the size of X, A and B erosion is still placed in A. After erosion action on image, the dilation action is done. The dilation action is defined by [16]:

$$A \oplus B = \left\{ X | [B_x \cap A] \subseteq A \right\} \tag{14}$$

A and B dilations are the collection of all X (es) that B_x and at least A overlap with a nonzero element. B_x is B symmetric around its own axis. And then, it transfers the symmetry of X. Finally, for extracting connected component, first, the holes are filled, and then, the area with eight connected neighbors is labeled as connected component. After creating connected component, the detected skin regions are labeled to obtain the outer boundary rectangle and pixel number of every region. Then, small regions that have pixels less than a given threshold, i.e., 350, will be removed. Figure 3d shows the sample result of this step.

3.4 Classification of Face and Non-face Candidates

As is clear from Fig. 3d, in the place of candidate face, there are several interconnected areas. First, the length of each of these candidates is obtained and candidates that have not geometry ratio of the face are considered as non-face. Then, for determining entire candidates as face or non-face, we use the existing of

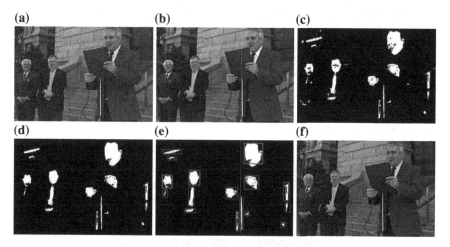

Fig. 3 Face detection gradually **a** entrance image, **b** denoised image, **c** detected skin region, **d** skin region after removing small region and applying morphological operations, **e** detected face and non-face region, where the face location is circled by green boxes and non-face location is circled by red boxes, and **f** final output on entrance image

eye region in candidate boundary. For detection of eye, we use the Viola and Jones detector. The detection approach by Viola and Jones introduced in 2001 has become one of the most popular real-time object detection frameworks. Original results were presented on the face detection problem, but the approach can easily be transferred to other domains [17]. The detector is basically a cascade of binary linear classifiers which are subsequently applied to the sliding window input. An example is passed through the cascade as long as it is positively classified by the current stage. Each stage consists of a certain number of weighted one-dimensional threshold classifiers that are fed with a single feature. During the stage-wise training, initially selected detection and false-positive rates are guaranteed to be met by each cascade stage which is trained using AdaBoost [18]. Thus, one is able to estimate the final performance given the number of stages. The training set for stage n is given by all positive examples and the false positives remaining after stage $n-1$, where those for the first stage are chosen randomly from the full images.

The real-time capability of the approach is mainly enabled by two properties: Most sliding windows are only evaluated by the first stages which contain few classifier features. The features offered during training are simple Haar-like (Fig. 4) filters which can be evaluated cheaply using a precalculated integral image of gray values. In fact, once the precalculation is done on the full image, responses of all basic types of Haar-like features (see Fig. 4) are computed by 5–8 additions/subtractions and a single division, independent of position and size. During detection, the sliding window was scaled to cover all possible sign sizes.

In order to achieve robustness toward intermediate-sized examples, positive samples were randomly scaled within the selected range. The same was done for translation, which is introduced during training to allow for larger step sizes of the sliding window. In this paper, we trained the Viola–Jones detector with a lot of eye images, such as the simple, the tilted, darken, and blurred with considering any distance and lighting condition that showed high accuracy in eye detection step. Figure 3e shows the result of this step for determining candidate as face or non-face by eye region judging. Also, Fig. 3f shows the detected faces on the entrance image where the face locations are circled by green boxes.

4 Practical Result

Our suggestive method has been done on Intel Core i3-2330 M CPU, 2.20 GHz with 2-GB RAM under MATLAB environment. Figure 5 shows the face of worked system. Performance of the skin detection and face detection stage is described in the next subsections.

Fig. 4 Basic types of Haar wavelet features used for the Viola–Jones detector [18]

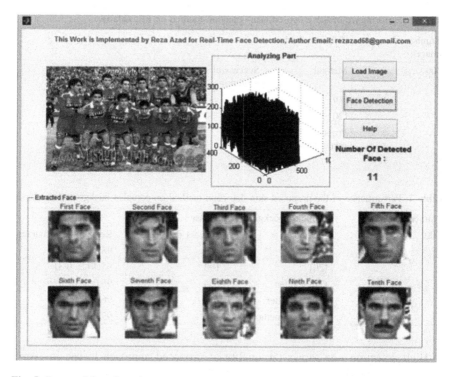

Fig. 5 Proposed face detection system

4.1 Performance of the Skin Detection Stage

In our experiments, statistical models of skin colors are estimated through histogram-based approach using a subset of FEI database, in which 200 images are used for training. Afterward, we applied our skin detection approach on FEI face database. The FEI face database is a Brazilian face database that contains a set of face images taken between June 2005 and March 2006 at the Artificial Intelligence Laboratory of FEI in São Bernardo do Campo, São Paulo, Brazil. In Table 1, our method is compared with color probabilistic method, represented in [19] and hybrid color space-based technique mentioned in [20] that both of them used this database for evaluation of their works. Further, our proposed method is compared with wavelet-based and Gabor-based methods, on the complex background images [20], and the result is detailed in Table 2. The accuracy rate formula that we used is mentioned in [19]. Equation below shows the accuracy rate:

$$\text{Accuracy} = 100 - (\text{False Detection Rate} + \text{False Dismissal Rate}) \quad (15)$$

Table 1 Performance of the proposed skin detection method on FEI database

Number of images	FEI face database		
400	Method	Successful skin region detection	Recognition accuracy
	[19]	400	97.10 ± 0.12
	[20]	400	99.25 ± 0.2
	Proposed method	400	99.40 ± 0.21

Table 2 Performance of the proposed skin detection method on complex background database

Number of images	Complex background database		
100	Method	Success detection	Recognition accuracy
	[19]	100	98.10 ± 0.19
	[20]	100	95.40 ± 0.31
	Gabor filter	100	98.0 ± 0.20
	Wavelet filter	100	96.0 ± 0.10
	Proposed method	100	99.0 ± 0.1

High detection rate shows the quality of proposed approach to use in every applications, which are needed a skin detection stage. Low complexity in computation and time is some of other advantages of the proposed approach. Figure 6 shows the sample output of skin detection stage on complex background database.

4.2 Performance of the Face Detection Stage

For evaluation of our face detection method on noisy images, 200 noisy images were generated in our office environments. Results of face detection stage are

Fig. 6 Sample output of the skin detection stage **a** original images and **b** detected skin region in each image

Table 3 Performance of the proposed face detection stage on noisy images

Number of images	Noisy face database		
200	Method	Successful face detection	Percent efficiency
	Proposed method	198	99 %

Table 4 Performance of the proposed face detection stage on FEI face database

Number of images	FEI face database		
400	Method	Successful face detection	Percent efficiency
	[21]	394	98.50 %
	[19]	397	99.25 %
	Proposed method	399	99.75 %

Fig. 7 Sample results on FEI, complex background, and noisy image databases

depicted in Table 3. Further, in Table 4, our face detection method is compared with color probabilistic method represented in [19] and morphological-based technique mentioned in [21] on FEI frontal face database.

In addition, our face detection algorithm is robust against parameters such as human pose variation, noise, and lightning condition. Figure 7 shows some of the detection results, where the human faces are circled by green boxes. From the examples, we may see that our algorithm can detect the human face with various sizes, positions, and noising from various backgrounds.

5 Conclusions

In this research, we studied the effect of denoising preprocessing step and skin color fusion model for real-time face detection. Phase-preserving denoising of images is used to preprocess the input images. In the mentioned method for face detection process, firstly, skin pixels were modeled by using supervised training, and at the run time, the skin pixels were detected using optimal boundary conditions. Further, by using the Jones detector, eye location has been extracted to get the face from the skin region. In the experimental results, we applied the proposed method on FEI, complex background, and noisy image database, and we achieved high recognition rate in comparison with other competing methods. In addition, experimental results have demonstrated our method robust in successful detection of skin and face regions even with variant lighting conditions and poses.

Acknowledgment This work is supported by the Shahid Rajaee Teacher Training University, Tehran, Iran (No. 22970060-9). The authors would like to thank Dr. Alok Kumar Jagadev, for his great advice and kindness.

References

1. Yang, M., Kriegman, D., Ahuja, N.: Detecting faces in images: a survey. IEEE Trans. Pattern Anal. Mach. Intell. **24**(1), 34–58 (2002)
2. Scheirer, W., Anthony, M., Nakayama, K., Cox, D.: Perceptual annotation: measuring human vision to improve computer vision. IEEE Trans. Pattern Anal. Mach. Intell. **99**(1), 1–20 (2014)
3. Azad, R., Azad, B., Kazerooni, I.T.: Optimized method for real-time face recognition system based on PCA and multiclass support vector machine. Adv. Comput. Sci.: Int. J. **2**(5), 126–132 (2013)
4. Abeer, A., Mohamad, A., Woo, W.L., Dlay, S.S.: Multi-linear neighborhood preserving projection for face recognition. Pattern Recogn. **47**(1), 544–555 (2014)
5. Fuzhen, H., Houqin, B.: Identity authentication system using face recognition techniques in human-computer interaction. In: 32nd Chinese Control Conference, pp. 3823–3827 (2013)
6. Shmaglit, L., Khryashchev, V.: Gender classification of human face images based on adaptive features and support vector machines. Opt. Mem. Neural Netw. **22**(4), 228–235 (2013)
7. Bae, G., Kwak, S., Byun, H., Park, D.: Method to improve efficiency of human detection using scale map. IEEE Electron. Lett. **50**(4), 265–267 (2014)
8. Wang, D., Ren, J., Jiang, J., Ipson, S.S.: Skin Detection from Different Color Spaces for Model-Based Face Detection. In: International Conference on Advanced Intelligent Computing Theories and Applications, pp. 487–494 (2008)
9. Bongjin, J., Inho, C., Daijin, K.: Local transform features and hybridization for accurate face and human detection. IEEE Trans. Pattern Anal. Mach. Intell. **35**(6), 423–1436 (2013)
10. Kalbkhani, H., Shayesteh, M.G., Mousavi, S.: Efficient algorithms for detection of face, eye and eye state. IET Comput. Vision **7**(3), 184–200 (2013)
11. Aiping, C., Lian, P., Yaobin, T., Ning, N.: Face Detection Technology Based on Skin Color Segmentation and Template Matching. In: IEEE Second International Workshop on Education Technology and Computer Science, pp. 708–711 (2010)
12. Pang, Y., Zhang, K., Yuan, Y., Wang, K.: Face detection technology based on skin color segmentation and template matching. IEEE Trans. Cybern. **99**(1), 1–20 (2014)
13. Anvar, S.M.H., Wei-Yun, Y., Khwang, T.E.: Multiview face detection and registration requiring minimal manual intervention. IEEE Trans. Pattern Anal. Mach. Intell. **35**(10), 1–20 (2013)
14. Kovesi, P.: Phase Preserving denoising of images. Digit. Image Comput. Tech. Appl. (1999)
15. Palus, H.: Representation of Color Images in Different Color Spaces. The Color Image Processing Handbook London (1998)

16. Azad, R., Shayegh, H.R.: New Method for Optimization of License Plate Recognition system with Use of Edge Detection and Connected Component. In: IEEE International Conference on Computer and Knowledge Engineering, pp. 21–25 (2013)

17. Azad, R.: View-independent traffic speed sign detection and recognition system for intelligent driving system. IJMECS 6(3), 31–37 (2014)

18. Viola, P., Jones, M..: Rapid Object Detection Using a Boosted Cascade of Simple Features. In: IEEE Conference on Computer Vision and Pattern Recognition, pp. 1–9 (2001)

19. Azad, R., Davami, F.: A robust and adaptable method for face detection based on color probabilistic estimation technique. Int. J. Res. Comput. Sci. 3(6), 1–7 (2013)

20. Azad, R., Shayegh, H.R.: Novel and tuneable method for skin detection based on hybrid colour space and colour statistical features. Int. J. Comput. Inf. Technol. 1(3), 211–221 (2013)

21. Patil, H.Y., Bharambe, S.V., Kothari, A.G., Bhurchandi, K.M.: Face Localization and its Implementation on Embedded Platform. In: 3rd IEEE International Advance Computing Conference, pp. 741–745 (2013)

Investigation of Full-Reference Image Quality Assessment

Dibyasundar Das and Ajit Kumar Nayak

Abstract The errors in imaging system create distorted image which affect the human perception of the image. This degradation in quality of image can be evaluated by image quality assessment (IQA) methods. To evaluate human subjectivity, the techniques need to follow the method of human visual system (HVS). The processing of image in extra cortical region of human brain is still unknown. Many attempts have been made to give an IQA algorithm that follows the philosophy of human observations. A brief experimental study of these philosophies has been made in this paper, which would help researchers to develop much improved IQA techniques in future.

Keywords Image quality assessment (IQA) · Subjective score · Objective score · SRCC · KRCC · SSIM · IWSSIM · Neural network

1 Introduction

Image quality assessment (IQA) is a process of giving a quality score to the image so as to determine how much the image has been degraded from its perceived perfect image. A perfect image in this context is an image that a human brain perceives to be perfect. In the physical world, no device has been perfectly designed to match that perception. But with advancement of technology, the methods to reach that perfect image are being attempted. As the technical revolution is at its peak, many advance and improved image acquisition, compression,

D. Das (✉) · A.K. Nayak
SOA University, Bhubaneswar, Odisha, India
e-mail: dibyasundarit@gmail.com

A.K. Nayak
e-mail: ajitnayak2000@gmail.com

© Springer India 2015
L.C. Jain et al. (eds.), *Intelligent Computing, Communication and Devices*,
Advances in Intelligent Systems and Computing 309,
DOI 10.1007/978-81-322-2009-1_50

transmission, processing, and reproduction techniques have been introduced. IQA provides the base to benchmark, monitor, and optimize these techniques.

The best quality score can be obtained by a human observer as only brain can perceive the perfect image. This scoring processing of IQA through human is called subjective evaluation and the mean value for an image is taken as mean opinion score (MOS), which is considered to be the quality score of the image [1]. But this process is time consuming, costly and not always radically possible. For this reason, machines need to evaluate the quality assessment, which is called objective evaluation. But the problem with objective process is that human visual system (HVS) still remains a big mystery. Without understanding HVS, a proper objective method that can predict the human perceived quality is not possible.

In paper, Sect. 2 describes about how to evaluate an objective IQA, Sect. 3 describes about different approaches to full-reference IQA methods, Sect. 4 gives the implementation results of different algorithms and finally the study has been concluded on Sect. 5.

2 Evaluation Procedure for Objective IQA Techniques

There are three types of objective evaluations namely [2]

(i) Full-reference: Where a better quality image is available for comparison.
(ii) Reduced-reference: Where concise information about perfect image is available.
(iii) No-reference: Where no prior know about image is given).

To evaluate the objective techniques, there are a few publically available IQA databases [8–10] and the description is as in Table 1.

All these databases come with MOS value of the image which has been evaluated and considered to be the base for objective score comparison. The range of the MOS may be 0–5 or 0–1 depending upon database. Similarly, the objective evaluation system may have different ranges as per their procedure. To compare different systems of values, the following methods are used.

(iv) Spearman's Rank Order Correlation Coefficient (SRCC):

Table 1 Description of subjective image quality databases

Database	Real images	Type of distortion	Distorted image
TID2013	25	24	3000
IVC	10	4	185
LIVE	29	5	779
A57	3	6	54

$$\text{SRCC} = 1 - \frac{6 \times \sum_{i=1}^{n} d_i^2}{n(n^2 - 1)} \tag{1}$$

where n is the number of image getting evaluated, d_i is the difference in rank of ith image in subjective score and objective score.

(v) Kendall's Rank Order Correlation Coefficient (KRCC):

$$\text{KRCC} = \frac{n_c - n_d}{0.5 \times n(n - 1)} \tag{2}$$

where n_c is the number of concordant pairs in dataset nd is the number of disconcordant pairs. A pair of image "x" and "y" having x_i, y_i as subjective score and x_j, y_j as objective score, and is said to be concordant if for $x_i > x_j$, $y_i > y_j$ must satisfy.

These rank ordering methods as described in Eqs. (1) and (2) can be replaced by some nonlinear mapping techniques which provide options to map the objective score to a nonlinear dimension and then find the rank ordering. These techniques give good results under fair assumption of systems. This discussion has not included those as for their ambiguities.

As explained above, there are three types of IQA techniques, but this paper has narrowed is study to full-reference IQA only.

3 Approaches to Full-Reference IQA

The knowledge about HVS is very limited. So, only assumptions have been made to understand the visual processing of human eye. First approach is error-based IQA. In this method, the assumption is made that human eye can detect error and same type of error will create same type of effect in human brain. Among the methods applied to model it, the most famous one are mean square error (MSE) [1] and peak signal to noise ratio (PSNR) [2].

3.1 Mean Square Error

It is the mean of the square of the error produced. From the historical perspective of signal processing, it is the most common error detecting strategy. For an image, it can be expressed as stated in Eq. (3).

$$MSE = \frac{1}{m \times n} \sum_{i=1}^{m} \sum_{j=1}^{n} x(i,j) - y(i,j) \tag{3}$$

where x is original image and y is distorted image with size (m, n).

3.2 Peak Signal to Noise Ratio

It is the second common method in signal processing world. It is based on MSE and finds the log relation of the MSE to the peak signal. For an image, it can be expressed as in Eq. (4).

$$PSNR = 10 \times \log\left(\frac{L^2}{MSE}\right) \tag{4}$$

where L is the peak signal and it is determined as per the imaging system. In this experimental study, gray scale is used for which its value would be 255.

The error-based methods are simple to use and they have a clear physical meaning. So, they are assumed to be the most useful tools. But they do not consider the spatial position of the pixel. Until the image errors are same, the position of error does not matter. These tend to give many type of error the same rating, which can be expressed from the following example:

In Fig. 1 image "a" is the original image. Images from "b" to "h" are introduced with different types of errors. As from images it can be seen that the quality of images are different but MSE value is same as 300.1899 and PSNR as 23.3568. So, it can be concluded that human eye not only responsive to error but also to the position of the error.

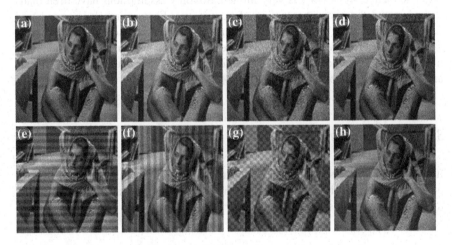

Fig. 1 Comparison of MSE for distortion quality

To comprehend this process, the philosophy of structural similarity [3] comes into play. It suggests that human eye is responsive to the structure of the image rather than error. That leads to the idea that the quality measurement must be made based on the structural disorder made by error introduction. For this purpose, the following methods have been used.

3.3 Structural Similarity Index

This structural similarity index (SSIM) [3] method includes the properties such as luminance, contrast and structure comparison. This can be expresses as Eq. (5).

$$\text{SSIM} = \frac{\left(2\mu_x\mu_y + C_1\right)\left(2\sigma_{xy} + C_2\right)}{\left(\mu_x^2 + \mu_y^2 + C_1\right)\left(\sigma_x^2 + \sigma_y^2 + C_2\right)} \tag{5}$$

where $C_1 = (K_1 L)^2$ and $C_2 = (K_2 L)^2$ the values of K_1 and K_2 as determined by philosophical studies are 0.01 and 0.03 [4]. But problem with SSIM is that it fails to measure the badly blurred images [5].

3.4 Information Content Weighting SSIM

All the above methods are the part of distortion measurement stage. IQA can be expressed as two stage process where stage one is distortion measurement stage which refers to the measurement of local distortion map and stage two is pooling which refers to the mapping of the error to the region of interest detection. It has been seen that optimal perceptual weight is directly proportional to local information content. So information content weighting can be incorporated with different error maps, which can enhance their performance [6]. As SSIM is better than the other methods, it is incorporated with SSIM to give information content weighting SSIM (IWSSIM).

Despite of performing well, these mathematical models are incapable to comprehend the processing of image in extra cortical regions and the human response to different error types which seemed to be same by computational analysis [7]. Here, a good approximation can be made with machine learning techniques. Following are some of the famous machine learning rule that can be used for IQA.

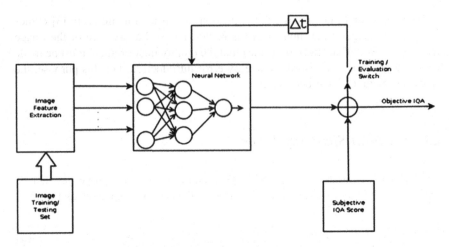

Fig. 2 Model for IQA learning in neural network

3.5 MultiLayer Perceptron (MLP) with Back Propagation Learning

The multilayer neural network can learn any nonlinear system efficiently and tolerant to errors. The multilayer ANN can comprehend complex human thinking by its learning methodology and back propagation learning is one of them. The process of back propagation learning on IQA can be expressed as in Fig. 2. The feature extraction process involves extracting features that have visual significance such as luminance, structure, edges (Hari's operator), directional change of intensities (with Sobel's operator) and change in frequency domain. The efficacy of this method is, it needs no prior knowledge of any mathematical relation of these features to human eye. The significance can be observed and verified on experimental level and then formulate a mathematical formula for the system. Here, experimental results are shown in Sect. 4 for the features luminance, structure and contrast taken together, which follow the results of SSIM. Figure 2 shows the learning process in an ANN with feature extraction and subjective value.

3.6 Functional Link ANN

The problem with multilayer back propagation methodology is that it needs more computation power as the number of hidden layer increases. To overcome this problem, the easiest way is to convert the nonlinear problem to linear problem by mapping it to a higher dimensional space. Functional link ANN (FLANN) used this concept to convert the multilayer network to single layer model. FLANN can summarize the nonlinear learning process to its underlying regression surface much faster with help of expansion function.

4 Experimental Result

4.1 Experimental Setup

The images of the database as described in Table 1 are converted to gray scale images. All the algorithms discussed in Sect. 3 have been implemented along with the SRCC and KRCC on MATLAB. The experimental results on IVC, A57 and LIVE datasets are shown in Tables 2, 3 and 4.

Table 2 Results of IVC dataset

	KRCC	SRCC
MSE	−0.52794	−0.69082
PSNR	0.52794	0.69082
SSIM	0.71277	0.88854
IWSSIM	0.7235	0.90024
C-L-S with BP	0.70012	0.83022

Table 3 Result A57 dataset

	KRCC	SRCC
MSE	−0.43007	−0.61763
PSNR	0.43007	0.61763
SSIM	0.60629	0.80666
IWSSIM	0.6358	0.8364
C-L-S with BP	0.5868	0.7886

Table 4 The result for LIVE dataset

	KRCC	SRCC
MSE	−0.6789	−0.8768
PSNR	0.6789	0.8768
SSIM	0.7846	0.9357
IWSSIM	0.8038	0.9447
C-L-S with BP	0.7794	0.9352

From the result, we can see that PSNR and MSE rank ordering magnitude is same but the negative indication in MSE suggests that ordering is in reverse order i.e., in all other case, larger value indicates better image but in case of MSE the larger value indicates worst quality.

5 Conclusion and Future Work

In Full-reference IQA, the error-based methods do not correlate to human eye so the philosophy of structural feature of image gives a better approach to human subjectivity. Again the result confirms the fact that IQA can be improved with pooling strategies. But all these methods are limited to human thinking of systems. So, the neural network models give a better approach to find new visual relevant features without changing the method of evaluation which can help to understand human visual observation and subjectivity.

Further studies are needed to find the proper feature that can be used in neural network to give better result and to reduce the computational time of the neural network.

References

1. Zhou, W., Bovik, A.C.: Mean squared error: love it or leave it? A new look at signal fidelity measures. Sig. Process. Mag. IEEE **26**(1), 98–117 (2009)
2. Hore, Alain, and Djemel Ziou. Image quality metrics: PSNR vs. SSIM, *Pattern Recognition (ICPR), 2010 20th International Conference, IEEE*, 2010
3. Wang, Z., Bovik, A.C., Sheikh, H.R., Simoncelli, E.P.: Image quality assessment: from error visibility to structural similarity. Image Proces. IEEE Trans. **13**(4), 600–612 (2004)
4. Zhou, W., Simoncelli, E.P., Bovik, A.C.: Multiscale structural similarity for image quality assessment. In: Conference Record of the Thirty-Seventh Asilomar Conference, vol. 2. IEEE (2003)
5. Chen, G.-H., et al.: Edge-based structural similarity for image quality assessment. In: Acoustics, Speech and Signal Processing, 2006. ICASSP 2006 Proceedings. 2006 IEEE International Conference, vol. 2. IEEE (2006)
6. Wang, Z., Li, Q.: Information content weighting for perceptual image quality assessment. Image Process. IEEE Trans. **20**(5), 1185–1198 (2011)
7. Wang, Z., Bovik, A.C., Lu, L.: Why is image quality assessment so difficult? In: Acoustics, Speech, and Signal Processing (ICASSP), 2002 IEEE International Conference, vol. 4. IEEE (2002)
8. IVC Database: http://www2.irccyn.ec-nantes.fr/ivcdb/IVC_SubQualityDB.zip (2014). Accessed 12 Mar 2014
9. Live Database: http://live.ece.utexas.edu/research/quality/release2/databaserelease2.zip (2014). Accessed 12 Mar 2014
10. A57 Database: http://foulard.ece.cornell.edu/dmc27/vsnr/a57_db.zip (2014). Accessed 12 Mar 2014

Image Compression Using Edge-Enhancing Diffusion

Lipsa Behera, Bibhuprasad Mohanty and Madhusmita Sahoo

Abstract Image compression technique minimizes the size in bytes of a graphics file without degrading the quality of the image to an unacceptable visual level. This piece of work deals with the method of image compression by preserving the edge information intact as the human visual system is much sensitive to these information. This is done by the use of Perona–Malik method for diffusion where the whole image is smoothened but the edge. Terming this as edge-enhancing diffusion (EED), we apply two established coding techniques, namely singular value decomposition (SVD) and set partitioning in hierarchical trees (SPHIT). Even though the above encoding schemes enjoy more superiority and advantages in their respective domain, extensive simulation for the proposed diffusion platform provides still better results in terms of PSNR and visual quality.

Keywords Perona–Malick diffusion · Partial differential equations · Image compression · Nonlinear diffusion · SVD · SPHIT

1 Introduction

Image compression addresses the problem of reducing the amount of data required to represent the digital image. Image compression demands an effective coding technique to meet the trade-off between bandwidth and storage capacity with the quality of reconstructed image. In natural images, most scenes are relatively blurred in the areas between edges, while large discontinuities occur at edge locations. Thus, the information between edges may be redundant and may be subjected to increased compression. However, the human visual system is highly

L. Behera · B. Mohanty (✉) · M. Sahoo
Department of ECE, ITER (Faculty of Engineering), Siksha O Anusandhana University, Bhubaneswar, India
e-mail: bibhumohanty@soauniversity.ac.in

© Springer India 2015
L.C. Jain et al. (eds.), *Intelligent Computing, Communication and Devices*,
Advances in Intelligent Systems and Computing 309,
DOI 10.1007/978-81-322-2009-1_51

457

sensitive to edges, and hence, an edge-preserving-based image compression technique can produce intelligible images at high compression ratios.

Diffusion-based image compression is well suited for extremely high compression rates and provides fast and compact coding. There are two types of diffusion: isotropic and anisotropic. The first nonlinear diffusion filter has been proposed by Perona and Malik in 1987 and is being widely used for intra-region smoothing and inter-region diffusion [1]. Its subsequent modifications [2–5] provide a great deal of support to the image processing community. The PM equation is a potential tool for image segmentation, noise removal, edge detection, and image enhancement. As far as the edge enhancement is concerned, the low-pass filtering and linear diffusion method further blurs the edge. On the other hand, using high-pass filter gives rise to numerically unstable computational method. [6] Hence, to overcome these two conflicts, anisotropic diffusion is proposed [7]. This is a more stable diffusion where the conduction coefficient of the equation is a function of the image gradient. This is popularly known as edge-enhancing anisotropic diffusion (EED).

The rest of this paper is organized as follows. In Sect. 2, we describe partial derivative-based EED interpolation techniques and briefly review the two encoding schemes, namely SPIHT and SVD, used popularly for their established superiority over other encoding scheme. In Sect. 3, the proposed methodology is detailed. Section 4 includes results and analysis. We have concluded the paper in Sect. 5 with a brief summary of work in future direction.

2 Brief Review on PM Model-Based Edge-Enhancing Diffusion and Coding Scheme

There are many partial derivative equations (PDEs)-based diffusion available in the literature [2, 5]. It is broadly classified as Linear and Nonlinear diffusion. Biharmonic, Tri-harmonic, and isotropic diffusions are the examples of linear type diffusion, whereas anisotropic diffusion is categorized under nonlinear diffusion. Let us assume that the image is considered as an array f. One of the most widely used methods for smoothing f is to apply the homogeneous linear diffusion process to it. The process is represented mathematically as [1, 2]:

$$\partial_t u = \nabla u$$

where u is the diffused image such that

$$u(x, 0) = f(x).$$

$$u(x, t) = \begin{cases} f(x) & (t = 0) \\ (K_{\sqrt{2t}} * f)(x) & (t > 0) \end{cases} \tag{1}$$

where K_σ denotes a Gaussian with standard deviation σ

$$K_\sigma(x) := \frac{1}{2\pi\sigma^2} \cdot \exp\left(-\frac{|x|^2}{2\sigma^2}\right). \tag{2}$$

For biharmonic smoothing $\partial_t u = \nabla^2 u$, and for triharmonic smoothing $\partial_t u = \nabla^3 u$. Linear diffusion dislocates edges when moving from finer to coarser scales. So structures which are identified at a coarse scale do not give the right location and have to be traced back to the original image. In practice, relating dislocated information obtained at different scales is difficult and bifurcations may give rise to instabilities. Average absolute error (AAE) between interpolated image and original image is a standard evaluation criterion for PDE-based image processing [5]. Experimentally, it has been verified that AAE of PM model-based EED is minimum, so it is better than other PDEs. This process can reduce noise and preserve edges.

Perona and Malik proposed a numerical method for selectively smoothing digital images. In anisotropic diffusion, flow is not only proportional to the gradient, but is also controlled by a function $g(|\nabla u|)$. Regions with low $|\nabla u|$ are plains. By choosing a high diffusion coefficient, the noise can be reduced. Regions with high $|\nabla u|$ can be found near edges. In order for those edges to be preserved, a low diffusion coefficient is chosen accordingly. This leads to the function, $[0, \infty] \rightarrow [0,1]$, $g(0) = 1$, $\lim s \rightarrow \infty \ g(s) = 0$ which is monotonically decreasing.

In order to obtain a complete system, which is needed to solve the equations, boundary conditions also have to be defined. As the Neumann boundary condition is common in image processing, $\nabla u = 0$ on $\partial\Omega$ is used. This ensures that there is no flow across the boundary and the overall brightness is thereby preserved. It is defined as

$$\frac{\partial u(x, y, t)}{\partial t} = \text{div}[g(\|\nabla u\|)\nabla u] \tag{3}$$

Where t is the time parameter, $u(x, y, 0)$ is the original image, and $\nabla u(x, y, t)$ is the gradient version of image at time "t".

$$u(0, x) = u_0(x) \tag{4}$$

$$g(|\nabla u|) = \frac{1}{1+|\nabla u|^2/\lambda^2} \tag{5}$$

λ is always greater than 0.

In the above equation, g is smooth non-increasing function with $g(0) = 1$, $g(x) \geq 0$, and $g(x)$ tending to zero at infinity. The idea is that the smoothing process obtained by the equation is "conditional"; that is, if $\nabla u(x)$ is large, then diffusion will be low, and therefore, the exact localization of the "edges" will be kept. If $\nabla u(x)$ is small, then the diffusion will tend to smooth still more around x.

Thus, the choice of g corresponds to a sort of thresholding which has to be compared to the thresholding of $|\nabla u|$ used in the final step of classical theory.

Perona and Malik discretized their anisotropic diffusion equation as

$$u_{t+1}(s) = u_t(s) + \frac{\lambda}{|\eta_s|} \sum_{p \in \eta_s} g_k(|\nabla u_{s,p}|) \nabla u_{s,p} \tag{6}$$

S denotes the pixel position in the discrete 2D grid, t denotes the iteration step, g is the conduction function, and λ is the gradient threshold parameter that determines the rate of diffusion. λ is a scalar quantity which determines the stability, and it is usually less than 0.25. η denotes the spatial neighborhood of pixel (x, y). $\eta_s = [N \ S \ E \ W]$, where N, S, E, and W are the north, south, east, and west neighbors of pixel S. η is equal to 4 (except for the image borders). The symbol ∇u is now representing a scalar defined as the difference between neighboring pixels to each direction.

Gradient ∇u in four different directions can be calculated as follows

$$\begin{aligned}
\nabla u_N(x, y) &= u(x, y - 1, t) - u(x, y, t) \\
\nabla u_s(x, y) &= u(x, y + 1, t) - u(x, y, t) \\
\nabla u_E(x, y) &= u(x + 1, y, t) - u(x, y, t) \\
\nabla u_W(x, y) &= u(x - 1, y, t) - u(x, y, t)
\end{aligned} \tag{7}$$

This model has some drawbacks such as if the image is noisy, with white noise, for example, then the noise introduces very large oscillations of the gradient ∇u. Thus, the conditional smoothing introduced by the Perona and Malik model will not help, since all these noise edges will be kept.

2.1 Singular Value Decomposition

Singular value decomposition (SVD) is a very useful method to represent a matrix as a product of matrices [8]. Assuming the image (I) to be a matrix of $m \times n$, one can write, $I = U\Sigma VT$, where U and V are orthogonal matrices (i.e., AT $= A-1$) and Σ is the diagonal matrix of $m \times n$ ("0" except its main diagonal) and is the "key" to the decomposition. The elements on the diagonal of Σ (Ki) are the singular value of I, and those values are arranged in decreasing order such that the 1st value is "big," and thereafter, the size decreases according to Ki $= \Sigma \ i, i > \Sigma \ i + 1, i + 1 = K \ i + 1$. This feature facilitates the objective of compression by allowing less storage space than the original image [9]. This is not necessarily the best way to compress image, as the reconstructed image is hardly recognizable for very small value of K. However, as more and more singular values are included, the quality of the image gets better. For detailed discussion on SVD, readers may refer to [10].

2.2 Set Partitioning in Hierarchical Trees

SPIHT is an embedded coding technique. In embedded coding algorithms, encoding of the same signal at lower bit rate is embedded at the beginning of the bit stream for the target bit rate. Effectively, bits are ordered in importance. This type of coding is especially useful for progressive transmission using an embedded code; an encoder can terminate the encoding process at any point. SPIHT algorithm is based on following concepts [11]: (1) ordered bit plane progressive transmission, (2) set partitioning sorting algorithm, (3) spatial orientation trees, and (4) ability to code for exact PSNR. For detailed description, the readers are referred to [11, 12].

3 The Proposed Scheme

Anisotropic diffusion is carried out on an image, and then, the diffused image is encoded using SVD or SPHIT encoding scheme, and then, the compressed image is transmitted as bit stream. At the receiver end, reverse operation is carried out. The following is the system overview (Fig. 1) to carry out the proposed image compression scheme. The EED scheme is a preprocessing scheme before the encoder in the transmitter side. By preprocessing the original image, we have smoothened the whole image but the edge. This process of smoothening introduces more redundancy. On transforming the image to the frequency domain, the energy is compacted, and hence, fewer number of bits are required to code the original one. On the decoder side, the reverse operation takes place. The output of the

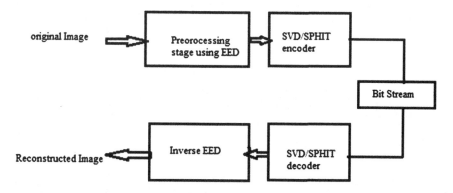

Fig. 1 Proposed SVD/SPHIT-based image codec

decoded image is once again subjected to the method of inverse operation of EED (post-processing) to approximate the lost data.

4 Results and Discussion

We will present some of the results of our implementation with MATLAB in this section. Extensive simulation is carried for over more than 32 images with different attributes. Results of experiments are depicted below with grayscale "Lena" and "Pepper" image for their established popularity. Pepper image contains more details than Lena image. All simulations are done with Pentium IV, 2.4 GHz,

Fig. 2 PSNR for LENA with SVD codec

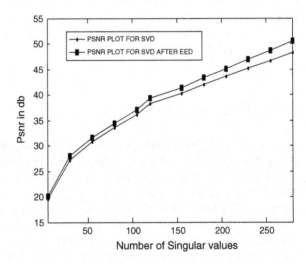

Fig. 3 PSNR for PEPPER with SVD codec

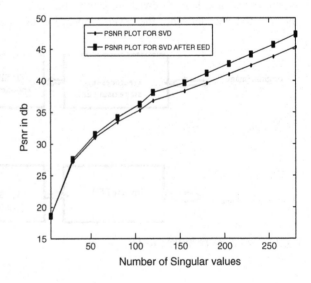

4 GB RAM. The quality of reconstruction is characterized by the well-known PSNR value calculated between the original and reconstructed image.

The following plots (Figs. 2 and 3) for PSNR (in y-axis) for different singular values (in x-axis) establish the fact that the edge-enhancing diffusion (EED)-based scheme as applicable to the SVD codec of Fig. 1 improves the quality of the reconstructed image. It also indicates that as the singular values increase, the PSNR increases irrespective of fact that whether diffusion is applied or not. But, with the application of diffusion, there is 1–2.3 dB increase of PSNR for Lena image and that for Pepper image is 0.7–1.9 dB. The small variation is due to the fact that the Pepper image contains large number of edge as compared to Lena. So, the redundancy reduction in Lena is larger than Pepper. The plot in Figs. 4 and 5 is for the SPIHT codec as depicted in Fig. 1. SPIHT being a wavelet-based codec, the transformed coefficient is more compact in energy and concentrated in much fewer planes than had it not been the wavelet transformed. One can see that there is a wide variation of PSNR (in y-axis) with different bit rate (in x-axis). The PSNR

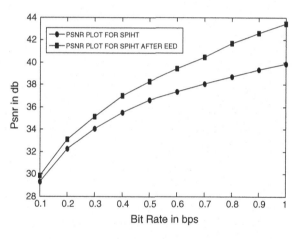

Fig. 4 PSNR for LENA with SPIHT codec

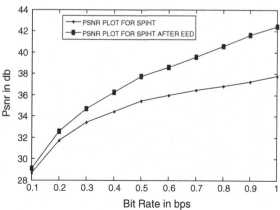

Fig. 5 PSNR for PEPPER with SPIHT codec

improvement is more than 4.3 dB for Lena image at 1 bit per pixel (bpp) and 3.7 dB for Pepper.

5 Conclusion

In this piece of work, an EED-based scheme is discussed for the purpose of still image compression and its inverse diffusion is proposed for the purpose of image reconstruction. Though the reconstructed image enjoys some superiority in terms of PSNR, it lacks the proper inverse operation as envisaged in the encoding stage. The application of wavelet transform may increase the speed of the codec and facilitate its hardware implementation. It is because of the fact that the preprocessed image is a smooth version of the original one except at the point of edges and hence bit plane coding scheme using wavelet transform will require less time for analysis as well as synthesis of the coefficients.

References

1. Perona, P., Malik, J.: Scale-space and edge detection using anisotropic diffusion. In: Proceedings of IEEE Computer Society Workshop on Computer Vision, pp. 16–22 (Nov 1987)
2. Perona, P., Malik, J.: Scale-space and edge detection using anisotropic diffusion. IEEE Trans. Pattern Anal. Mach. Intell. 12(7), 629–639 (1990)
3. Sapiro, G.: From active contours to anisotropic diffusion: relations between basic PDEs in image processing. In: Proceedings ICIP, Lausanne, Switzerland (Sep 1996)
4. Shah, J.: A common framework for curve evolution, segmentation, and anisotropic diffusion. In: Proc CVPR, San Francisco, CA, pp. 136–142 (June 1996)
5. Barash, D., Comaniciu, D.: A common framework for nonlinear diffusion, adaptive smoothing, bilateral filtering and mean shift. Image Vis. Comput. 22(1), 73–81 (2004)
6. Canny, J.: A computational approach to edge detection. IEEE Trans. Pattern Anal. Mach. Intell. PAMI-8, 679–698 (1986)
7. Black, M.J., Sapiro, G., Marimont, D.H., Heeger, D.: Robust anisotropic diffusion. IEEE Trans. Image Process. 7(3), 421–432 (1998)
8. Linear Algebra and its Application, 3rd edn. In: Lay, D.C. Addison-Wesley Publishing Co., Boston (2002)
9. Numerical Analysis. In: Saucer, T. George Mason University, Pearson Education Inc, Pearson (2006)
10. An investigation into using SVD as a method of image compression, http://www.haroldthecat.f2s.com/project
11. Said, A., Pearlman, W.A.: A new, fast and efficient image codec based on set partitioning in hierarchical trees. IEEE Trans. Circ. Syst. Video Technol. 6, 243–250 (1996)
12. Kim, B.-J., Xiong, Z., Pearlman, W.A.: Low bit-rate scalable video coding with 3-D set partitioning in hierarchical trees (3-D SPIHT). IEEE Trans. Circ. Syst. Video Technol. 10(8), 1374–1387 (2000)

Comparative Analysis of Edge Detection Techniques for Extracting Blood Vessels in Diabetic Retinopathy

Sunita Sarangi, Arpita Mohapatra and Sukanta Kumar Sabut

Abstract Diabetic retinopathy (DR) is one of the serious complications caused by diabetes. It damages the small blood vessel of the retina, which leads to loss of vision. Accurate extraction of retinal blood vessels is an important task in computer-aided diagnosis of DR. The edge detection technique has been greatly benefited in interpreting the information contents in the retinal blood vessel. The preprocessing of retinal image may help in detecting the early stage of symptoms in DR. In this article, we compared the results of Sobel and Canny edge operators for finding the abnormalities in retinal blood vessels. The Canny operator is found to be more accurate in detecting even tiny blood vessels compared to Sobel operator for an affected diabetic retinal image. The peak signal-to-noise ratio (PSNR) is also found to be high in the Canny operator.

Keywords Diabetic retinopathy · Edge detection · Sobel · Canny operator

1 Introduction

Diabetes has emerged as a major healthcare problem worldwide. The Indian Council of Medical Research estimated that around 65.1 million patients affected by diabetes and, by 2030, India will have 101.2 million diabetic people [1]. The high incidence is due to genetic susceptibility plus adoption of a high-calorie, low-activity lifestyle. Nearly 1 million dies due to diabetes every year [2]. Diabetic retinopathy (DR) damages the small blood vessel of the retina, which leads to loss of

S. Sarangi (✉) · A. Mohapatra · S.K. Sabut
Department of Electronics and Instrumentation Engineering, Institute of Technical
Education and Research, Siksha 'O' Anusandhan University, Bhubaneswar, Odisha, India
e-mail: sunitasarangi@soauniversity.ac.in

S.K. Sabut
e-mail: sukantsabut@soauniversity.ac.in

© Springer India 2015
L.C. Jain et al. (eds.), *Intelligent Computing, Communication and Devices*,
Advances in Intelligent Systems and Computing 309,
DOI 10.1007/978-81-322-2009-1_52

vision [3]. Current research indicates that at least 90 % of new diabetic cases could be reduced by proper treatment and monitoring of the eyes in early stages [4, 5].

Retinal blood vessels can be automatically visualized by image processing with very high resolution for clinical scenarios [6]. Edge detection techniques are used for identifying sharp discontinuities in an image by preserving the important structural properties [7]. The 2D matched-filter detection method retains the computational simplicity of retina compared to the Sobel operator [8]. It is found that MF-FDOG by using both the matched-filter (MF) and the first-order derivative of the Gaussian (FDOG) distinguishes better vessel structures [9]. The substantial improvement is obtained in retinal vessel by extracting centerline combining with matched-filter responses and vessel boundary measures [10]. The automatic tracking algorithms are used to detect the vascular tree and local vessel parameters such as vessel edge points and the vessel width in retinal images [11].

In this paper, we have implemented different edge detection techniques and compared their performances in diabetic retinal images. The angle histogram also performed for different edged detectors, and computed peak signal-to-noise ratio (PSNR) values are compared.

2 Methodology

Blood vessels are extracted for the identification of damages in DR. Retinal images are collected from an eye institute and the image processing operations to detect blood vessels (shown in Fig. 1). A healthy image and an affected DR image are selected from ten sample retinal images (shown in Fig. 3a, b). The RGB images are converted to grayscale to strengthen the appearance of blood vessels. Initially, the original image was scaled to enhance the contrast of image and the Sobel and Canny operators were performed to detect blood vessels by using MATLAB 11.0.

2.1 Median Filter

It is a nonlinear digital filtering technique that is used to remove noise in pre-processing step to improve the image in edge detection. It removes noise while preserving edges for a given fixed window size. For a grayscale input image with intensity values $x_{i,j}$, the two-dimensional median filter is defined as

$$y_{i,j} = \text{median}(x_{i+r,j+s})$$
$$(r,s) \in w$$

where w is a window over which the filter is applied.

Fig. 1 Flowchart of blood
vessel extraction of retinal
image

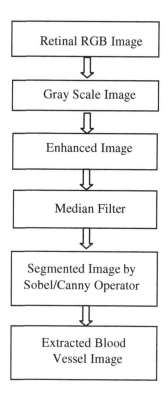

2.2 Sobel Operator

The operator consists of a pair of 3×3 convolution kernels (Fig. 2) and is designed to respond maximally to edges running vertically and horizontally relative to the pixel grid. The kernels can be applied separately to the input image, to produce separate measurements of the gradient component in each orientation (call these G_x and G_y). These can then be combined together to find the absolute magnitude of the gradient at each point and the orientation of that gradient.

The gradient magnitude is given by: $|G| = \sqrt{G_x^2 + G_y^2}$; the approximate magnitude computed by $|G| = |G_x| + |G_y|$ and the angle of orientation by: $\theta = \tan^{-1}\left(\frac{G_y}{G_x}\right)$.

2.3 Canny Operator

The Canny method can detect edges with noise suppressed at the same time [12]. The raw images were smoothed by a canny detector convolving with a Gaussian filter with $\sigma = 4$. Compute the gradient of image $g(x, y)$ by convolving it with the

$$G_x = \begin{bmatrix} -1 & -2 & -1 \\ 0 & 0 & 0 \\ 1 & 2 & 1 \end{bmatrix} \quad ; \quad G_y = \begin{bmatrix} -1 & 0 & 1 \\ -2 & 0 & 2 \\ -1 & 0 & 1 \end{bmatrix}$$

Fig. 2 Masks used for Sobel operator

first derivative of Gaussian masks $f(x, y)$. The Gaussian filter size [3 3] is used to smooth an image using a weighted mean based on the 3D Gaussian point spread function.

The circularly symmetric 3D Gaussian equation centered at the origin has the form of: $f(x,y) = \frac{1}{2\pi\sigma^2} e^{-\frac{x^2+y^2}{2\sigma^2}}$.

Compute the gradient of $g(x, y)$ by using any of the gradient operators to get:

$$G(x,y) = \sqrt{g_x^2(x,y) + g_y^2(x,y)}$$

The Canny edge detection algorithm is known to many as the optimal edge detector.

$$g(x, y) = G_\sigma(x, y) * f(x, y)$$

where $G_\sigma(x, y)$ is Gaussian filter.

2.4 Angle Histogram

The angle histogram shows the distribution of values according to their numeric range. The distribution of theta in 20 angle bins or less determines the angle of each bin from the origin and the length of each bin reflects the number of elements in theta.

2.5 Evaluation Parameters

The objective measures PSNR that is used to evaluate the intensity changes of an image between the original and the enhanced images. PSNR can be computed as

$$PSNR = 10 \log_{10}\left(\frac{255^2}{MSE}\right),$$

between the original–affected and the output images, where mean square error (MSE) is given by

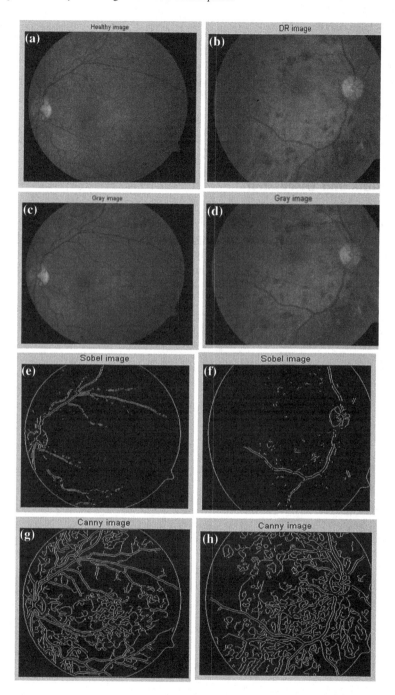

Fig. 3 The simulation results of healthy image and DR image. **a** Healthy image, **b** DR image, **c** gray image, **d** gray image, **e** Sobel image, **f** Sobel image, **g** Canny image, and **h** Canny image

Table 1 Edge detection analysis

Parameters	Healthy Sobel	Healthy Canny	DR Sobel	DR Canny
PSNR	11.7816	11.7917	10.0033	10.0119
MSE	0.004348	0.004338	0.006548	0.006535

$$\text{MSE} = \frac{1}{mn} \sum_{i=1}^{m} \sum_{j=1}^{n} \|Ia(i,j) - Io(i,j)\|^2$$

3 Results and Discussion

A healthy image and a DR image were chosen from ten sample images. Sobel and
Canny edge detection techniques were applied to the affected image according to
Fig. 1, and the results are presented in Fig. 3a–h. The PSNR and MSE parameters
were calculated between the grayscaled image and Sobel, Canny operator's output

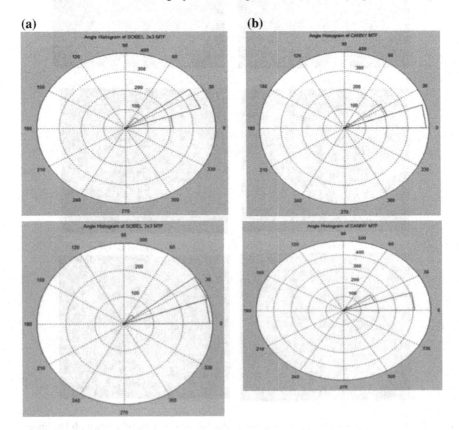

Fig. 4 Angle histogram of healthy and DR images. **a** Sobel operator, **b** Canny operator

images. The simulation result shows that the blood vessels extracted from the retinal images are with better PSNR values in Canny operator than the Sobel operator (Table 1).

The quality of edge detection results was evaluated subjectively by angular. Less attenuation has shown in Canny compared to Sobel operator (Fig. 4). The PSNR value is also high in Canny operator. We observed that both edge detection techniques help in detecting the retinal blood vessels but the Canny operator performs better as it could detect even very small variation of intensity in the tiny retinal blood vessels.

4 Conclusion

In this paper, we have implemented the Sobel and Canny operators in retinal images for detecting the blood vessels. The result indicates that the Canny operator performs better compared to Sobel operator for detecting even tiny blood vessels of retina. We also performed the angular histogram on retinal image for finding the distribution of values according to numeric range. The calculated PSNR value of Canny operator is high that indicates better results for analysis of image in DR. We conclude that the Canny operator performed better than the Sobel operator, since it produced the continuous edges, which is helpful for identifying the blockages in the tiny blood vessels. Thus, it will be easier for ophthalmologist for identifying the pathological blood vessels in retina for diagnosing the DR. Our future work focused on calculating the sensitivity and accuracy of the edge detectors in DR images.

Acknowledgements The authors thank the doctors and administrator of L.V. Prasad Eye Institute for providing the retinal images used in this work.

References

1. Basu, S., Vellakkal, S., Agrawal, S., Struckler, D., Popkin B., Ebrahim, S.: Averting obesity and type 2 diabetes in India through sugar-sweetened beverage taxation: an economic-epidemiologic modelling study. PLoS Med. **11**, (2014)
2. Jason, G.: India's Diabetes Epidemic Cuts Down Millions Who Escape Poverty, Bloomberg. Retrieved 8 Jun 2012
3. Diabetes and Eye Disease, http://www.ncbi.nlm.nih.gov. Retrieved on 21 Apr 2012
4. Taylor, H.R., Keeffe, J.E.: World blindness: a 21st century perspective. Br. J. Ophthalmol. **85**, 261–266 (2001)
5. Lee, S.J., McCarty, C.A., Keeffe, J.E.: Costs of mobile screening for diabetic retinopathy: a practical framework for rural populations. Aust. J. Rural Health **8**, 186–192 (2001)
6. Dougherty, G.: Image analysis in medical imaging: recent advances in selected example. Biomed. Imaging Intervention J. **6**, 1–9 (2010)

7. Ramalho, M., Curtis, K.M.: Edge detection using neural network arbitration. In: 5th Int. Conf. on Image Processing and its Applications, pp. 514–518 (1995)
8. Chaudhuri, S., Chatterjee, S., Katz, N., et al.: Detection of blood vessels in retinal images using 2D matched filters. IEEE Trans. Med. Image **8**, 263–269 (1989)
9. Zhang, B., Zhang, L., Karray, F.: Retinal vessel extraction by matched filter with first-order derivative of Gaussian. Comput. Biol. Med. **40**, 438–445 (2010)
10. Sofka, M., Stewart, C.V.: Retinal vessel centerline extraction using multi-scale matched filters, confidence edge measures. IEEE Trans. Med. Imaging **25**, 1531–1546 (2006)
11. Yin, Y., Adel, M., Bourennane, S.: Automatic segmentation and measurement of vasculature in retinal fundus images using probabilistic formulation. Comput Math Methods Med: 1–16 (2013)
12. Canny, J.: A computational approach to edge detection. IEEE Trans. Pattern Anal. Mach. Intell., PAMI-8(6), 679–698 (1986)

Realization of Different Algorithms Using Raspberry Pi for Real-Time Image Processing Application

Mrutyunjaya Sahani and Mihir Narayan Mohanty

Abstract As automated system is an efficient one which decreases the malfunctions and error, industries having more demand to adopt such systems day by day. To enhance this capability, image processing-based system has a major role. But to design an embedded system with the facilities, it is difficult and challengeable task. Another issue for the use interfacing of such systems should be simple so that the system can be user friendly. To develop this system with various facilities, this piece of work has been attempted. Hence in this paper, authors showcase the attempt to realize the algorithms based on image processing for different applications. The work is developed in Raspberry pi development board with python 2.7.3 and OpenCV 2.3.1 platform. The result shows for the fully automated without any user intervention. The prime focus is on python image library (PIL) in addition to numpy, scipy, and matplotlib form a powerful platform for scientific computing.

Keywords ARM1176JZF-S · Image processing · Raspberry pi · Python image library · Raspbian-Wheezy

1 Introduction

Real-time image processing techniques are important not only in terms of improving productivity, but also in reducing operator errors associated with visual feedback delay. For implementing the different image processing algorithm, we have used the Raspberry Pi (shown in Fig. 1) which is a single-board computer developed by Cambridge University. The Pi has been extremely popular among

M. Sahani (✉) · M.N. Mohanty
ITER, Siksha 'O' Anusandhan University, Odisha, India
e-mail: mrutyunjayasahani@soauniversity.ac.in

M.N. Mohanty
e-mail: mihirmohanty@soauniversity.ac.in

© Springer India 2015
L.C. Jain et al. (eds.), *Intelligent Computing, Communication and Devices*,
Advances in Intelligent Systems and Computing 309,
DOI 10.1007/978-81-322-2009-1_53

Fig. 1 Raspberry pi board (model B)

the academic fraternity due to its low cost. The model B of the Pi ships with 512 Mb of RAM, 2 USB ports, and an Ethernet port. It packs an ARM1176JZF-S 700 MHz processor, Video Core IV GPU into the Broadcom BCM2835 System on Chip which is cheap, powerful, and also low on power. The Pi has HDMI support and has an SD card slot for booting up due to lack of BIOS and a persistent memory [1]. Python is one of the primary languages supported by the Pi; hence, we have used the python image library (PIL) which adds powerful image processing functions to the Python interpreter. There are many different operating systems supported by the Raspberry Pi. We have chosen Raspbian-Wheezy which can be downloaded from the University of Cambridge website. Raspbian-Wheezy is a freely available version of Debian Linux which has been customized to run on the Pi. It has been designed to make use of the Pi's floating point hardware architecture, thus enhancing performance. The program designed by me is completely automatic, and there is no need for user intervention. The program displays the processed image on the screen for comparison.

This chapter is organized as follows. Section 1 introduces the work, and Sect. 2 deals with the system configuration. Section 3 proposes the implementation method that follows the result. Finally, Sect. 4 concludes this piece of contribution.

2 System Configuration

The basic hardware requirements are a Raspberry Pi board (model B), a USB hub, a USB keyboard, and mouse. An HDMI-enabled display device is also required along with a micro-USB charger to power the board. An SD card is required to boot the Pi.

2.1 Preparation of SD Card for Raspberry Pi

Raspberry Pi lacks a very important feature, i.e., BIOS. So the Pi always needs a SD card loaded with a snapshot of an operating system to boot up. After downloading the image, we must prepare a snapshot of the image on the SD card for the Pi to boot [2].

2.2 Network Configuration

Network configuration should be done for the Pi to enable installation of different packages, dependencies and to regularly update and upgrade the operating system.

2.3 List of Packages

Following packages are to be installed for implementing the proposed model. Installation commands have been listed below.

(a) sudo apt-get install camorama
(b) sudo apt-get install python-dev
(c) sudo apt-get install libjpeg62-dev libpng12-dev
(d) sudo apt-get install python-matplotlib
(e) sudo apt-get install python-numpy python-scipy
(f) sudo apt-get install python-imaging python-tk

2.4 OpenCV Installation

OpenCV is used to implement different algorithms in the python environment to make the system automatic [3]. Before final installation of OpenCV, following dependencies have to be installed.

1. Menu → Accessories → LX Terminal → this opens a terminal window in RPi graphical interface. Then, we use following commands.

   (a) sudo apt-get install build-essential
   (b) sudo apt-get install cmake
   (c) sudo apt-get install pkg-config
   (d) sudo apt-get install libpng12-0 libpng12-dev libpng++dev libpng3
   (e) sudo apt-get install zlib1g-dbg zlib1g zlib1g-dev
   (f) sudo apt-get install libpnglite-dev libpngwriter0-dev libpngwriter0c2

(g) sudo apt-get install pngtools libtiff4-dev libtiff4 libtiffxx0c2 libtiff-tools

(h) sudo apt-get install libjpeg8 libjpeg8-dev libjpeg8-dbg libjpeg-progs

(i) sudo apt-get install ffmpeg libavcodec-dev libavcodec52 libavformat52 libavformat-dev

(j) sudo apt-get install libgstreamer0.10-0-dbg libgstreamer0.10-0 libgstreamer0.10-dev

(k) sudo apt-get install libxine1-ffmpeg libxine-dev libxine1-bin

(l) sudo apt-get install libunicap2 libunicap2-dev

(m) sudo apt-get install libdc1394-22-dev libdc1394-22 libdc1394-utils

(n) sudo apt-get install libv4l-0 libv4l-dev

(o) sudo apt-get install python-numpy

(p) sudo apt-get install libpython2.7 python-dev python2.7-dev

(q) sudo apt-get install libgtk2.0-dev pkg-config

2. OpenCV was downloaded from the Internet using the following command. Wgethttp://sourceforge.net/projects/opencvlibrary/files/opencv-unix/2.3.1/ OpenCV2.3.1a.tar.bz2/download.

3. Then, we make, install, and configure the OpenCV installed by using following commands.

(a) cmake -D CMAKE_BUILD_TYPE=RELEASE –D MAKE_INSTALL_ PREFIX = "/usr/local/lib"–D BUILD_NEW_PYTHON_SUPPORT=ON –D BUILD_ EXAMPLES = ON

(b) sudo make

(c) sudo make install

Now OpenCV-2.3.1 is installed and ready to run on the Raspberry Pi. The installation takes a very long time to complete.

3 Proposed Method for System Realization

3.1 Geometrical Transformation

Geometrical transformation refers to modification of the spatial relationship of the pixels in an image [4]. Some or all of the pixels in the original image are mapped to a new coordinate in the modified image. Mapping is usually of two types, forward mapping and inverse mapping. Inverse mapping is more efficient than forward mapping and produces better results. Affine transforms are one of the most commonly used transforms and can be used to achieve cropping, flipping, and rotation as shown in Fig. 2. Affine transforms can be achieved by applying the following technique:

(a)   **(b)**   **(c)**   **(d)**   **(e)**

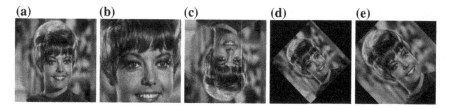

Fig. 2 **a** Original image. **b** Magnified image. **c** Inverted image. **d** Rotation with resizing. **e** Rotation without resizing

$$
\begin{bmatrix} x' \\ y' \\ 1 \end{bmatrix} = \begin{bmatrix} m1 & m2 & m3 \\ n1 & n2 & n3 \\ 0 & 0 & 1 \end{bmatrix} \times \begin{bmatrix} x \\ y \\ 1 \end{bmatrix}
$$

Here, x' and y' are the modified pixels locations, and x and y are the original image locations. By altering the values of the constants in the middle matrix, the image can be rotated, flipped, and cropped [5]. If we wish to perform multiple operations, then the resultant matrix is the product of all the required matrices. For determining the intensity levels, we use interpolation techniques. Usually, bilinear is preferred over the nearest neighbor approach of interpolation.

3.2 Smoothing and Gray-level Slicing

Smoothing operation is usually applied for removal of noise and blurring. Smoothing tends to replace the pixel values with the average of the values in the neighborhood of that pixel. They are also known as low-pass filters or averaging filters. Random noise usually consists of sharp changes in intensity levels, so this filter is a great tool for noise reduction, but since edges are also represented by sharp changes in intensity values so they have the undesirable effect of blurring the edges.

For image enhancement, some applications may need us to focus on certain intensity levels instead of the entire range to highlight the region of interest. There are usually two approaches in this case. In the first approach, the intensity levels within certain desired range are enhanced as required and the remaining intensity levels are suppressed. In the second approach, the intensity levels within the desired range are enhanced and remaining intensity levels are left as it is shown in Fig. 3.

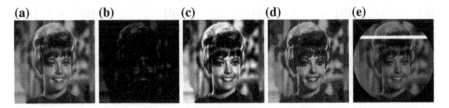

Fig. 3 **a** Original image. **b** Darkened image. **c** Brightened image. **d** Contour plot. **e** Sliced image

3.3 Sharpening and Denoising

Sharpening is done to enhance details and edges in an image. Among the various options available, we will restrict ourselves to sharpening by Laplacian operator. The matrix A represents the Laplacian mask considering only the horizontal and vertical directions. The matrix B represents the Laplacian mask considering horizontal, vertical as well as diagonal directions.

$$A = \begin{bmatrix} 0 & 1 & 0 \\ 1 & -4 & 1 \\ 0 & 1 & 0 \end{bmatrix} \quad B = \begin{bmatrix} 1 & 1 & 1 \\ 1 & -8 & 1 \\ 1 & 1 & 1 \end{bmatrix}$$

In case of median filter, we take a mask of the desired size and apply it to each pixel in the image. On applying the mask, we replace the original intensity value by the median (by arranging in ascending order and selecting the middle value) of the neighborhood intensity values. For denoising applications, median filter has greater advantage over box filter or Gaussian filter since it tends to retain the sharpness and contrast of the image after processing it which is shown in Fig. 4. Median filter is particularly very effective in case of salt and pepper noise.

3.4 Edge Detection

Industry is using edge detection technique to automate the process for increasing efficiency by eliminating defects or malfunctioning of any system. An edge may be

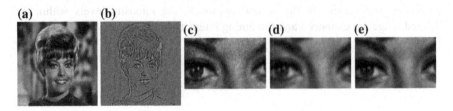

Fig. 4 **a** Original image. **b** Sharpened image. **c** Noisy image. **d** Gaussian filter. **e** Median filter

(a) (b) (c) (d) (e)

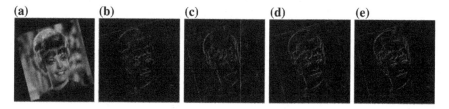

Fig. 5 **a** Original image. **b** Sobel along *x*. **c** Sobel along *y*. **d** Sobel. **e** Canny

defined as a high-frequency component in an image *f*(*x*, *y*) or a set of continuous pixels in an image in which there is a considerable change in some physical aspect of the image. Such discontinuities in intensity values of an image are detected by using first- and second-order derivatives as shown in Fig. 5. There is several edge detection algorithms [2] implemented in digital image processing like Sobel, Prewitt, Roberts, and Canny [6], but all of them can be categorized into two categories, gradient and Laplacian.

4 Conclusion

Matplotlib, Numpy, Scipy, PIL, and OpenCV were installed successfully on Raspberry Pi development board. It was found that the algorithm developed for the Raspberry Pi executes successfully for different image processing applications and gives very colorful images. Therefore, we conclude that the Raspberry Pi module can easily replace with a host processor for any kind of real-time image processing applications.

References

1. Schmidt, M.: Pragmatic Raspberry Pi, 1st edn. The Pragmatic Programmers, LLC., USA (2012)
2. Tank, J.K., Patel, V.: Edge detection using different algorithms in raspberry pi. Int. J. Sci. Res. Dev. **1**(4), 984–986 (2013)
3. Solem, J.F.: Programming Computer Vision with Python, 1st edn. O'Reilly Media Inc., USA (2012)
4. Gonzalez, R.C., Woods, R.E.: Digital Image Processing, 3rd edn. Pearson Prentice Hall, New Jersy (2008)
5. Vlasic, D.: Geometric image transformation. Internet: http://groups.csail.mit.edu/graphics/classes/6.837/F01/Lecture07/Slide01.html. 30 Oct 2013
6. Canny, J.: A computational approach to edge detection. IEEE Trans. Pattern Anal. Mach. Intell. **PAMI-8**(6), 679–698 (1986)

Edge Preserving Region Growing
for Aerial Color Image Segmentation

Badri Narayan Subudhi, Ishan Patwa, Ashish Ghosh
and Sung-Bae Cho

Abstract Many image segmentation techniques are available in the literature. One of the most popular techniques is region growing. Research on region growing, however, has focused primarily on the design of feature extraction and on growing and merging criterion. Most of these methods have an inherent dependence on the order in which the points and regions are examined. This weakness implies that a desired segmented result is sensitive to the selection of the initial growing points and prone to over-segmentation. This paper presents a novel framework for avoiding anomalies like over-segmentation. In this article, we have proposed an edge preserving segmentation technique for segmenting aerial images. The approach implicates the preservation of edges prior to segmentation of images, thereby detecting even the feeble discontinuities. The proposed scheme is tested on two challenging aerial images. Its effectiveness is provided by comparing its results with those of the state-of-the-art techniques and the results are found to be better.

Keywords Image segmentation · Region growing · Edge detection · Thresholding · Aerial image

B.N. Subudhi (✉) · A. Ghosh
Machine Intelligence Unit, Indian Statistical Institute, Kolkata 700108, India
e-mail: subudhi.badri@gmail.com

A. Ghosh
e-mail: ash@isical.ac.in

I. Patwa
Department of Instrumentation and Control Engineering, National Institute of Technology,
Tiruchirappalli 620015, India
e-mail: ishan.patwa@gmail.com

S.-B. Cho
Soft Computing Laboratory, Department of Computer Science, Yonsei University,
262 Seongsanno, Seodaemun-Gu, Seoul 120-749, South Korea
e-mail: sbcho@cs.yonsei.ac.kr

© Springer India 2015
L.C. Jain et al. (eds.), *Intelligent Computing, Communication and Devices*,
Advances in Intelligent Systems and Computing 309,
DOI 10.1007/978-81-322-2009-1_54

481

1 Introduction

Image segmentation may be defined as a process of dividing an image into a number of non-overlapping meaningful homogenous regions. It has numerous important applications including object detection, object representation, robotics, and land cover analysis. Segmentation of aerial and satellite image is one of the most important tasks in computer vision. Few popular existing segmentation techniques include thresholding-based, statistical-based, and region-based segmentations [1].

In thresholding-based segmentation scheme, an image is divided into a number of segments by defining some threshold value. Thresholding is a well-known and simple approach for segmentation. As the computational complexity is low, thresholding-based schemes are considered for real-time computer vision systems. A simple approach of determining the threshold value/s is by analyzing the peaks and valleys of the histogram of the image frame [2]. A hierarchical satellite image segmentation technique is studied by Peak and Tag [3]. However, since thresholding-based segmentation scheme depends on the gray-level information of a scene, segmentation result obtained from this scheme provides disconnected segmented regions rather than a connected one.

Markov random field (MRF), a well-known statistical model, provides a convenient way to model contextual features [4] of an image such as image gray values, edge, and color. A robust use of MRF theory in segmentation of satellite image is proposed by Berthod et al. [5], where MRF is used with three deterministic relaxation methods and are found to be much faster. However, it may be noted that MRF-based segmentation scheme inherently depends on the assumption about the data distribution, and hence, accuracy may degrades in real-life scenes.

In this regard, it is observed that region-based segmentation technique is one of the popular image segmentation techniques. Region growing technique is simple and can segment different image pixels that have similar properties to form large regions or objects/background [6]. Research in region growing methods of image segmentation has focused either on the designing of feature extraction and growing/merging criterion [7] or on algorithmic efficiency and accuracy [8]. A modification of region growing technique for image segmentation can also be found for aerial image application [9].

In the proposed scheme, we have provided a novel aerial image segmentation technique using edge preserving region growing technique. The approach implicates the preservation of edges in the segmentation of aerial images. In the proposed scheme, we have considered the advantage of local histogram of the considered aerial image. From the local histogram, we computed three contrast features: local contrast (L), region ratio (R) and edge potential (E_p). From these three features, we compute an edge function at each pixel location of the image [7]. The computed edge function of the complete image is thresholded with Pun entropy [2] to obtain the seeds for region growing. The region growing scheme is followed to create the segmentation map of the considered image. Since the considered segmentation scheme relies on the edge function, the segmentation output gives an edge preserved segmented map rather than over-segmented result. The proposed scheme is tested on two challenging aerial images. Its effectiveness is provided by comparing its results with those of the conventional region growing, edge preserving mean-shift, and MRF-based segmentation techniques, and the results are found to be better.

2 Proposed Edge Preserving Region-Based Segmentation

The proposed scheme is an efficient application of the idea floated in Kim et al. [7] for aerial image segmentation. In the proposed scheme, initially we plot the local histogram at each pixel location considering a window/mask around it. From each local histogram, we compute an edge potential function. Then, the computed edge strengths are thresholded with Pun entropy [2]-based thresholding scheme. The edge functions which are below the selected threshold are considered as seed points; and region growing scheme is started from the above seed points to segment the input image. In the subsequent parts, we have clearly described each block.

2.1 Local Histogram Creation and Calculation of Edge Potential

We have implemented a histogram-based edge detection algorithm for our work. In this regard, initially for each input aerial image, a window/mask of size $n \times n$ (where n is considered to be 7, 9,... and is selected according to the size of the image) is chosen.

In the above process, we assume that the local histogram is bimodal and we calculate the major valley and two major peaks on both sides of the major valley. The use of these three parameters of the histogram help in calculating three contrast measures that gives the edge information of the pixel at that location. The considered contrast measures are as follows:

2.1.1 Local Contrast

Local Contrast can be defined as the variation in the intensity of the desired region from the intensity of its neighborhood. The contrast is one of the good descriptors of the edge in the scene. We calculate a local contrast between two or more distinct regions based on the image histogram as follows: v_i is considered to be the major valley point, and L_1 and L_2 are the intensity level at the maximum peak on the left side and right side of v_i, respectively. We may calculate the local contrast at any pixel location by considering L_1, L_2, and v_i, as in [7],

$$L = L_2 - L_1. \tag{1}$$

2.1.2 Region Ratio

Distinctiveness between the two regions depends on the area of the two regions, and this effect is termed as aerial effects. The ratio of these two regions also gives variation of one region from another. For any bimodal distributed probability density function (pdf), the region ratio (r) is defined as the ratio of the minimum area to the maximum area. This can be defined as [7]

$$r = \frac{R_1}{R_2}, \tag{2}$$

where R_1 is the area having less size and R_2 is the area having more size (from *pdf*).

2.1.3 Edge Potential

Similarly, we have computed the local edge potential function as in [7]

$$E_p = \text{Min}\{h_1, h_2\} - h_{v_i} \tag{3}$$

where h_1 is the maximum peak on the left side of the local major valley, h_2 is the maximum on the right side of the local major valley, and h_{vi} is the probability at the local minimum v_i. The edge potential gives variation of the majority peaks from its desired or important region defined over small neighborhood. This also describes the contrast of a region in an image.

2.1.4 Edge Function

We define an edge function for the analyzed parameters L, r, and E_p to obtain the edge information using the defined parameters as follows [7];

$$E = w_1 \frac{L_i}{L_{max}} + w_r \frac{r_i}{r_{max}} + w_e \frac{E_{p_i}}{E_{p_{max}}}; \tag{4}$$

where w_1, w_r, and w_e are the weighing factors and are set to 0.33. The other parameters are considered to be

$$L_{\max} = 255, \; r_{\max} = 1 \text{ and } E_{p_{\max}} = 0.5.$$

The image obtained by using the above-described algorithm is found to produce prominent edges with less noise as compared to the results obtained by Sobel and Laplacian edge detection techniques.

2.2 Seed Point Selection

For any color image, we take the maximum value among the computed edge information E (from the three R, G, and B planes) as the edge response at the given image. However, the image obtained by the process may contain noises, i.e., unwanted edges, thus in order to enhance edge features the image is segmented using region-based segmentation. In this work, all the pixels are quantified by the predefined quantification level with the acquired edge information map.

In this regard, we initially set a proper threshold value T by Pun's entropic thresholding [2]. The pixels edge values less than T are considered as the initial seed points for region growing.

2.3 Region Growing Segmentation

In the region growing algorithm, we start off with the selection of seed points. The seed points are selected with some predefined criteria. These seed points act as the initial points of different regions available in the considered. As the region growing process starts, the neighboring pixels of a seed point from a particular region are tested for homogeneity and are added to a that region. In this way, each pixel in that image is assigned to a particular region in the segmented image. After region growing is over, region merging is performed; different regions of the image are merged to a single region with some similarity criterion.

3 Results and Discussion

The proposed scheme is tested on different images; however, for space constraint, we have provided results on two benchmark aerial images. To validate the proposed scheme, results obtained by it are compared with those of the region growing [1], mean-shift, [10], and MRF-based segmentation [5].

The first example considered in our experiment is San Diego North Island NAS data set, a 512×512 size image, and is obtained from SIPI image database. The

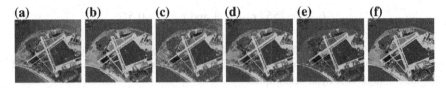

Fig. 1 **a** Original image, **b** ground-truth image, **c** segmentation using region growing scheme, **d** segmentation using mean-shift scheme, **e** segmentation using MRF scheme, **f** segmentation using proposed scheme

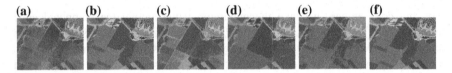

Fig. 2 **a** Original image, **b** ground-truth image, **c** segmentation using region growing scheme, **d** segmentation using mean-shift scheme, **e** segmentation using MRF scheme, **f** segmentation using proposed scheme

original image is shown in Fig. 1a and the corresponding manually segmented ground-truth image is shown in Fig. 1b. The conventional region growing technique has produced a segmented map as shown in Fig. 1c, where it is observed that many details in the image are missed and many places are over-segmented into multiple classes. The result obtained by the mean-shift segmentation technique is shown in Fig. 1d. It is observed from this result that many parts are over-segmented and most regions in the image are split into multiple classes. Hence, there is a higher misclassification error. The result obtained by the MRF-based segmentation scheme is shown in Fig. 1e. Due to large noise and blurred boundary, many false alarms are obtained in the image. The result obtained by the proposed scheme is shown in Fig. 1f. From this result, we can summarize that the results obtained by the proposed scheme have provided a better segmentation map than the other considered techniques.

The second aerial image considered in our experiment is Stockton date set, a 1,024 × 1,024 size image and is obtained from SIPI image database. The original image is shown in Fig. 2a, and the corresponding ground-truth image is shown in Fig. 2b. The segmentation map obtained by the region growing, mean-shift, and MRF-based techniques are shown in Fig. 2c–e, where it is observed that many grass regions are over-segmented and falsely segmented. Similarly, few regions on the upper right side of the images are misclassified into another class. The result obtained by the proposed scheme (Fig. 2f) has provided a better segmentation map with improved boundary.

To provide a quantitative evaluation of the proposed scheme, we have provided a *ground-truth*-based performance measures. For evaluating the accuracy of the proposed segmentation scheme, we have used the pixel by pixel comparison of the

Table 1 Misclassification error

Approaches used	Region growing	Mean-shift	MRF	Proposed
San Diego North Island NAS	11,243	7,431	9,723	**3,558**
Stockton	14,557	8,729	11,483	**5,841**

ground-truth images with the obtained spatial segmentation results. This measure is also called *number of misclassified pixels*. The *number of misclassified pixels* obtained by different scheme is tabulated in Table 1 and reveals that the proposed scheme gives a better result as compared to the other considered techniques.

4 Conclusion and Future Work

In this article, we have proposed an edge preserving region growing technique for segmentation of aerial images. The proposed scheme is demonstrated by segmentation of two-color aerial image segmentation. The results obtained by the proposed scheme are found to be good. The accuracy of the proposed scheme is evaluated by comparing the results obtained by it with those of the conventional region growing, mean-shift, and spatio-contextual MRF-based segmentation techniques, and results are found to be better. It is observed that the time taken by the proposed scheme is comparable with that of the conventional region growing technique and mean-shift technique and is very less than that of the MRF-based segmentation technique. It is also observed that in a less illuminated and blurred environment, it fails to give satisfactory results. In this regard, we would like to develop some algorithm that takes care of illumination invariant edge features to highlight and improve the segmentation accuracy.

References

1. Gonzalez, R.C., Woods, R.E.: Digital Image Processing. Pearson Education, Singapore (2001)
2. Pun, T.: Entropic thresholding. A new approach. Comput. Graph. Image Process. **16**, 210–236 (1981)
3. Peak, J., Tag, P.: Segmentation of satellite weather imagery using hierarchical thresholding and neural networks. J. Appl. Meteorol. **33**, 605–616 (1994)
4. Ghosh, A., Subudhi, B.N., Ghosh, S.: Object detection from videos captured by moving camera by fuzzy edge incorporated markov random field and local histogram matching. IEEE Trans. Circuits Syst. Video Technol. **22**(8), 1127–1135 (2012)
5. Berthod, M., Kato, Z., Yu, S., Zerubia, J.: Bayesian image classification using markov random fields. Image Vis. Comput. **14**, 285–295 (1996)
6. Zucker, S.W.: Region growing: childhood and adolescence. Comput. Graph. Image Process. **5**, 382–399 (1976)

7. Kim, B.-G., Shim, J.-I., Park, T.-J.: Unsupervised video object segmentation and tracking based on new edge features. Pattern Recogn. Lett. **25**, 1731–1742 (2004)
8. Hu, X., Tao, C.V., Prenzel, B.: Automatic segmentation of high-resolution satellite imagery by integrating texture, intensity, and color features. Photogram. Eng. Remote Sens. **71**(12), 1399–1406 (2005)
9. Hojjatoleslami, S.A., Kittler, J.: Region growing: a new approach. IEEE Trans. Image Process. **7**, 1079–1084 (1998)
10. Comaniciu, D., Meer, P.: Mean shift: a robust approach toward feature space analysis. IEEE Trans. Pattern Anal. Mach. Intell. **24**(5), 603–619 (2002)

ILSB: Indicator-Based LSB Steganography

Prasenjit Das and Nirmalya Kar

Abstract The age-old LSB steganography is one of the easiest and most commonly used data-hiding techniques. Although higher payload delivery can be achieved, it has poor acceptance due to the perceivable impact it leaves on the cover media. Our proposed algorithm strengthens the LSB encoding by not only minimizing the perceivable distortion but also making original message reconstruction by an attacker highly impossible, due to vast key-space. The bit replacement process is guided by a selected indicator color channel, and the embedding sequence is controlled by an indicator pattern table value indexed by the secret message bits. For efficient extraction at the receiver end, the indicator and other metadata are hidden inside the same cover in the form of a header. Encryption of data and header by RC4 cipher adds another layer of security.

Keywords Information security · LSB · Indicator · *Stego*Rect · Header · Metric of distortion · LSB enhancement · Neighborhood · Pixel difference · PoV

1 Introduction

Image steganography techniques exploit several features of digital image to increase the amount of payload significantly by means of simple modifications that preserve the perceptual content of the underlying cover image [1]. LSB steganography is one of the many spatial domain steganography techniques available in many variations.

Techniques involving RGB bitmap images as cover media use single- or multi-channel hiding, RNG or color cycle methods [2], etc. Gutub et al. [3] describes the

P. Das (✉) · N. Kar
Department of Computer Science and Engineering,
National Institute of Technology Agartala, Agartala, India
e-mail: pj.cstech@gmail.com

N. Kar
e-mail: nirmalya@nita.ac.in

© Springer India 2015
L.C. Jain et al. (eds.), *Intelligent Computing, Communication and Devices*,
Advances in Intelligent Systems and Computing 309,
DOI 10.1007/978-81-322-2009-1_55

pixel indicator technique where the LSBs of one color channel are used to determine the channel to store data. Reference [4] discuses a technique to use color intensity to decide the number of bits to store in each pixel. In triple-A [5] technique, selection of color channel and number of bits introduced in each color is randomized. Max-Bit algorithm [6] hides data in random pixels after measuring the intensity. Hossain et al. [7] proposed three efficient steganographic methods that utilize the neighborhood information to estimate the amount of data to be embedded into an input pixel of cover image. In [8], a shared key-based algorithm is used in which the color channel (green or blue) used to hide information bit decided by the secret key. The LSB substitution compatible steganography (LSCS) [9] method has the equal capacity to that of the conventional LSB embedding.

2 Proposed Algorithm

The proposed algorithm hides 8 bit data in a single pixel of cover image inside some rectangular areas named *Stego*Rect by replacing the LSBs of red, green, and blue color. The entire process has 2 parts—embedding and extraction algorithm. The cover image contains both data and metadata (length of data (3 bytes) and header information with its maximum length spread over 657 pixels). So

$$\text{Length of secret data}(L) + 3 + 657 < \text{cover\_image\_size}$$

where cover_image_size = cover_image_width $\times$ cover_image_height
   For text secret message, L = number of characters [as 1 byte/character].
   For image secret message, L = secret_image_width $\times$ secret_image_height $\times$ 3

Embedding Algorithm It has the following steps.

- Indicator selection: In a pixel, any 1-color channel is chosen as indicator and gets 2 bits replaced. The indicator bits decide the pattern table, which decides the embedding sequence of the remaining 6 bits in the other 2 channels.
- Indicator data position selection: The 2 indicator bits can be picked from 3 different parts of the secret data byte: front (first two bits), middle (4th and 5th bit), and rear (last two bits). The remaining (3 + 3) bits are chosen in right-circular fashion.
- *Stego*Rect selection: Every *Stego*Rect information is stored in a 39-bit data structure which has the following members: left (11 bits), top (10 bits), width (9 bits), height (9 bits). Maximum 128 *Stego*Rects can be selected.
- Prepare *Stego*Data and encrypt: The *Stego*Data along with its 2 parts—secret data and its length (L = 3 bytes = 24 bits)—is encrypted by RC4 cipher. For secret image, we allocate 12 bits for both width and height.
- Embed *Stego*Data: The secret data embedding operation can be performed on both text and image data by almost similar process explained in the form of a flowchart in Fig. 1a. Byte values for R, G, and B channels taken sequentially.

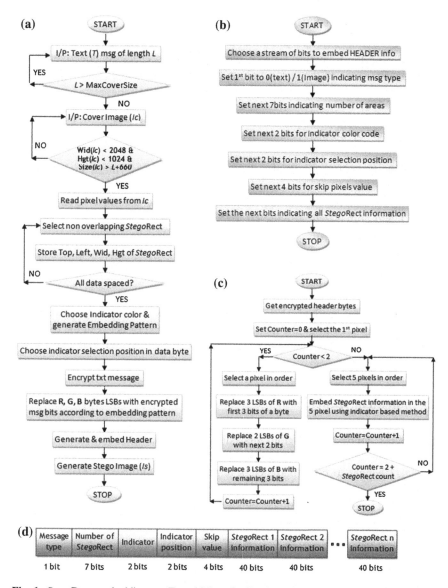

Fig. 1 *Stego*Data embedding. **a** Text hiding. **b** Header generation. **c** Header embedding. **d** Header format

- Header generation and embedding: The header contains several information, such as message type, indicator, indicator position, *Stego*Rect count, and all StegoRect information. At the end of data embedding, it is generated and hidden at the top of cover using the flowcharts in Fig. 1b–c.

3 Experimental Results

We conducted the following tests on our algorithm to check its effectiveness of hiding against some of the very common to most sophisticated attacks till date.

3.1 Metrics of Distortion

To assess the quality, we calculated the following metrics—mean squared error (MSE), peak signal-to-noise ratio (PSNR), average difference (AD), structural content (SC), normalized cross-correlation (NCC), Laplacian mean squared error (LMSE), and normalized absolute error (NAE) (see Table 1).

3.2 Visual Attack

LSB enhancement method [10] is applied to check any visually perceivable change between the carrier and the stegogramme. Difference between the two LSB planes is barely distinguishable even with 71.98 % payload (see Fig. 2).

3.3 Statistical Attacks

Histograms Analysis Histogram analysis on luminance channel shows that our algorithm preserves the general shape of the histogram with 63.4 % payload (see Fig. 3a). Changes to other parameters (e.g., relative entropy = 0.0038) are also negligible.

Neighborhood Histogram Neighborhood histogram [11] is performed with 80 % payload. It shows that change in neighbors count is negligible, although the frequency hike is remarkable (see Fig. 3b).

Table 1 Metrics of distortion values at different payloads

Payload (%)	MSE	PSNR	AD	SC	NCC	LMSE	NAE
10	0.26861	53.83960	0.00032	0.99998	1.000001	0.00191	0.001094
30	0.77070	49.26194	−0.00073	0.99997	0.999995	0.00576	0.003190
50	1.27329	46.08152	−0.00467	1.00005	0.999987	0.01046	0.010142
70	1.76303	41.66821	−0.00174	0.99997	0.999957	0.01409	0.008228
80	2.02025	39.07676	−0.00512	0.99991	0.999978	0.01619	0.009422

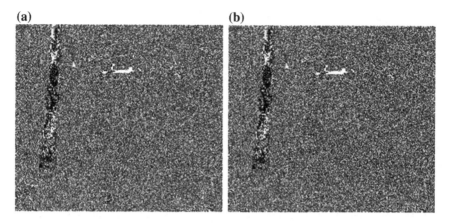

Fig. 2 LSB plane. **a** Original cover. **b** Stegogramme with 71.98 % payload

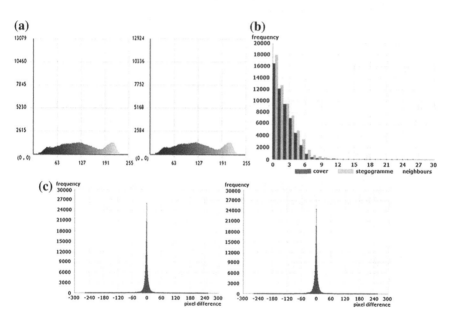

Fig. 3 Histograms of **a** luminance, **b** neighborhood, **c** difference image

Difference Image Histogram We analyzed the difference image histograms [12] with and without 40 % payload. Results show that no visible pairs of values (PoVs) are identified in the histogram, and it retains the natural Gaussian shape by maintaining the slope of bars in each range (see Fig. 3c).

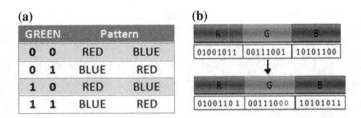

Fig. 4 Message hiding example. **a** Sample pattern table. **b** Bit replacement

4 Example of Message Hiding

Let 01100101 be the secret data byte. Selected indicator = green, indicator data position = middle. 2 LSBs of green is replaced by middle 2 bits, i.e. 00. Green colors' pattern table entry for '00' is red blue. Next 3 bits '101' are inserted in LSB of red. Next 3 bits '011' in right-circular fashion goes in LSB of blue (see Fig. 4).

5 Conclusion

In this paper, we introduced a new image steganography using LSB encoding with two primary objectives in mind: the technique to provide the maximum possible payload and the embedded data must be imperceptible to the observer. We achieved a capacity of 8 bpp, and also the distortions made to the image are as minimal as possible. To achieve best security, it is crucial to select a proper cover and indicator. Automation of the indicator selection process based on image features will be the scope of our future work.

References

1. Raja, K.B., Chowdary, C.R., Venugopal, K.R., Patnaik, L.M.: A secure image steganography using lsb, dct and compression techniques on raw images. In: 3rd International Conference on Intelligent Sensing and Information Processing (ICISIP), pp. 171–176 (2005)
2. Bailey, K., Curran, K.: An evaluation of image based steganography methods using visual inspection and automated detection techniques. Multimedia Tools Appl. **30**(1), 55–88 (2006)
3. Gutub, A., Ankeer, M., Abu-Ghalioun, M., Shaheen, A., Alvi, A.: Pixel indicator high capacity technique for RGB image based steganography. In: 5th IEEE International Workshop on Signal Processing and its Applications (WoSPA), University of Sharjah, Sharjah, U.A.E (2008)
4. Parvez, M.T., Gutub, A.A.: RGB intensity based variable-bits image steganography. In: Proceedings of 3rd IEEE Asia-Pacific Services Computing Conference (APSCC), Yilan, Taiwan (2008)

5. Gutub, A., Al-Qahtani, A., Tabakh, A.: Triple-A: Secure RGB image steganography based on randomization. In: ACS International Conference on Computer Systems and Applications, pp. 400–403. Rabat, Morocco (2009)

6. Bagchi, N.: Secure BMP image steganography using dual security model (I.D.E.A, image intensity and bit randomization) and max-bit algorithm. Int. J. Comput. Appl. 1(21), 18–22 (2010)

7. Hossain, M., Haque, S.A., Sharmin, F.: Variable rate steganography in gray scale digital images using neighborhood pixel information. In: Proceedings of 12th International Conference on Computer and Information Technology (ICCIT 2009), Dhaka, Bangladesh (2009)

8. Karim, S.M.M., Rahman, M.S., Hossain, M.I.: A new approach for LSB based image steganography using secret key. In: 14th International Conference on Computer and Information Technology (ICCIT 2011), Dhaka, Bangladesh (2011)

9. Sun, H.M., Wang, K.H.: A LSB substitution compatible steganography. In: IEEE region 10 conference (TENCON 2007) (2007)

10. Westfeld, A., Pfitzmann, A.: Attacks on steganographic systems. In: Information Hiding, LNCS, vol. 1768. Springer, Heidelberg (1999)

11. Westfeld, A.: Detecting low embedding rates. In: F.A.P. Petitcolas (ed.), Information Hiding. 5th International Workshop, pp. 324–339. Springer, Berlin (2003)

12. Zhang, T., Ping, X.: Reliable detection of LSB steganography based on the difference image histogram. In: IEEE International Conference on Acoustics, Speech, and Signal Processing (ICASSP 03), vol. 3 (2003)

Odia Running Text Recognition Using Moment-Based Feature Extraction and Mean Distance Classification Technique

Mamata Nayak and Ajit Kumar Nayak

Abstract Optical character recognition (OCR) is a process of automatic recognition of character from optically scanned documents for the purpose of editing, indexing, searching, as well as reduction in storage space. Development of OCR for an Indian script is an active area of research today because the presence of a large number of letters in the alphabet set, their sophisticated combinations, and the complicated grapheme's they formed is a great challenge to an OCR designer. We are trying to develop the OCR system for Odia language, which is used as official language of Odisha (formerly known as Orissa). In this paper, we attempt to recognize the vowels, consonants, matras, and compound characters of running Odia script. At first, the given scanned text is segmented into individual Odia symbols, then, extract corresponding feature vectors, using two-dimensional moments and Hough transform (based on topological and geometrical properties), which are used to classify and recognize the symbol. We found that the proposed model can recognize up to 100 % running test having no touched characters.

Keywords Optical character recognition · Odia language · Matras · Juktakhyara · Image processing · Feature extraction · Recognition

1 Introduction

Optical character recognition (OCR) system decreases the barrier of the keyboard interface between man and machine to a great extent. It helps in office automation with huge saving of time and human effort. Overall, it performs automatic

M. Nayak (✉) · A.K. Nayak
Siksha 'O' Anusandhan University, Bhubaneswar, India
e-mail: mamatanayak@soauniversity.ac.in

A.K. Nayak
e-mail: ajitnayak@soauniversity.ac.in

© Springer India 2015
L.C. Jain et al. (eds.), *Intelligent Computing, Communication and Devices*,
Advances in Intelligent Systems and Computing 309,
DOI 10.1007/978-81-322-2009-1_56

extraction of text from an image. Research in OCR is popular for its various applications potentials in banks, library automation, post offices, defense organizations, and language processing. For the first time, OCR was realized as a data processing approach, with particular applications for the business world. Currently, PC-based systems are commercially available to read printed documents of single font with very high accuracy and documents of multiple fonts with reasonable accuracy.

Many research works have been done for various types of Indian languages such as Bangla, Devnagari, Urdu, Telugu, Gujrati, etc. [1, 2]. A survey of tools used for recognition of character for Indian languages is described in detail [3]. However, to the best of our knowledge, very few research reports are available for Odia language. Recognition of the Odia character is done by performing segmentation of each line into three zones as following to the Bangla script [4, 5], but it cannot handle a larger variety of touching characters, which occur fairly often in images obtained from inferior-quality printed material. Support vector machine technique is used to design Odia corps design [6, 7] that deals with only clean machine printed text with minimum noise. The organization of the rest of this paper is as follows: In Sect. 2, we have discussed properties of Odia script and different recognition approached used for other languages. Sect. 3 presents the details of proposed approach and algorithm description, Sect. 4 shows the testing and result discussions, and finally, in Sect. 5, the paper is concluded.

2 Review of the Previous Approaches

2.1 Properties

The Odia script is derived through various transformations from the ancient Brahmi script. We explain here some properties of the Odia script that makes easier to recognize the characters. The Odia alphabet set consists of 58 numbers of basic characters (i.e., 12 independent vowels +36 consonant characters +10 digits) identical to the names for corresponding characters in other scripts like Devanagari and Bengali. As like other Indian scripts, the vowels take its dependent form whenever it followed by consonants called as matras. These are appears in left, right, bottom, or top of a consonant. Figure 1 shows place of the matras with the character 'Ka.'

କା କି କୀ କୁ କୂ କେ କୋ କୈ କୌ

Fig. 1 All matras of Odia script with character 'Ka'

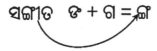

Fig. 2 A Juktakhyra of Odia script

Similarly, consonants take a compound form whenever it followed by another vowel or consonant called as Juktakhyra in Odia language. These characters can also be combined with the matras to generate a meaningful word as shown in Fig. 2. There are a total of 110 number Juktakhyar supported by the script those are formed as a combination of at most three characters. As like other Indian scripts writing style of Odia script is from left to right and also concept of upper/lower case is absent in it. Many of other characteristics of the Odia script have been explained nicely [7].

2.2 Recognition Techniques

The solution to the problem of text recognition has been attempted by many approaches. Some of them are: template matching approach, feature extraction approach, and training Tesseract OCR.

Template matching approach is one of the most simplistic approaches. According to this approach, a test pattern is compared with the existing template patterns by measuring the degree of similarities and recognizes the character that produces high score. It works effectively with the standard fonts, but needs large memory space to store template. This technique gives very poor result for the Indian languages because of its complex shape.

According to feature extraction approach, a set of values are generated for each type of character called as feature vector. These feature vectors are generated based on geometrical or structural properties (such as relative position, number of joints, number of end points, aspect ratio, etc.) or image transformation (such as discrete cosine transformation, discrete wavelet transformation) or stroke- and curvature-based feature, etc. Then, based on the feature vector, characters are classified into various groups using different techniques such as mean distance measure, artificial neural network, support vector machine, etc. Thereafter, in the recognition phase, again feature vector of the test characters is generated and needs to match [8, 9].

Another way of designing an Odia OCR is by using Tesseract, an open source OCR engine that is considered as one of the most accurate FOSS OCR engines. It was originally developed by Hewlett-Packard from 1985 until 1995 and is currently maintained by Google. It has already been designed to recognizing English, Italian, French, German, Spanish, and Dutch, and many more, as well as for few Indian languages such as Bengali, Tamil, Telugu, and Malayalam. Similarly,

Tesseract can be made to recognize other scripts if the engine can be trained with the requisite data. We train the Tesseract engine to recognize the Odia script, whereas the difficulty with it is font specific and difficult to train the compound characters [10]. Thus, Odia running script recognition remains a highly challenging task for OCR developer.

3 Proposed Approach

We studied various research papers on existing OCR of Indic languages to gain better knowledge, techniques, and solutions. Image processing toolbox of Matlab provides a set of functions that extend the capability to develop and analyze the result. All these studies helped us in clarifying our target. Broadly, the development procedure of OCR is divided into two phases named as: training phase and testing phase.

There are five basic steps followed to complete these two phases: image acquisition, preprocessing (noise reduction, binarization, segmentation), feature vector generation, classification and recognition, and post-processing.

Image Acquisition: In this step, images of Odia characters in .tiff, .jpg (with no LZW compression) format are created using the Kalinga font style. It can also be obtained by using a scanner. A part of it is shown in Fig. 3. Then, the image is imported to a graphical user interface (GUI) which is designed using Matlab.

Preprocessing: In this process, the image is first converted into grayscale (two dimensional) image from RGB (three dimensional) images. As practically, any scanner is not perfect; thus, the scanned image may have some noise. So, to get a noise-free image, the generated image is further processed to remove the objects having pixel values less than a threshold value. The obtained image is to be saved for further processing. Now, the elements of the matrix of the generated image value ranges from 0 to 255, which is complicated for further processing. Therefore, the generated grayscale images are converted into two tone image by performing binarization (i.e., values of 0–255 are replaced by 0 or 1). Then, the position of the object (i.e., the character in the image) is found out and is cropped as an image. Character segmentation is a two stage process in which at first, each line is removed as sub-images and then the individual characters are segmented from it. Image segmentation plays a crucial role in character recognition. Here, we consider that the scanned image consists of multiple lines. Thus, it first needs to extract each line and then each character. The steps followed for segmentation are explained below. Figures 4 and 5 show the segmentation of a line in a multiline training document and extract each symbol separately by using the 8 connected component analysis.

କା କି କୀ କେ କୋ କୌ   କୁ କୃ ଣ୍ଡ ଞ୍ଜ ରୁ ନ୍ତ ଣ୍ଟ

Fig. 3 Few characters of Odia training image

Fig. 4 Text line extraction

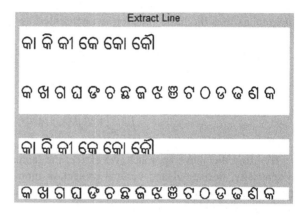

Fig. 5 Character extraction

Algorithm to extract line(Input: Image matrix [Img]m×n , Output: Line Segments)

Step1: $Sum_i \leftarrow \sum_{j=1}^{to\ n}$ [Img]i,j

Step2:   maxPixelVal $\leftarrow$ max(Sum_i) for all i= 1... m

Step3:   Initialize r1 $\leftarrow$ -1, r2 $\leftarrow$ -1

Step4:   For each i= 1 ... m

Step4.a: if Sum_i = =maxPixel Val

                    r1 $\leftarrow$ -1 and Increment i

Step4.b: if Sum_i = = maxPixelValue

                    r2 $\leftarrow$ -1 and Increment i

Step4.c: if r1 and r2 $\leftarrow$ -1

Step4.c.i: Extract pixel values from r1 to r2 and draw the picture

Step4.c.ii: Reinitialize r1 $\leftarrow$ -1, r2 $\leftarrow$ -1

Step4.d: Continue Step 4

Feature vector generation: The purpose of feature extraction is to extract properties of the characters that help to identify it uniquely. In this approach, we use geometric and orthogonal moments [11]. The general two-dimensional $(p + q)$th-order moments of an image are defined as given below:

$$\iint_{R^2} \emptyset_{p,q}(x,y)f(x,y)\,dxdy \tag{1}$$

$\emptyset(x,y)$ is the basic function and $f(x, y)$ is the image intensity.

Geometrical moments are the projection of the image function onto the monomials thus the basic function is represented as

$$\emptyset_{p,q}(x,y) = x^p y^q c \tag{2}$$

Therefore, the geometric moment is defined as

$$\iint_{R^2} (x^p y^q)f(x,y)\,dxdy \tag{3}$$

By using the above approach, we extract the features of each character such as eccentricity, horizontal and vertical projection moments, horizontal and vertical projection moments, horizontal projection and vertical projection skewness, and horizontal and vertical projection Kurtosis.

Algorithm to extract feature of each character (I/O: A character image Img, A vector)

Step 1: Based on the above techniques extract features of each
Step 2: Assign a class label to each of the image
Step 3: Repeat the Step 2 for all characters
Step 4: Save the feature vector in a file

We use integer values as the class labels for the training characters, few of which are given in below:

Along with the class label, the features of each character are stored in a data file as sown in the Fig. 6.

Classification and Recognition: This step is the heart of the system. It refers to the grouping of character based on the vector generated to simplify the recognition. As explained earlier, there are many techniques used for classification. However, in our approach, at first, we need to find the group of characters having the same class label, then calculate the mean of all values for each feature, and then normalized the values (i.e., mapping of values within 0–1) to represent that class. Character recognition system is the base for many different types of applications in various fields, and many of which we use in our daily life. To recognize a character, its feature vector is calculated and normalized. After that, the distance among the normalized feature vectors of the test character with each trained character is measured as given below:

$$D = \min_{(i=1:n)}(\text{Sum of elements}(T_r - T_s))$$

T_r, T_s: *Normalized feature vector of trained and test characters, n: Number of classes*

Fields	Eccentricity	Orientation	⊞I ⊟I ⊟'	HPMoment	VPMoment	HCenter	VCenter	HPSkewness	VPSkewness	HPKurtosis	VPKurtosis	Class
1	0.5059	-79.4642	36x.. 36x.. 31x..	[374;7057;3.831...	[374;5868;2.90...	18.8690	15.6898	-0.0130	0.0011	0.0051	0.0047	11
2	0.9961	90	35x.. 35x.. [33...	[101;1818;9554;...	[101;202;66;0;66]	18	2	0	0	0.0179	0.0152	101
3	0.5059	-79.4642	36x.. 36x.. 31x..	[374;7057;3.831...	[374;5868;2.90...	18.8690	15.6898	-0.0130	0.0011	0.0051	0.0047	11
4	0.9752	-6.2894e-16	7x2.. [8;.. 26x..	[98;368;258.122...	[98;1323;5.436...	3.7551	13.5000	0.0200	0	0.0236	0.0183	102
5	0.5059	-79.4642	36x.. 36x.. 31x..	[374;7057;3.831...	[374;5868;2.90...	18.8690	15.6898	-0.0130	0.0011	0.0051	0.0047	11
6	0.9849	-85.8987	36x.. 36x.. [3;...	[128;2042;1.411...	[128;928;484;-...	15.9531	7.2500	0.0275	-0.1468	0.0137	0.0413	103
7	0.8043	-86.0962	37x.. 37x.. 23x..	[300;6076;3.834...	[300;3534;1.36...	19.7273	11.4740	-0.0103	0.0050	0.0058	0.0053	104

Fig. 6 Feature vector of each character

The trained character is assigned to the class with minimum distance. The algorithm we use to classify and recognize the Odia character is explained below: *Algorithm for classification and recognition*

Input: Normalized feature vector file generated in training, testing phase.
Output: Generate class level of the test character

Step 1: Sort the trained characters with respect to their class label and find the mean feature value for each type of features.
Step 2: The generated mean feature vectors are normalized and stored.
Step 3: As explained earlier, the test data image is preprocessed and its normalized feature vector is computed and is stored for further processing.
Step 4: Compute the distance between the normalized feature vectors of the testing character with each training character.
Step 5: Assign the character to the class with least distance.

Post-processing: After recognition the character, it needs to convert the generated Unicode output to its corresponding character and write into a text file. However, in Odia script, one character may consist of a combination of more than one glyph, so it needs to rearrangement of the generated Unicode output to get the correct result.

4 Testing and Results

To accomplish the task of recognition for running Odia test, using moment-based feature extraction and mean distance classification techniques, we develop a tool using Matlab. The developed tool can be used to recognize the characters of scanned copy of an Odia document. The program is developed in such a manner that the same module can be used for training as well as testing purposes. In fact, the testing was done at two levels: individual letter level and paragraph level. At first, we perform testing our classifier at character level on scanned copy of typed character Odia document without having touched characters, and the recognition is excellent with the rate of recognition approximately 100 % few of which are shown in the Table 1.

The result can also be analyzed by representing the comparison of graph for two characters consisting of more than one glyph. Here, we used the feature vector of

Table 1 Recognized class labels for the testing Odia characters

Original Character	Classified values				
ଚୋଲା	107	11	101	22	101
ପୂଜକ	31	105	18	11	
ଗଣେଶ	13	107	25	41	
କଣ୍ଠା	11	509	101		

Table 2 Class labels used for the training Odia characters

କ	ଖ	ଗ	ଘ	ଙ	ଚ	ଛ	ଜ	ଝ	ଞ	ଟ	ଠ	ଡ
11	12	13	14	15	16	17	18	19	20	21	22	23
ା	ି	ୀ	ୁ	ୂ	ୃ	େ	ୈ	ୋ	ୖ	ୖ	ୈ	ୄ
101	102	103	104	105	106	107	108	109	110	111	112	113
କ	ଖ	ଗ	ଘ	ଙ	ଚ	ଛ	ଜ	ଝ	ଞ	ଟ	ଠ	ଡ
501	502	503	504	505	506	507	508	509	510	511	512	513

two characters କଟ·ା and କଟ·ୀ, both having combinations of three glyphs. As in both characters the first two glyphs are same, they belongs to their corresponding class levels (i.e., 107, 11) as given in the Table 2. However, the third glyph is different (i.e., 101, 109) as shown in Fig. 8. Similarly, we can draw figures to compare the features for glyphs of each character, which concludes that each glyph has recognized correctly (Fig. 7).

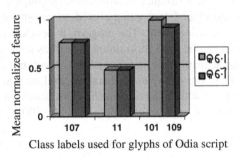

Fig. 7 Feature vector of the Odia script glyphs

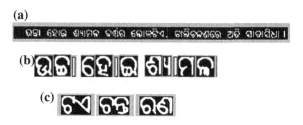

Fig. 8 Test file (**a**), generated output (**b**), and incorrect output (**c**)

Then, paragraph-level testing was performed for a scanned copy of an ancient running document, and we analyzed that it recognizes the matra and juktakhyra, but not able to recognize when two characters are touched to each other as shown in the Fig. 8a, b, c.

5 Conclusion

OCR technology provides fast, automated data capture. Proper scanning of a document, good form design, sufficient data validation, and targeted manual review deliver accurate results with huge savings over manual processes. In this work, we could recognize the composite characters as well as conjugate modifiers of Odia script. This is the first report attempt to recognize Odia complex character, as of our knowledge. The tool we have developed gives about 95 % accurate result for scanned documents with the constraint that no two conjugative characters are touched to each other at a common pixel. So, the error of 5 % is for touched characters that leads to extent this work in future. The result achieved is good, but every work has a limitation. Our efforts are continued to increase the effectiveness of the result.

References

1. Pal, U., Chaudhuri, B.B.: Indian script character recognition: a survey. J. Pattern Recogn. **37**, 1887–1899 (2004)
2. Dongre, V.J., Mankar, V.H.: A review of research on Devnagari character recognition. Int. J. Comput. Appl. (0975–8887) **12**(2), 8–14 (2010)
3. Kumar, M.P., Ravikiran, S.S., Nayani, A., Jawahar, C.V., Narayanan, P.J.: Tools for developing OCRs for Indian scripts. CVIT, pp. 1–6 (2011)
4. Jayadevan, R., Kolhe, S.R., Patil, P.M., Pal, U.: Offline recognition of Devanagari script: a survey. IEEE Trans. Syst. Man Cybern **41**(6), 2011 (2011)
5. Chaudhuri, B.B., Pal, U., Mitra, M.: Automatic recognition of printed Oriya script. Special Issue Sadhana, Printed in India **27**(1), 23–34 (2002)

6. Mohanty, S., Behera, H.K.: A complete OCR development system for Oriya Script. In: Proceeding of SIMPLE RC-ILTS-Oriya, vol. 4 (2004)
7. Mohanty, S., Bebartta, H.N.D.: A novel approach for Bilingual (English–Oriya) script identification and recognition in a printed document Sangh. Int. J. Image Process. (IJIP) **4**(2), 175–191 (2010)
8. Pall, U., Wakabayashi, T., Kimura, F.: A system for off-line Oriya handwritten character recognition using curvature feature. In: 10th International Conference on Information Technology (ICIT), IEEE Computer Society, pp. 227–229 (2007)
9. Meher, S., Basa, D.: An intelligent scanner with handwritten Odia character recognition capability. In: Fifth International Conferrence On Sensing Technology, IEEE Computer Society, pp. 53–59 (2011)
10. Nayak, M., Nayak, A.K.: Odia characters recognition by training tesseract OCR engine. International Conference in Distributed Computing and Internet Technology (ICDCIT-2014), published in Int. J. Comput. Appl. (0975–8887), pp. 25–30 (2013)
11. Sridevi, N., Subashini, P.: Moment based feature extraction for classification of handwritten ancient Tamil document. Int. J. Emerg. Trends Eng. Dev. **7**(2), 106–115 (2012)

Keyword Extraction from Hindi Documents Using Statistical Approach

Aditi Sharan, Sifatullah Siddiqi and Jagendra Singh

Abstract Keywords of a document give us an idea about its important points without going through the whole text. In this paper, we propose an unsupervised, domain-independent, and corpus-independent approach for automatic keyword extraction. The approach is general and can be applied to any language. However, we have tested the approach on Hindi language. Our approach combines the information contained in frequency and spatial distribution of a word in order to extract keywords from a document. Our work is specially significant in the light that it has been implemented and tested on Hindi which is a resource poor and underrepresented language.

Keywords Keyword extraction · Spatial distribution · Standard deviation · Frequency · Hindi

1 Introduction

Our need to quickly sift through large amount of textual information is growing on a daily basis at an overwhelming rate. This task can be made easier if we have a subset of words (keywords) which can provide us with the main features, concept, theme, etc., of the document. Keyword extraction is also an important task in the field of text mining. Keywords can be extracted either manually or automatically but the former approach is very time-consuming and expensive. Automatic approaches for keyword extraction can be language-dependent or statistical

A. Sharan (✉) · S. Siddiqi · J. Singh
Jawaharlal Nehru University, New Delhi, India
e-mail: aditisharan@gmail.com

S. Siddiqi
e-mail: sifatullah.siddiqi@gmail.com

J. Singh
e-mail: jagendrasngh@gmail.com

© Springer India 2015
L.C. Jain et al. (eds.), *Intelligent Computing, Communication and Devices*,
Advances in Intelligent Systems and Computing 309,
DOI 10.1007/978-81-322-2009-1_57

approaches, which are language independent. Most of the work in this area has been done in English language. Other languages are underrepresented due to scarcity of domain-specific resources and non-availability of data, especially labeled data. Now due to availability of large amount of textual data in different languages, there is a growing need for developing keyword extraction techniques in other languages as well.

There are many approaches by which keyword extraction can be carried out, such as supervised and unsupervised machine learning, statistical methods, and linguistic ones. Statistical methods for the extraction of keywords from documents have certain advantages over linguistic-based approaches such as the same approach can be applied to many different languages without the need to develop different set of rules each time for a different language. However, most of the statistical approaches are based on the corpus statistics. An obvious disadvantage of such methods is that they cannot be applied in the absence of availability of corpus. There are some other disadvantages as well of corpus-oriented keyword extraction methods. Keywords which occur in many documents of the corpus are less likely to be statistically discriminating, and most of the corpus-oriented approaches typically work on single words which are used in different contexts with different meanings. As availability of the corpus is a severe limitation in resource poor languages, it is important to focus on document-based approaches which can extract keywords from single document. Such methods are especially suitable for live corpuses such as news and technical abstracts. **Therefore, there is a need to find approaches which are unsupervised, domain independent, and corpus independent.**

Some work has been done for extracting keywords using statistical measures. Among one of the frequently used statistical measures to judge the importance of a particular word in the document is the frequency of a word in the document [1]. The intuition behind this measure is that the more important a word is to a document, the more number of times it should occur in the document. But in a document, most of the words having highest frequencies are stopwords (the, and, of, it, etc.). It was proposed in [2] to select the intermediate frequency words as keywords and discard the high- and low-frequency words as stopwords. Tf-idf measure [3] gives much better results than frequency score, but it requires a corpus and cannot be used on a single document. Shannon's entropy measure [4] was used to extract keywords from literary text in conjunction with randomly shuffled text. Spatial distribution [5] of words in the document was utilized to estimate the importance of a word in the document.

In this paper, we present a statistical approach for keyword extraction from single documents based on frequency and next nearest neighbor analysis of words in the document. We present an algorithm for the same. The algorithm has been tested on a famous Hindi novel "Godan" by Munshi Premchand.

2 Framework

This paper is based on the observation that both the frequency and spatial distribution of a word play an important role in evaluating its importance. Therefore, we have tried to develop an integrated approach which combines the information contained in frequency and spatial distribution of a word in order to extract keywords from a document. Our approach has strong theoretical background as well as empirical evidence. It works better than simple frequency-based approach as well as better than the approach based on spatial distribution only.

It has been observed that in a long text, the words with middle range of frequency are more important, because very high-frequency words are stopwords, whereas very low-frequency words are not important. We start with removing very low-frequency words.

Our spatial distribution-based approach is based on the observation that occurrence pattern for important words (keywords) should be different from that of non-important words. For a non-important word, its word distribution pattern should be random in the text and significant clustering should not be observed, whereas for a keyword, its distribution pattern should indicate some level of clustering because a keyword is expected to be repeated more often in specific contexts or portions of text.

To estimate the degree of clustering or randomness in the text for different words, we can use the standard deviation which measures the amount of dispersion in a data series. One such data series which can be analyzed is the successive differences between positions of occurrence of the word in the text. In other words, if a word W occurs N times in the document at positions $X_1, X_2, X_3, \ldots, X_N$, then successive differences are (X_2-X_1), (X_3-X_2), ... (X_N-X_{N-1}). Representing the intermediate difference series for word W as S_W we have,

$$S_W = \{(X_2 - X_1), (X_3 - X_2), (X_4 - X_3), \ldots (X_N - X_{N-1})\}$$

For a word W with frequency N in the text, we have $N-1$ elements in the series S_W.

To eliminate the effect of frequency on the standard deviation analysis of different words, it is convenient to normalize the standard deviation (σ_W) of series S_W with its corresponding mean (μ_W), so that normalized standard deviation of series S_W is $\hat{\sigma}_W$. Higher values of generally represent more pronounced clustering or lesser random behavior which is what we expect for important words.

Figures 1 and 2 show two words from the document (Godan) with similar frequencies, while the first word is a keyword, the other is a stopword. The clustering is evident in Fig. 1, while in Fig. 2, it can be easily seen that distribution is random throughout and no significant clustering is observed. As observed, spatial distribution can be used for differentiating between a keyword and a non-keyword.

Fig. 1 Spatial distribution of the keyword "होरी" with frequency 623 in the document

Fig. 2 Spatial distribution of stopword "गया" with frequency 677 in the document

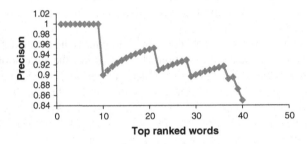

Fig. 3 Precision curve for *top* 40 words ranked on *decreasing* standard deviation. X-axis represents the ranks of words. The *curve* shows the number of keywords found in *top N* words ranked on $\hat{\sigma}$

2.1 Algorithm

1. Generate the list of unique words in the document which forms the vocabulary.
2. Calculate the frequency (f) of each unique word in the document.
3. Eliminate low-frequency words.
4. Generate the dataset of next nearest neighbor distances series S_W for each unique word in the document.
5. Calculate the mean (μ_W) and standard deviation (σ_W) of dataset S_W for each unique word.
6. Normalize the σ_W with μ_W.
7. Rank the resulting words list in order of decreasing normalized standard deviation ($\hat{\sigma}_w$).
8. Select the words with highest standard deviation as keywords.

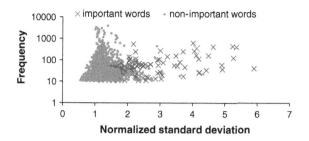

Fig. 4 Plot of frequency versus standard deviation of words in the document. *Frequency* has been plotted on logarithmic scale, whereas *normalized standard deviation* is plotted on linear scale. It can be seen that for $\hat{\sigma} > 3$, most of the words are important words

Table 1 *Top* ranked words from the document and their standard deviation scores

Keyword	Frequency (f)	$\hat{\sigma}$
चौधरी	40	5.899899
मेहता	404	5.35546
सलियि	125	5.26475
मालती	440	5.237255
नोहरी	57	4.902299
गोबर	414	4.633352
मिर्जा	159	4.490566
हीरा	113	4.190991
धनिया	355	4.125608
गाय	164	4.112118
खन्ना	262	4.050105
होरी	623	4.015021
संपादक	45	3.760877
झुनिया	249	3.703393
गोविंदी	79	3.604878
ओंकारनाथ	52	3.528352
भोला	155	3.48224
तंखा	54	3.42949
सोना	146	3.217998
मातादीन	74	3.085875

Table 2 Some stopwords from the document and their standard deviation scores

Stopword	Frequency (f)	$\hat{\sigma}$
हैं	1417	1.627264
था	1735	1.498282
है	3914	1.390569
ने	1841	1.289441
वह	1519	1.276871
नहीं	2306	1.255755
तो	3001	1.201025
की	2532	1.124765
न	1971	1.120748
भी	1711	1.091106
हो	2021	1.090015
को	2012	1.082808
के	2684	1.080228
का	1994	1.074362
और	3113	1.064317
ही	1241	1.063947
पर	1577	1.060343
कर	3053	1.027002
में	3311	1.022517
से	2640	0.997308

3 Experiment

We performed our experiment on a novel "Godan" of Hindi language by Premchand. Some document statistics are as follows:

Total number of words in document = 167,707.

Total number of unique words in the document = 11,160.

Number of words with frequency greater than 10 = 1,565.

Words with frequency lesser than or equal to 10 were removed from consideration.

The proposed algorithm was implemented. The result is shown through a precision curve for top ranked 40 words in Fig. 3. It can be easily visualized that top ranked 10 words are all keywords of "Godan."

The result can be better understood with the help of Tables 1 and 2 and Fig. 4. Table 1 gives the list of top ranked words on $\hat{\sigma}$. Though many words have large frequency differences between them, but they have much closer values of standard deviation $\hat{\sigma}$. Table 2 gives a list of stopwords used in the document. Although these words have quite larger frequencies than the important words in the document, their $\hat{\sigma}$ values are lower compared to the latter. Thus, $\hat{\sigma}$ works better in differentiating between important and non-important words. Figure 4 shows the distribution of words of "Godan" on frequency and $\hat{\sigma}$ where important and non-important words with frequency greater than 10 are labeled separately.

4 Conclusion

In this paper, we have proposed a novel document-based statistical approach to extract important words from Hindi literary document. Our approach hybridizes the information from both the frequency value and spatial distribution of words in the document. Standard deviation of nearest next neighbor distances of the word in the document was used as a discriminating factor. It was found that higher values of standard deviation generally correspond to the important words in the document. Considering that our approach is unsupervised, domain independent, and corpus independent, the results are quite motivating. Further, it was also observed that most of the important words which were extracted were named entities (NEs). Thus, a further research area which can be explored is the application of statistical methods to extract important named entities from literary texts.

References

1. Salton, G., Buckley, C.: Weighting approaches in automatic text retrieval. Inf. Process. Manage. **24**(5), 513–523 (1988)
2. Luhn, H.P.: A statistical approach to mechanized encoding and searching of literary information. IBM J. Res. Dev. **1**(4), 309–317 (1957)
3. Jones, K.: A statistical interpretation of term specificity and its application in retrieval. J. Documentation **28**(1), 11–21 (1972)
4. Herrera, J.P., Pury, P.A.: Statistical keyword detection in literary corpora. Eur. Phys. J. B. **63**(1), 135–146 (2008)
5. Ortuño, M., Carpena, P., Bernaola-Galván, P., Muñoz, E., Somoza, A.M.: Keyword detection in natural languages and DNA. Europhys. Lett. **57**, 759–764 (2002)

Development of Odia Language Corpus from Modern News Paper Texts: Some Problems and Issues

Bishwa Ranjan Das, Srikanta Patnaik and Niladri Sekhar Dash

Abstract In this paper, we have tried to describe the details about the strategies and methods we have adapted to design and develop a digital Odia corpus of newspaper texts. We have also attempted to identify the scopes of its utilization in different domains of Odia language technology and applied linguistics. The corpus is developed with sample news reports produced and published by some major Odia newspapers published from Bhubaneswar and neighboring places. We have followed several issues relating to text corpus design, development, and management, such as size of the corpus with regard to number of sentences and words, coverage of domains and sub-domains of news texts, text representation, question of nativity, determination of target users, selection of time span, selection of texts, amount of sample for each text types, method of data sampling, manner of data input, corpus sanitation, corpus file management, and problem of copyright. The digital corpus is basically in machine readable format, so that the text becomes easy to process very quickly. We presume that the corpus we have developed will come to a great help to look into the present texture of the language as well as to retrieve various linguistic data and information required for writing a modern grammar for Odia with close reference to its empirical identity, usage, and status. The electronic Odia corpus that we have generated can also be used in various fields of research and development activities for Odia.

B.R. Das (✉) · S. Patnaik
Department of Computer Science and Information Technology, Institute of Technical Education and Research, Siksha'O' Anusandhan University, Khandagiri, Bhubaneswar, India
e-mail: biswadas.bulu@gmail.com

S. Patnaik
e-mail: patnaik_srikanta@yahoo.co.in

N.S. Dash
Linguistic Research Unit, Indian Statistical Institute, 203, Barrackpore Trunk Road, Baranagar, Kolkata, India
e-mail: ns_dash@yahoo.com

© Springer India 2015
L.C. Jain et al. (eds.), *Intelligent Computing, Communication and Devices*,
Advances in Intelligent Systems and Computing 309,
DOI 10.1007/978-81-322-2009-1_58

515

Keywords Corpus · Odia · Newspaper · Sentence · Word · Text representation ·
Time span · File management · Copyright

1 Introduction

The development of corpora for Odia language is a challenging task for the research
and development activities. A corpus, developed in digital form, can help us in
research works in various forms and types. Also, it provides basic resource for
developing many text processing tools and techniques, which make research into
language easy and user-friendly. The application of the Odia newspaper corpus can
be visualized in several ways for retrieving information to be used in research and
development activities in various field of Odia linguistics. It can also help scholars to
realize the empirical and referential value of corpus as well as can inspire them to
initiate projects for corpus development in Odia and other Indian languages.

The work of corpus generation in the Indian language is first started at individual level nearly 40 years ago. The first corpus developed in Indian language is
the *Kolhapur Corpus of Indian English* (KCIE) by Shastri and his team of Shivaji
University, Kolhapur, India. This corpus is made from written texts of Indian
English. It will be helpful to all Indian research scholars for different research area
in linguistics. Detailed methodology and some of the salient features of corpus
generation are described with some references to Bengali language corpus [1].

In this paper, we basically focus to generation of corpus in electronic form in Odia.
The Odia corpus is collected from different sources such as Odia newspaper, printed
books, documents of government publication, and press materials. The whole Odia
corpus generation is divided into two different ways according to nature of work and
level of participation of scholars. In this paper, the structure of Odia corpus is
described in some detail keeping its possible use in various domains of linguistics and
language technology, particularly writing a descriptive grammar for Odia. The
electronic form of Odia corpus data uses in various fields of research and development activities in today's world. The digital corpus basically is in the form of
machine readable format, so it becomes easy to process very large amount of text
very quickly. Developing Odia corpus is used for different field of computer
including machine translation, language processing, speech processing, and text
analysis.

People of Odisha and different territory can use this corpus for their research
work. Today, Odia corpus linguistic is closely connected to the use of "computer
readable" text for language processing/study. The language study means to know
the term of grammatical units such as subject, verb, object, noun, and pronoun, and
phrases to explain the structure of the language and to write and speak any sentence
without any mistake. The objective of the grammar of a language study means to
write or speak the whole sentences without any error. How computer system
understand a particular language without any grammatical mistake? The digital

corpus is basically designed and developed by human being to process the grammar without making any mistake and retrieved the text by the system very easily.

2 Issues of Odia Corpus Development

There are several issues related to corpus design, development, and management. The written text corpus development involves size of corpus, text representation, question of nativity, determination of target users, selection of time span, selection of documents, collection of documents, collection of text documents, method of data sampling, manner of data input, corpus sanitation, corpus file management, and problem of copyright.

2.1 Size of Corpus

Size of the corpus means how big it is? It belongs to total number of words different words to be taken into a corpus. Two main factors of a size of corpus are as follows:

  i. The kind of query that is anticipated from users.
  ii. The methodology the use to study the data.

It involves the decision of how many categories we like to keep in the corpus, how many samples of texts we need to put into each category, and how many words we shall keep in each sample. The bigger size of the corpus is more reliable for research. There is no maximum size. We will begin with the kind of figures found in general reference corpora, but the principles are the same, no matter how large or small the corpus happens to be. To relate the kind of query to the size of the corpus, it is best to start with a list of the "objects" that you intend to study; the usual objects are the physical word forms or objects created by tags, such as lemmas. Then, try them out on one of the corpora that are easy to interrogate, such as the million-word corpora on the ICAME CD-ROM [2]. The Brown group of corpora are helpful here, because they have been proofread and tagged and edited over many years, and with a million words, the sums are easy.

2.2 Text Representation

It is always better to emphasis in the quality of data or text samples. That means data should be proportionately represented from all possible domains of language use within a corpus. In other words, the overall size of corpus and balance need to be set against the diversity of source texts for achieving proper text representation. Any large collection of text cannot said to be a proper corpus if the large collection of texts is not in a proper text format for any kind of generalization.

2.3 Question of Nativity

It is a talking about native users and non-native users. Any monitor corpus produced by the native users should get priority over the texts produced by the non-native users. Because the main aim of a monitor corpus is to represent the language, which can be consider as an ideal form for all kind of works in linguistics and language technology. Our basic objective for building a corpus is to know the naturally occurring language, in order to see what exactly occur and what does not exactly occur.

2.4 Identification of Target Users

Various types of corpus user and their needs of types of corpus are described for his/her research work. Each researcher has specific requirements; a corpus has to be designed accordingly. A person who works in tagging, processing and parsing, and word sense disambiguation will require annotated, monitor, written, spoken, and general corpus. The types of corpus users and their needs with regard to the type of corpus are mentioned below in Table 1.

2.5 Selection of Time Span

To capture the features of a language or a variety reflected within a specific time span. A corpus should attempt to capture a particular period of time with a clear time indicator. Odia corpora are collected between the year of 2012 and 2013. Data are collected from Odia newspaper The Samaja only. The data will be sufficiently present the nature and character of the use of the language within the time span. This database will provided faithful information about the changes taking place within this period.

Table 1 Types of corpus users and their used corpus

Target users	Corpus
NLP and language technology (LT) people	General, monitor, parallel, spoken, aligned corpus
Sociolinguistics	General, written, speech, monitor corpus
Information retrieval specialists	General, monitor and annotated corpus
Tagging, processing, and parsing specialists	Annotated, monitor, written, spoken, general corpus
Word sense disambiguation specialists	Annotated, monitor, written, spoken, general corpus

2.6 Selection of Texts Type

A corpus will contain a collection of texts taken from all possible branches of human knowledge. The writings of highly reputed authors as well of little known writers are of equal importance. The collection of source materials may be collected from different sources, but here, we collected corpus from only Odia newspaper. A corpus gathers texts from various sources and disciplines where individuality safeguards to corpus against any kind of skewed text representation. The Odia corpus contains texts from literature, arts, commerce, science, and sports – some major domains of news circulated in standard Odia newspapers. The following list (Table 2) records the amount of news texts data collected from different domains for constructing the present corpus.

2.7 Method of Data Sampling

Data should be selected and sampled according to the researcher. Sampling of data can be random, regular, or selective. Data sampling can be done according to the types of data, i.e., historical data, sports data, tourism data, spiritual data, science related data, and art and commerce data. Data also can be sampling frame based on the nature of study. Brown corpus is one of the important examples.

2.8 Method of Data Input

Data are collected from different sources such as electronic sources (news paper, journals, magazines, books, etc.) if these are found in electronic form, Web sites (web pages, Web sites, and home pages), emails (electronic typewriting), machine reading of text (optical character recognition), and manual data input (typing text in computer).

Table 2 The text type included in the Odia corpus

Text types	Total number of words	Percentage of words (%)
Sports	88,930	09
Finance	1,26,327	12
Tourism	1,01,933	10
Politics	82,573	08
Agriculture	1,03,086	10
Total	6,74,352	49

2.9 Hardware Requirement

For developing a corpus, we use a personal computer, script processor, monitor, one keyboard, and a multilingual printer. Text files are developed with the help of the TC installed in the PC.

2.10 Management of Data Files

It involves various related tasks such as holding, storing, processing, screening, and retrieving information from the corpus, which require utmost care and sincerity on the part of the corpus creators. Once corpus is stored into computer, it should maintain frequently and add new things if required. If error is occurred, then it to be corrected, modification to be done, and some adjustment to be made.

3 Corpus Sanitation

Corpus sanitation means purify the entire corpus. There are some errors which may occur at the time of corpus entry manually, i.e., omission of character, addition of character, repetition of character, substitution of character, and transposition of character [3, 145–160].

4 Corpus Processing

Various corpus processing techniques are statistical analysis, concordance, lexical collocation, keyword search, local word grouping, lemmatization, morphological processing and generation, chunking, word processing, parts of speech tagging, annotation, and parsing.

4.1 Frequency Count

The frequency of occurrence of letter, syllable, morph, word, or a phrase in a text can be counted using software. For Odia language, word frequency can be studied only after the removing the suffix and prefix from the root.

4.2 Keyword in Context

Keyword in context (KWIC) means the occurrence of particular word used in corpus processing. The word under searching appears at the center of each line, with space on both the directions. A KWIC index is formed by sorting and aligning the words within an article title to allow each word (except the stopwords) in titles to be searchable alphabetically in the index. It was a useful indexing method for technical manuals before computerized full-text search became common.

4.3 Local Word Grouping

Local word grouping is another method of corpus processing in which attempt is made to locate and identify a group of words which often occur together to indicate a unique pattern of lexical combination that carries special lexico-grammatical significance in understanding the syntactic-semantic functions of words used in a piece of text. Word order is important for determining semantic analysis. Local word grouping is achieved by defining regular expressions for the word groups.

4.4 Morphological Analysis

Morphological analysis means finding the root word of a particular word in the corpus. It includes processing of single word units, double word units, and multiword units. Morphology is the identification, analysis, and description of the structure of a given language's morphemes and other linguistic units, such as root words, affixes, parts of speech, intonations and stresses, or implied context.

4.5 Part of Speech Tagging

The method of part of speech (POS) tagging is applied on texts to identify part of speech of each and every word used in the corpus. There are some well-defined hierarchical schemas for tagging words for English and other languages. Recently, the Bureau of Indian Standard (BIS) has proposed a POS tagset for tagging words in Odia corpus. We are planning to use this tagset on the present Odia corpus to develop the Odia tagged corpus for future application.

4.6 Lemmatization

Dash [4, 5] The process of grouping together the different inflected forms of word can be analyzed as a single item. In other words, there are many words in form of infected used in the sentence. Lemmatization is an algorithmic process of determining the lemma for a particular word.

5 Conclusion

The Odia corpus which was developed is very accurate and reliable for research work. It helps the researcher and scholars to do their research work based on natural language processing. The size of the corpus at present for Odia is very small. It needs to increase the size of corpus near about 200 millions words. Very large amount of corpus gives the accuracy more for a specific research domain. The use of corpora is not limited to certain area of linguistic research or to certain hypotheses. We can just as well compile a corpus of text and compare it to a reference corpus.

References

1. Dash, N.S.: Corpus Linguistics and Language Technology: with Reference to Indian Languages. Mittal Publications, New Delhi (2005)
2. Hofland, K.: Concordance programs for personal computers. In: Johansson, S., Stenström, A.-B. (eds.) English Computer Corpora: Selected Papers and Research Guide, pp. 283–306. Mouton de Gruyter, Berlin (1991)
3. Dash, N.S.: Techniques of text corpus processing. In: Mohanty, P., Reinhard, K. (eds.) Readings in Quantitative Linguistics, pp. 81–115. Indian Institute of Language Studies, New Delhi (2008)
4. Dash, N.S.: Corpus Linguistics: an Introduction. Person Education-Longman, New Delhi (2008)
5. Hunston, S.: Corpora in Applied Linguistics. Cambridge University Press, Cambridge (2002)

A Rule-Based Concatenative Approach to Speech Synthesis in Indian Language Text-to-Speech Systems

Soumya Priyadarsini Panda and Ajit Kumar Nayak

Abstract Several text-to-speech (TTS) systems are available today for languages such as English, Japanese, and Chinese, but still Indian languages are lacking behind in terms of good quality synthesized speech. Even though almost all Indian languages share a common phonetic base, till now a usable TTS system for all official Indian languages is not available. Also the existing speech synthesis techniques are found to be less effective in the scripting format of Indian languages. Considering the intelligibility of speech production and increasing memory requirement for Indian language TTS systems, in this paper we have proposed a rule-based concatenative technique for speech synthesis in Indian languages. It is being compared with the existing technique and the results of our experiments show our technique outperforms the existing technique.

Keywords Speech synthesis · Text-to-speech system · Natural language processing · Concatenative synthesis · Indian languages

1 Introduction

Recent research in the area of speech and language processing enables machines to speak naturally like humans. Speech synthesis deals with artificial production of speech and a text-to-speech (TTS) system in this aspect converts natural language text into its corresponding spoken wave form or speech. The intelligible speech synthesis systems have a widespread area of application in human–computer interaction systems such as talking computer systems and talking toys. Speech

S.P. Panda (✉) · A.K. Nayak
Siksha 'O' Anusandhan University, Bhubaneswar, India
e-mail: sppanda.cse@gmail.com

A.K. Nayak
e-mail: ajitnayak2000@gmail.com

© Springer India 2015
L.C. Jain et al. (eds.), *Intelligent Computing, Communication and Devices*,
Advances in Intelligent Systems and Computing 309,
DOI 10.1007/978-81-322-2009-1_59

synthesis, combined with speech recognition, allows for interaction with mobile devices via natural language processing interfaces [1].

India is a multilingual country where there are 23 constitutionally recognized languages spoken and written by people. A TTS system in Indian languages may benefit people in listening to different mythological documents available in the regional languages. For people with speech or reading disabilities, it could be a way to articulate the typed expressions or to listen to written documents. A TTS system in the regional languages may help people to learn different languages by listening to the pronunciation of the words. Even though the importance of a TTS system in Indian language is much higher, till now a usable TTS system for all official Indian languages is not available.

The usability of a TTS system depends on the quality of synthesized speech, where the quality is measured by two parameters: intelligibility and naturalness [2]. While intelligibility refers to the system's intelligence in producing speech, naturalness refers to how closely the speech sounds like a human generated speech. Out of a number of available speech synthesis techniques [3], the concatenative technique is used widely due to the highly natural quality in speech production. However, the quality depends on the size of the speech database. As the size of the stored data is reduced, desired segments are not always available in the stored data, and audio discontinuity may result.

Considering the quality of the produced speech and increasing memory requirements in concatenative technique for Indian language TTS systems, in this paper we have proposed a rule-based concatenative technique. The main focus here was to reduce the size of the speech database by storing only the basic sound units from which all other sound units may be derived. The experimental results show the effectiveness of our proposed technique in terms of intelligibility of the technique and naturalness of the produced speech.

The remainder of the paper is organized as follows. In the next section, we discusses about some related work in this area. Section 3 describes the details about our proposed model and the rule-based concatenative technique for speech production. The experimental methodology and result analysis for our technique is given in Sect. 4, showing the effectiveness of this technique in producing quality speech. Section 5 concludes the discussion, explaining the findings of our experiments and the future directions of this work, where further work can be undertaken.

2 Related Work

To overcome the major deficiency of concatenative technique with respect to the increasing memory requirements and audio discontinuities, a number of hybrid approaches are adopted by different researchers. A combination of the concatenative technique with statistical synthesis units is discussed in [4] to produce high quality speech. Another hybrid approach is discussed in [5] which integrate the rule-based formant synthesis and waveform concatenation techniques to produce waveform fragments dynamically. The authors in [6–8] proposed different models for developing TTS systems in different Indian languages. The dhvani—TTS system for Indian languages [9], uses a syllable-based concatenative speech synthesis technique to produce natural speech segments in 11 official languages of India requiring a large number of speech units to be stored in its database. However till date, a generalized speech synthesis technique is not available to produce high quality speech for all official languages of India satisfying user requirements.

3 Proposed Model

In this section, we discussed about the methodology of text-to-speech conversion in the proposed rule-based concatenative technique.

3.1 Text-to-Speech Conversion Methodology

As compared to English, the level of complexity for Indian languages in designing TTS systems is lesser in terms of pronunciation variation. Indian languages are highly phonetic; i.e., the pronunciation of new words can reliably be predicted from their written form, reducing the complexity of designing language processing component, as compared to English, where the pronunciations of words with similar written form has different pronunciation rules. Figure 1 shows the text-to-speech conversion process in our proposed technique and the steps are explained below.

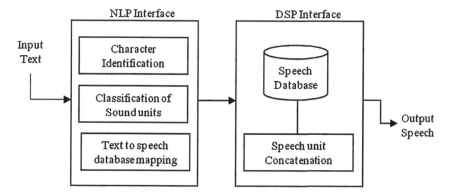

Fig. 1 Text-to-speech conversion process in Indian languages TTS systems

Fig. 2 Possible states of a character

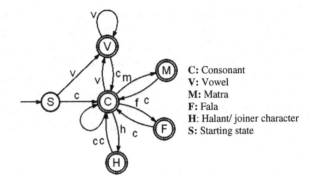

C: Consonant
V: Vowel
M: Matra
F: Fala
H: Halant/ joiner character
S: Starting state

3.1.1 Character Identification

The input to our system is text in any Indian language. The character sets in Indian languages may be divided into two broad categories as independent characters: vowels (V), consonants (C), and dependent characters: matra (M), fala (F), halant (H). In our methodology, we classify each character to be either in any one of the five states as shown in Fig. 2. The job of character identification phase is to identify the dependent and independent characters in a word for further processing.

3.1.2 Classification of Sound Units

In our methodology, we have considered the sound units to be broadly categorized into two types: V sounds and C sounds out of which all other sounds can be derived. For example the "matra" attached with a C is derived from the V sounds where as the "fala" attached to the same C is derived from the C sounds. The character "halant" act as a nonjoiner between any two C to produce the special characters. After the characters are identified this phase classifies them to their appropriate sound units.

3.1.3 Text-to-Speech Database Mapping

This phase maps the character to its appropriate V or C sound from which the unit is derived and passes it to the speech synthesizer so that the speech segment can be produced based on the rule-based concatenative technique.

3.1.4 Speech Database Creation

As compared to English, Indian languages have approximately twice as many V and C (total 56 characters instead of 26 in English) along with a number of possible special characters formed by combination of two or more characters.

Table 1 Speech units in database

Set of fundamental speech units						
\a	\o	\cha	\ttha	\tha	\pha	\lla
\aa	\ka	\chha	\dda	\da	\ba	\la
\ee	\kha	\ja	\ddha	\dha	\bha	\sha
\uu	\ga	\jha	\nna	\na	\ma	\ha
\ae	\gha	\tta	\ta	\pa	\ra	\ya

There are a number of Indian languages available, and the possible combination of characters is not limited. The desired segments may not be always available in the speech data. Therefore, the concatenative speech synthesis technique requires more number of speech segments to be stored in the speech database to produce natural speech.

In our approach, we have not limited the database to store all possible sound units; instead we create the database by storing only some fundamental speech units of the C and V sounds and derive all other sound units from these basic fundamental sound units using our proposed rule-based concatenative algorithm, reducing the memory requirements. For example, the sound unit "\ai" for the character "ai" in an Indian language may be derived from the combination of the units "\a" and "\ee," the sound unit "\rru" can be derived from "\ra," "\uu," etc. There are a number of special characters available in Indian language scripts which may be produced by combining the fundamental sound units. Table 1 shows the list of basic sounds required to produce any sound in our technique for an Indian language script. However, some more sound units may be added later on depending on the requirement of the language. All the speech segments stored in our database are *.wav* files recorded at 8,000 Hz in male voice.

3.1.5 Speech Unit Concatenation

The Text-to-speech database mapping step gives the appropriate database unit for any entered character. Depending on the type of the character, the final speech is produced by using our proposed rule-based concatenation algorithm discussed in the next subsection.

3.2 Algorithm

The Unicode standard [10] defines a range of values for each of the language, (e.g., for Odia −0B01–0B71, for Bengali: 0981–09FA, etc.). Based on the Unicode ranges, we have classified the characters to form 3 set of groups: consonant_set, vowel_set, and dependent_char_set. The rule-based concatenative algorithm checks a pair of characters each time to identify the type of characters in the input text based on their Unicode values. The type of characters in the input text determines how the final speech will be generated.

Rule-based Concatenative Algorithm

Open a blank output file (*f1.wav*)
1. While $i \leq$ word length
2. {
3. $U_1 =$ Unicode equivalent of char$_i$
4. $U_2 =$ Unicode equivalent of char$_{i+1}$
5.   If $U_1 \in$ [consonant_set] && $U_2 \in$ [vowel_set]
          a.   wavConcat(U_1, U_2)
          b.   i=i+2
6.   Else if $U_1 \in$ [consonant_set] && $U_2 \in$ [dependent_char_set]
          a. U_e=database equivalent of U_2
          b.   wavPortionConcat(U_1, U_e)
          c.   i=i+2
7.   Else if $U_1 \in$ [consonant_set] && $U_2 \in$ [consonant_set] || $U_1 \in$ [vowel_set]
          a.   singleWavConcat(U_1)
          b.   i=i+1;
8.   Else if i==length
          a.   singleWavConcat(U_1)
      }

If the character belongs to C_V category, the wavConcat () concatenates the whole wave pattern or the data in the .wav files for the pair to f1.wav for producing the final speech. If a character belongs to the C/V set, singleWavConcat() append the wav data for that character to f1.wav. If the Unicode belongs to dependent_char_set, after identifying the origin sounds from the database, the wavPortionConcat() concatenates, wave patterns (or the data in the .wav files) for the pair with portions amount from each to f1.wav for producing the final speech. We have fixed the portions to be 20 % from the independent character and 80 % from the dependent character. These percentages are fixed by performing experiments on the .wav files and analyzing the quality of the produced speech segment.

3.3 Example

The methodology for portion concatenation is same for all the words having dependent characters or special characters constitute of two or more characters. Figure 3 shows the wave patterns of "\ka" and "\ra" sounds in our database, and Fig. 4 shows the formation of the special character "\kra" from these two units. In producing words with independent characters like V or C the whole wave patterns for the pair is concatenated to produce the final speech segment.

Fig. 3 Wave pattern of \ka sound (*left*) \ra sound (*right*)

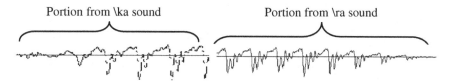

Fig. 4 Wave pattern of "\kra" after concatenating portions from "\ka" and "\ra" sounds

4 Experimental Methodology and Result Analysis

The proposed model and the speech synthesis algorithm are implemented in C/C++. Even though the rule-based concatenative technique may work for all the Indian languages, at an initial step of result analysis, we have considered the Odia language only. We have tested this technique for producing different types of words in Odia and our output is being compared with dhvani TTS system [9]. To analyze the performance, we have considered the storage requirement and execution time aspects along with the overall performance of the technique in producing speech. The result analysis of both the techniques is discussed below.

4.1 Storage Requirement

While the total number of speech samples in the speech database of dhvani is 800 requiring a memory size 1 Mb, our approach requires very less number of speech samples in the database (35 in total), requiring a memory of 235 Kb (23 % of dhvani). We store the speech samples in .wav format without further coding them to .gsm format like dhvani TTS system. Therefore, further decompression to the wave data is not required to produce the final speech in our approach.

4.2 Execution Time

To analyze the performance of our technique in terms of execution time compared to dhvani TTS system, we have prepared different text files containing 100 numbers of words from each category, V, VV, VCM, C, CC, CMC, CMV, and

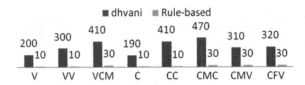

Fig. 5 Execution time for dhvani and rule-based technique

CFV. Figure 5 shows the execution times for both techniques in milliseconds for each category. The results of our experiments shows the exponential increase of execution time, due to the increase in number of decompression to the .gsm files in dhvani, while our approach shows relatively very less time in all the scenarios.

4.3 Subjective Measure for Speech Quality

We have performed listening tests to evaluate the proposed model. In these tests, we have evaluated the mean opinion score (MOS) [7] for a set of words in Odia from all 7 possible sound units (V, VC, VV, CV, CM, CF, and CHC). We have used a 5-point scale (1–5: very low to high) to analyze the performance of the proposed technique on the two individual subjective questions: the intelligibility of the technique in producing speech and naturalness of the produced speech. We have chosen 10 different listeners belonging to different classes such as student, layman, and academicians with no experience with TTS systems.

All the tests were performed with a headphone set. The only information the listeners are provided with is, they have to compare two speech synthesis techniques for producing speech for the same words with respect to the set of subjective questions. Figure 6 shows the MOS test results for our experiments for produced speech in both the approaches. The results show that our approach achieves relatively better or same results in different scenarios compared to the dhvani TTS system.

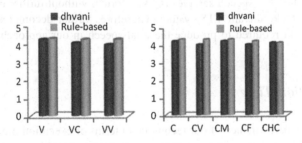

Fig. 6 MOS test results for starting vowel character (*left*) and starting consonant character (*right*)

Table 2 Comparative analysis of dhvani and rule-based concatenative technique

Technique	Storage requirement	Execution time	Naturalness	Intelligibility
dhvani	More	Exponential	High	Average
Rule-based	Less (23 % of dhvani)	Linear with low growth rate	High	High

5 Conclusions

In this paper, we describe a rule-based concatenative technique for speech synthesis in Indian language TTS systems. To analyze the performance of the technique, we have considered the storage requirement and execution time aspects along with the two quality measures intelligibility and naturalness. The subjective measure analysis shows the effectiveness of the proposed technique compared to dhvani in terms of intelligibility and naturalness. The summary of our experiments are given in Table 2. At the first step, we have implemented this concept for Odia language only. In our future work, we try to enhance the model to work for all possible characters in any Indian languages. Also some smoothening techniques may be used at the concatenation points to achieve better performance.

References

1. Feng, J., Ramabhadran, B., Hansel, J., Williams, J.D.: Trends in speech and language processing. In: IEEE Signal Processing Magazine (2012)
2. Alías, F., Sevillano, X., Socoró, J.C., Gonzalvo, X.: Towards high-quality next-generation text-to-speech synthesis: a multidomain approach by automatic domain classification. Audio Speech Lang. Process. IEEE **16**(7) (2008)
3. Tabet, Y., Boughazi, M.: Speech synthesis techniques. A survey. In: System, Signal Processing and their Applications, IEEE (2011)
4. Tiomkin, S., Malah, D., Shechtman, S., Kons, Z.: A hybrid text-to-speech system that combines concatenative and statistical synthesis units. Audio Speech Lang. Process. IEEE **19**(5), 1278–1288 (2011)
5. Hertz, S.R.: Integration of rule-based formant synthesis and waveform concatenation: a hybrid approach to text-to-speech synthesis. In: IEEE, pp. 87–90 (2002)
6. Mohanty, S.: Syllable based indian language text to speech system. IJAET **1**(2), 138–143 (2011)
7. Narendra, N.P., Rao, K.S., Ghosh, K., Vempada, R.R., Maity, S.: Development of syllable-based text to speech synthesis system in Bengali. IJST **14**(3), 167–181 (2011)
8. Bhakat, R.K., Narendra, N.P., Rao, K.S.: Corpus based emotional speech synthesis in Hindi. In: Pattern Recognition and Machine Intelligence. Springer LNCS, Heidelberg, vol. 8251, pp 390–395 (2013)
9. dhvani: A text to speech system for indian languages. http://dhvani.sourceforge.net
10. The Unicode Consortium. http://www.unicode.org

Author Index

© Springer India 2015
L.C. Jain et al. (eds.), *Intelligent Computing, Communication and Devices*,
Advances in Intelligent Systems and Computing 309,
DOI 10.1007/978-81-322-2009-1